A Passion for Mathematics

Works by Clifford A. Pickover

– – – – – –

The Alien IQ Test

Black Holes: A Traveler's Guide

Calculus and Pizza

Chaos and Fractals

Chaos in Wonderland

Computers, Pattern, Chaos, and Beauty

Computers and the Imagination

Cryptorunes: Codes and Secret Writing

Dreaming the Future

Egg Drop Soup

Future Health

Fractal Horizons: The Future Use of Fractals

Frontiers of Scientific Visualization

The Girl Who Gave Birth to Rabbits

Keys to Infinity

Liquid Earth

The Lobotomy Club

The Loom of God

The Mathematics of Oz

Mazes for the Mind: Computers and the Unexpected

Mind-Bending Visual Puzzles (calendars and card sets)

The Paradox of God and the Science of Omniscience

The Pattern Book: Fractals, Art, and Nature

The Science of Aliens

Sex, Drugs, Einstein, and Elves

Spider Legs (with Piers Anthony)

Spiral Symmetry (with Istvan Hargittai)

Strange Brains and Genius

Sushi Never Sleeps

The Stars of Heaven

Surfing through Hyperspace

Time: A Traveler's Guide

Visions of the Future

Visualizing Biological Information

Wonders of Numbers

The Zen of Magic Squares, Circles, and Stars

A Passion for Mathematics

- - - - -

Numbers, Puzzles, Madness, Religion, and the Quest for Reality

CLIFFORD A. PICKOVER

WILEY

John Wiley & Sons, Inc.

Published by John Wiley & Sons, Inc., Hoboken, New Jersey
Published simultaneously in Canada

Illustration credits: pages 91, 116, 137, 140, 142, 149, 150, 151, 157, 158, 159, 160, 162, 164, 167, 168, 169, 179, 199, 214, 215, 224, 225, 230, 274, 302, 336, 338, 341, 343, 345, and 348 by Brian C. Mansfield; 113, 114, 115, 145, 146, 332, 333, and 334 by Sam Loyd; 139 courtesy of Peter Hamburger and Edit Hepp; 141 by Stewart Raphael, Audrey Raphael, and Richard King; 155 by Patrick Grimm and Paul St. Denis; 165 and 166 from *Magic Squares and Cubes* by W. S. Andrews; 177 and 352 by Henry Ernest Dudeney; 200 by Bruce Patterson; 204 by Bruce Rawles; 206 by Jürgen Schmidhuber; 253 by Abram Hindle; 254, 255, 256, and 257 by Chris Coyne; 258 and 259 by Jock Cooper; 260 by Linda Bucklin; 261 by Sally Hunter; 262 and 263 by Jos Leys; 264 by Robert A. Johnston; and 266 by Cory and Catska Ench.

Design and composition by Navta Associates, Inc.

For general information about our other products and services, please contact our Customer Care Department within the United States at (800) 762-2974, outside the United States at (317) 572-3993 or fax (317) 572-4002.

Wiley also publishes its books in a variety of electronic formats. Some content that appears in print may not be available in electronic books. For more information about Wiley products, visit our web site at www.wiley.com.

Library of Congress Cataloging-in-Publication Data:

Pickover, Clifford A.
  A passion for mathematics : numbers, puzzles, madness, religion, and the quest for reality
/ Clifford A. Pickover.
    p. cm.
  Includes bibliographical references and index.
  ISBN-13 978-0-471-69098-6 (paper)
  ISBN-10 0-471-69098-8 (paper)
  1. Mathematics. I. Title

QA39.3.P53 2005
510—dc22   2004060622

Printed in the United States of America

10 9 8 7 6 5 4 3 2 1

Ramanujan said that he received his formulas from God. This book is dedicated to all those who find Ramanujan's π formulas beautiful to look at:

$$\frac{1}{\pi} = \sqrt{8} \sum_{n=0}^{\infty} \frac{(1103 + 26390n)(2n-1)!!(4n-1)!!}{99^{4n+2} 32^n (n!)^3}$$

$$\pi = \frac{5\sqrt{5}}{2\sqrt{3}} \left(\sum_{n=0}^{\infty} \frac{(11n+1)\left(\frac{1}{2}\right)_{Poch(n)} \left(\frac{1}{6}\right)_{Poch(n)} \left(\frac{5}{6}\right)_{Poch(n)}}{(n!)^3} \left(\frac{4}{125}\right)^n \right)^{-1}$$

(where *Poch(n)* refers to the Pochhammer notation described in chapter 2)

"Truly the gods have not from the beginning
revealed all things to mortals, but by long seeking,
mortals make progress in discovery."

—Xenophanes of Colophon (c. 500 B.C.)

— — — — —

"Every blade of grass has its angel
that bends over it and whispers, 'grow, grow.'"

—Talmudic commentary Midrash
Bereishis Rabbah, 10:6

Contents

In which we encounter religious mathematicians, mad mathematicians, famous mathematicians, mathematical savants, quirky questions, fun trivia, brief biographies, mathematical gods, historical oddities, numbers and society, gossip, the history of mathematical notation, the genesis of numbers, and "What if?" questions.

In which we encounter fascinating numbers and strange number sequences. We'll explore transcendental numbers, octonions, surreal numbers, obstinate numbers, cyclic numbers, Vibonacci numbers, perfect numbers, automorphic numbers, prime numbers, Wilson primes, palindromic primes, Fibonacci primes, Sophie Germain primes, Baxter-Hickerson primes, star-congruent primes, narcissistic numbers, amenable numbers, amicable numbers, p-adic numbers, large palindromes, factorions, hyperfactorials, primorials, palindions and hyperpalindions, exotic-looking formulas for ≠, the Golay-Rudin-Shapiro sequence, the wonderful Pochhammer notation, and famous and curious math constants (such as Liouville's constant, the Copeland-Erdös constant, Brun's constant, Champernowne's number, Euler's gamma, Chaitin's constant, the Landau-Ramanujan constant, Mills's constant, the golden ratio, Apéry's constant, and constants even more bizarre).

3. Algebra, Percentages, Weird Puzzles, and Marvelous Mathematical Manipulations   III

- - - - - - - - - - - - - - - -

In which we encounter treasure chests of zany and educational math problems that involve algebra, fractions, percentages, classic recreational puzzles, and various types of mathematical manipulation. Some are based on problems that are more than a thousand years old. Others are brand new. Get ready to sharpen your pencils and stretch your brains!

4. Geometry, Games, and Beyond   135

- - - - - - - - - - - - - - - -

In which we explore tiles, patterns, position problems, arrays, Venn diagrams, tic-tac-toe, other games played on boards, Königsberg bridges, catenaries, Loyd's and Dudeney's puzzles, Sherck's surface, magic squares, lituuses, inside-out Mandelbrot sets, the Quadratrix of Hippias, hyperspheres, fractal geese, Schmidhuber circles, Pappus's Arbelos, Escher patterns, and chess knights.

5. Probability: Take Your Chances   209

- - - - - - - - - - - - - - - -

In which we explore casinos, logic, guessing, decisions, combinations, permutations, competition, possibilities, games involving choice, monkeys typing Hamlet, *Benford's law, combinatorics, alien gambits, Rubik's Cubes, card shuffles, marble mazes, nontransitive dice, dangerous movements of air molecules, the board game of the gods, and the tunnels of death and despair.*

6. Big Numbers and Infinity   233

– – – – – – – – – – – – – – – – – –

In which we explore very large numbers, the edges of comprehension, infinity, the Funnel of Zeus, the infinite gift, the omega crystal, the Skewes number, the Monster group, Göbel's number, the awesome quattuordecillion, the Erdös-Moser number, Archimedes' famous "Cattle Problem," classic paradoxes, Gauss's "measurable infinity," and Knuth's arrow notation.

7. Mathematics and Beauty   251

– – – – – – – – – – – – – – – – – –

In which we explore artistic forms generated from mathematics—delicate fungi, twenty-second-century cityscapes, fractal necklaces and seed pods, alien devices, and a rich panoply of patterns that exhibit a cascade of detail with increasing magnifications.

Answers   267

References   383

Index   387

Acknowledgments

I thank Brian Mansfield for his wonderful cartoon diagrams that appear throughout the book. Over the years, Brian has been helpful beyond compare.

Numerous people have provided useful feedback and information relating to the solutions to my puzzles; these individuals include Dennis Gordon, Robert Stong, Paul Moskowitz, Joseph Pe, Daniel Dockery, Mark Nandor, Mark Ganson, Nick Hobson, Chuck Gaydos, Graham Cleverley, Jeffrey Carr, Jon Anderson, "Jaymz" James Salter, Chris Meyers, Pete Barnes, Steve Brazzell, Steve Blattnig, Edith Rudy, Eric Baise, Martie Saxenmeyer, Bob Ewell, Teja Krasek, and many more. I discussed some of the original puzzles in this book at my Pickover Discussion Group, located on the Web at groups.yahoo.com/group/CliffordPickover, and I thank the group members for their wonderful discussions and comments.

Many of the fancy formulas derived by the Indian mathematician Srinivasa Ramanujan come from Calvin Clawson's *Mathematical Mysteries*, Bruce Berndt's *Ramanujan's Notebooks* (volumes 1 and 2), and various Internet sources. With respect to some of the elegant prime number formulas, Clawson cites Paulo Ribenboim's *The Book of Prime Number Records*, second edition (New York: Springer, 1989), and *The Little Book of Big Primes* (New York: Springer, 1991). Calvin Clawson's *Mathematical Mysteries* and David Wells's various Penguin dictionaries provide a gold mine of mathematical concepts. A few problems in this book draw on, update, or revise problems in my earlier books, with information provided by countless readers who spend their lives tirelessly tackling mathematical conundrums.

Other interesting sources and recommended reading are given in the reference section. Numerous Web sites are proliferating that give comprehensive mathematical information, and my favorites include Wikipedia, the Free Encyclopedia (wikipedia.com), Ask Dr. Math (mathforum.org/dr.math/), The MacTutor History of Mathematics Archive (www.gap-system.org/~history/), The Published Data of Robert Munafo (www.mrob.com/pub/index.html), and MathWorld (mathworld. wolfram.com).

Introduction

"An equation means nothing to me unless it expresses a thought of God."

—Srinivasa Ramanujan (1887–1920)

The Ramanujan Code

- - - - - - - - - - - - - - - - -

"An intelligent observer seeing mathematicians at work might conclude that they are devotees of exotic sects, pursuers of esoteric keys to the universe."

—Philip Davis and Reuben Hersh, *The Mathematical Experience*, 1981

Readers of my popular mathematics books already know how I feel about numbers. Numbers are portals to other universes. Numbers help us glimpse a greater universe that's normally shielded from our small brains, which have not evolved enough to fully comprehend the mathematical fabric of the universe. Higher mathematical discussions are a little like poetry. The Danish physicist Niels Bohr felt similarly about physics when he said, "We must be clear that, when it comes to atoms, language can be used only as in poetry."

When I think about the vast ocean of numbers that humans have scooped from the shoreless sea of reality, I get a little shiver. I hope you'll shiver, too, as you glimpse numbers that range from integers, fractions, and radicals to stranger beasts like transcendental numbers, transfinite numbers, hyperreal numbers, surreal numbers, "nimbers," quaternions, biquaternions, sedenions, and octonions. Of course, we have a hard time thinking of such queer entities, but from time to time, God places in our midst visionaries who function like the biblical prophets, those individuals

1

who touched a universe inches away that most of us can barely perceive.

Srinivasa Ramanujan was such a prophet. He plucked mathematical ideas from the ether, out of his dreams. Ramanujan was one of India's greatest mathematical geniuses, and he believed that the gods gave him insights. These came in a flash. He could read the codes in the mathematical matrix in the same way that Neo, the lead character in the movie *The Matrix*, could access mathematical symbols that formed the infrastructure of reality as they cascaded about him. I don't know if God is a cryptographer, but codes are all around us waiting to be deciphered. Some may take a thousand years for us to understand. Some may always be shrouded in mystery.

In *The Matrix Reloaded*, the wise Architect tells Neo that his life is "the sum of a remainder of an unbalanced equation inherent in the programming of the matrix." Similarly, the great Swiss architect Le Corbusier (1887–1965) thought that gods played with numbers in a matrix beyond our ordinary reality:

> The chamois making a gigantic leap from rock to rock and alighting, with its full weight, on hooves supported by an ankle two centimeters in diameter: that is challenge and that is mathematics. The mathematical phenomenon always develops out of simple arithmetic, so useful in everyday life, out of numbers, those weapons of the gods: the gods are there, behind the wall, at play with numbers. (Le Corbusier, *The Modulor*, 1968)

A century ago, Ramanujan was *The Matrix*'s Neo in our own reality. As a boy, Ramanujan was slow to learn to speak. He seemed to spend all of his time scribbling strange symbols on his slate board or writing equations in his personal notebooks. Later in life, while working in the Accounts Department of the Port Trust Office at Madras, he mailed some of his equations to the renowned British mathematician G. H. Hardy. Hardy immediately tossed these equations into the garbage—but later retrieved them for a second look. Of the formulas, Hardy said that he had "never seen anything in the least like them before," and that some had completely "defeated" him. He quickly realized that the equations "could only be written down by a mathematician of the highest class." Hardy wrote in *Ramanujan: Twelve Lectures* that the formulas "must be true because, if they were not true, no one would have had the imagination to invent them."

Indeed, Ramanujan often stated a result that had come from some sense of intuition out of the unconscious realm. He said that an Indian goddess inspired him in his dreams. Not all of his formulas were perfect, but the avalanche of actual gems that he plucked from the mine of reality continues to boggle our modern minds. Ramanujan said that only in mathematics could one have a concrete realization of God.

Blood Dreams and God's Mathematicians

Repeatedly, [Western mathematicians] have been reduced to inchoate expressions of wonder and awe in the face of Ramanujan's powers—have stumbled about, groping for words, in trying to convey the mystery of Ramanujan."

—Robert Kanigel, *The Man Who Knew Infinity*, 1991

According to Ramanujan, the gods left drops of vivid blood in his dreams. After he saw the blood, scrolls containing complicated mathematics unfolded before him. When Ramanujan awakened in the morning, he scribbled only a fraction of what the gods had revealed to him.

In *The Man Who Knew Infinity*, Robert Kanigel suggests that the ease with which Ramanujan's spirituality and mathematics intertwined signified a "peculiar flexibility of mind, a special receptivity to loose conceptual linkages and tenuous associations. . . . " Indeed, Ramanujan's openness to mystical visions suggested "a mind endowed with slippery, flexible, and elastic notions of cause and effect that left him receptive to what those equipped with purely logical gifts could not see."

Before we leave Ramanujan, I should point out that many other mathematicians, such as Carl Friedrich Gauss, James Hopwood Jeans, Georg Cantor, Blaise Pascal, and John Littlewood, believed that inspiration had a divine aspect. Gauss said that he once proved a theorem "not by dint of painful effort but so to speak by the grace of God." For these reasons, I have included a number of brief pointers to religious mathematicians in chapter 1. I hope these examples dispel the notion that mathematics and religion are totally separate realms of human endeavor.

Our mathematical description of the universe forever grows, but our brains and language skills remain entrenched. New kinds of mathematics are being discovered or created all the time, but we need fresh ways to think and to understand. For example, in the last few years, mathematical proofs have been offered for famous problems in the history of mathematics, but the arguments have been far too long and complicated for experts to be certain they are correct. The mathematician Thomas Hales had to wait *five years* before expert reviewers of his geometry paper—submitted to the journal *Annals of Mathematics*—finally decided that they could find no errors and that the journal should publish Hale's proof, but only with a disclaimer saying they were not certain it was right! Moreover, mathematicians such as Keith Devlin have admitted (in the May 25, 2004, *New York Times*) that "the story of mathematics has reached a stage of such abstraction that many of its frontier problems cannot even be understood by the experts." There is absolutely no hope of explaining these concepts to a popular audience. We can construct theories and do computations, but we may not be sufficiently smart to comprehend, explain, or communicate these ideas.

A physics analogy is relevant here. When Werner Heisenberg worried that human beings might never truly understand atoms, Bohr was a bit more optimistic. He replied, "I think we may yet be able to do so, but in the process we may have to learn what the word *understanding* really means." Today, we use computers to help us reason beyond the limitations of our own intuition. In fact, experiments with computers are leading mathematicians to discoveries and insights never dreamed of before the ubiquity of these devices. Computers and computer graphics allow mathematicians to discover results long before they can prove them formally, thus opening entirely new fields of mathematics. Even simple computer tools, such as spreadsheets, give modern mathematicians power that Heisenberg, Einstein, and Newton would have lusted after. As just one example, in the late 1990s, computer programs designed by David Bailey and Helaman Ferguson helped to produce new formulas that related pi to log 5 and two other constants. As Erica Klarreich reports in the April 24, 2004, edition of *Science News*, once the computer had produced the formula, proving that it was correct was extremely easy. Often, simply *knowing* the answer is the largest hurdle to overcome when formulating a proof.

The Mathematical Smorgasbord

– – – – – – – – – – – – – – – – – –

As the island of knowledge grows, the surface that makes contact with mystery expands. When major theories are overturned, what we thought was certain knowledge gives way, and knowledge touches upon mystery differently. This newly uncovered mystery may be humbling and unsettling, but it is the cost of truth. Creative scientists, philosophers, and poets thrive at this shoreline.

—W. Mark Richardson, "A Skeptic's Sense of Wonder," *Science*, 1998

Despite all of my mystical talk about mathematics and the divine, mathematics is obviously practical. Mathematics has affected virtually every field of scientific endeavor and plays an invaluable role in fields ranging from science to sociology, from modeling ecological disasters and the spread of diseases to understanding the architecture of our brains. Thus, the fun and quirky facts, questions, anecdotes, equations, and puzzles in this book are metaphors for an amazing range of mathematical applications and notations. In fact, this book is a smorgasbord of puzzles, factoids, trivia, quotations, and serious problems to consider. You can pick and choose from the various delicacies as you explore the platter that's set before you. The problems vary in scope, so you are free to browse quickly among concepts ranging from Champernowne's number to the Göbel number, a number so big that it makes a trillion pale in comparison. Some of the puzzles are arranged randomly to enhance the sense of adventure and surprise. My brain is a runaway train, and these puzzles and factoids are the chunks of cerebrum scattered on the tracks.

Occasionally, some of the puzzles in this book will seem simple or frivolous; for example, Why does a circle have 360 degrees? Or, Is zero an even number? Or, What's the hardest license plate to remember? Or, Could Jesus calculate 30×24? However, these are questions that fans often pose to me, and I love some of these "quirkies" the best. I agree with the Austrian physicist Paul Ehrenfest, who said, "Ask questions. Don't be afraid to appear stupid. The stupid questions are usually the best and hardest to answer. They force the speaker to think about the basic problem."

In contrast to the "quirkies," some of the puzzles I pose in this book are so insanely difficult or require such an exhaustive search that only a computer hacker could hope to answer them, such as my problem "Triangle of the Gods" (see page 61).

These are questions with which I have challenged my geeky colleagues, and on which they have labored for hours and sometimes days. I think you'll enjoy seeing the results. Don't be scared if you have no chance in hell of solving them. Just enjoy the fact that intense people will often respond to my odd challenges with apparent glee. Most of the problems in the book are somewhere in-between the extremes of simplicity and impossibility and can be solved with a pencil and paper. Chapter 3 contains most of the problems that *teachers* can enjoy with students.

I will also tease you with fancy formulas, like those decorating the book from Ramanujan. Sometimes my goal is simply to delight you with wonderful-looking equations to ponder. Occasionally, a concept is repeated, just to see if you've learned your lesson and recognize a similar problem in a new guise. The different ways of getting to the same solution or concept reveal things that a single approach misses.

I've been in love with recreational mathematics for many years because of its educational value. Contemplating even simple problems stretches the imagination. The *usefulness* of mathematics allows us to build spaceships and investigate the geometry of our universe. Numbers will be our first means of communication with intelligent alien races.

Ancient peoples, like the Greeks, also had a deep fascination with numbers. Could it be that in difficult times numbers were the only constant thing in an ever-shifting world? To the Pythagoreans, an ancient Greek sect, numbers were tangible, immutable, comfortable, eternal—more reliable than friends, less threatening than Zeus.

Explanation of Symbols

- - - - - - - - - - - - - - - -

Non-Euclidean calculus and quantum physics are enough to stretch any brain, and when one mixes them with folklore, and tries to trace a strange background of multidimensional reality behind the ghoulish hints of the Gothic tales and the wild whispers of the chimney-corner, one can hardly expect to be wholly free from mental tension.

—H. P. Lovecraft, "Dreams in the Witch House," 1933

I use the following symbols to differentiate classes of entries in this book:

signifies a thought-provoking quotation.

signifies a mathematical definition that may come in handy throughout the book.

signifies a mathematical factoid to stimulate your imagination.

🔮 signifies a problem to be solved. Answers are provided at the back of the book.

These different classes of entries should cause even the most right-brained readers to fall in love with mathematics. Some of the zanier problems will entertain people at all levels of mathematical sophistication. As I said, don't worry if you cannot solve many of the puzzles in the book. Some of them still challenge seasoned mathematicians.

One common characteristic of mathematicians is an obsession with completeness—an urge to go back to first principles to explain their works. As a result, readers must often wade through pages of background before getting to the essential ingredients. To avoid this, each problem in my book is short, at most only a few paragraphs in length. One advantage of this format is that you can jump right in to experiment or ponder and have fun, without having to sort through a lot of verbiage. The book is not intended for mathematicians looking for formal mathematical explanations. Of course, this approach has some disadvantages. In just a paragraph or two, I can't go into any depth on a subject. You won't find much historical context or many extended discussions. In the interest of brevity, even the answer section may require readers to research or ponder a particular puzzle further to truly understand it.

To some extent, the choice of topics for inclusion in this book is arbitrary, although these topics give a nice introduction to some classic and original problems in number theory, algebra, geometry, probability, infinity, and so forth. These are also problems that I have personally enjoyed and are representative of a wider class of problems of interest to mathematicians today. Grab a pencil. Do not fear. Some of the topics in the book may appear to be curiosities, with little practical application or purpose. However, I have found these experiments to be useful and educational, as have the many students, educators, and scientists who have written to me. Throughout history, experiments, ideas, and conclusions that originate in the play of the mind have found striking and unexpected practical applications.

A few puzzles come from Sam Loyd, the famous nineteenth-century American puzzlemaster. Loyd (1841–1911) invented thousands of popular puzzles, which his son collected in a book titled *Cyclopedia of Puzzles*. I hope you enjoy the classics presented here.

Cultivating Perpetual Mystery

– – – – – – – – – – – – – – – –

"Pure mathematics is religion."

—Friedrich von Hardenberg, circa 1801

A wonderful panoply of relationships in nature can be expressed using integer numbers and their ratios. Simple numerical patterns describe spiral floret formations in sunflowers, scales on pinecones, branching patterns on trees, and the periodic life cycles of insect populations. Mathematical theories have predicted phenomena that were not confirmed until years later. Maxwell's equations, for example, predicted radio waves. Einstein's field equations suggested that gravity would bend light and that the universe is expanding. The physicist Paul Dirac once noted that the abstract mathematics we study now gives us a glimpse of physics in the future. In fact, his equations predicted the existence of antimatter, which was subsequently discovered. Similarly, the mathematician Nikolai Lobachevsky said that "there is no branch of mathematics, however abstract, which may not someday be applied to the phenomena of the real world."

A famous incident involving Murray Gell-Mann and his colleagues demonstrates the predictive power of mathematics and symmetry regarding the existence of a subatomic particle known as the Omega-minus. Gell-Mann had drawn a symmetric, geometric pattern in which each position in the pattern, except for one empty spot, contained a known particle. Gell-Mann put his finger on the spot and said with almost mystical insight, "There is a particle." His insight was correct, and experimentalists later found an actual particle corresponding to the empty spot.

One of my favorite quotations describing the mystical side of science comes from Richard Power's *The Gold Bug Variations*: "Science is not about control. It is about cultivating a perpetual condition of wonder in the face of something that forever grows one step richer and subtler than our latest theory about it. It is about reverence, not mastery."

Today, mathematics has permeated every field of scientific endeavor and plays an invaluable role in biology, physics, chemistry, economics, sociology, and engineering. Math can be used to help explain the structure of a rainbow, teach us how to make money in the stock market, guide a spacecraft, make weather forecasts, predict population growth, design buildings, quantify happiness, and analyze the spread of AIDS.

Mathematics has caused a revolution. It has shaped our thoughts. It has shaped the way we think.

Mathematics has changed the way we look at the world.

This introduction is dedicated to anyone who can decode the following secret message.

```
.. / - .... .. -. -. -.- / - .... .-. - / ...- . .-. -.-- /
..-. . .-- / --- ..-. / -.-- --- ..- /   .-- .. .-. .-.. /
      -... --- - .... . .-. / - --- /
         -.. . -.-. --- -.. . /
            - .... .. ... /
         ... . -.-. .-. . - /
      -- . ... ... .- --. . . .-.-.-
```

"I am the thought you are now thinking."

—Douglas Hofstadter, *Metamagical Themas*, 1985

Numbers,
History, Society,
and People

I N WHICH WE ENCOUNTER RELIGIOUS MATHEMATICIANS, MAD MATHEMATICIANS, famous mathematicians, mathematical savants, quirky questions, fun trivia, brief biographies, mathematical gods, historical oddities, numbers and society, gossip, the history of mathematical notation, the genesis of numbers, and "What if?" questions.

> Mathematics is the hammer that shatters
> the ice of our unconscious.

Ancient counting. Let's start the book with a question. What is the earliest evidence we have of humans counting? If this question is too difficult, can you guess whether the evidence is before or after 10,000 B.C.— and what the evidence might be? (See Answer 1.1.)

- - - - - -

Mathematics and beauty. I've collected mathematical quotations since my teenage years. Here's a favorite: "Mathematics, rightly viewed, possesses not only truth, but supreme beauty— a beauty cold and austere, like that of sculpture" (Bertrand Russell, *Mysticism and Logic*, 1918).

The symbols of mathematics. Mathematical notation shapes humanity's ability to efficiently contemplate mathematics. Here's a cool factoid for you: The symbols + and −, referring to addition and subtraction, first appeared in 1456 in an unpublished manuscript by the mathematician Johann Regiomontanus (a.k.a. Johann Müller). The plus symbol, as an abbreviation for the Latin *et* (and), was found earlier in a manuscript dated 1417; however, the downward stroke was not quite vertical.

- - - - - -

Mathematics and reality. Do humans invent mathematics or discover mathematics? (See Answer 1.2.)

Math beyond humanity. "We now know that there exist true propositions which we can never formally prove. What about propositions whose proofs require arguments beyond our capabilities? What about propositions whose proofs require millions of pages? Or a million, million pages? Are there proofs that are possible, but beyond us?" (Calvin Clawson, *Mathematical Mysteries*).

- - - - - -

The multiplication symbol. In 1631, the multiplication symbol × was introduced by the English mathematician William Oughtred (1574–1660) in his book *Keys to Mathematics*, published in London. Incidentally, this Anglican minister is also famous for having invented the slide rule, which was used by generations of scientists and mathematicians. The slide rule's doom in the mid-1970s, due to the pervasive influx of inexpensive pocket calculators, was rapid and unexpected.

- - - - - -

Mathematics and the universe. Here is a deep thought to start our mathematical journey. Do you think humanity's long-term fascination with mathematics has arisen because the universe is constructed from a mathematical fabric? We'll approach this question later in the chapter. For now, you may enjoy knowing that in 1623, Galileo Galilei echoed this belief in a mathematical universe by stating his credo: "Nature's great book is written in mathematical symbols." Plato's doctrine was that God is a geometer, and Sir James Jeans believed that God experimented with arithmetic. Isaac Newton supposed that the planets were originally thrown into orbit by God, but even after God decreed the law of gravitation, the planets required continual adjustments to their orbits.

Math and madness. Many mathematicians throughout history have had a trace of madness or have been eccentric. Here's a relevant quotation on the subject by the British mathematician John Edensor Littlewood (1885–1977), who suffered from depression for most of his life: "Mathematics is a dangerous profession; an appreciable proportion of us goes mad."

- - - - - -

Mathematics and murder. What triple murderer was also a brilliant French mathematician who did his finest work while confined to a hospital for the criminally insane? (See Answer 1.3.)

- - - - - -

Creativity and madness. "There is a theory that creativity arises when individuals are out of sync with their environment. To put it simply, people who fit in with their communities have insufficient motivation to risk their psyches in creating something truly new, while those who are out of sync are driven by the constant need to prove their worth. They have less to lose and more to gain" (Gary Taubes, "Beyond the Soapsuds Universe," 1977).

Mathematicians and religion. Over the years, many of my readers have assumed that famous mathematicians are not religious. In actuality, a number of important mathematicians were quite religious. As an interesting exercise, I conducted an Internet survey in which I asked respondents to name important mathematicians who were also religious. Isaac Newton and Blaise Pascal were the most commonly cited religious mathematicians.

In many ways, the mathematical quest to understand infinity parallels mystical attempts to understand God. Both religion and mathematics struggle to express relationships between humans, the universe, and infinity. Both have arcane symbols and rituals, as well as impenetrable language. Both exercise the deep recesses of our minds and stimulate our imagination. Mathematicians, like priests, seek "ideal," immutable, nonmaterial truths and then often venture to apply these truths in the real world. Are mathematics and religion the most powerful evidence of the inventive genius of the human race? In "Reason and Faith, Eternally Bound" (December 20, 2003, *New York Times*, B7), Edward Rothstein notes that faith was the inspiration for Newton and Kepler, as well as for numerous scientific and mathematical triumphs. "The conviction that there is an order to things, that the mind can comprehend that order and that this order is not infinitely malleable, those scientific beliefs may include elements of faith."

In his *Critique of Pure Reason*, Immanuel Kant describes how "the light dove, cleaving the air in her free flight and feeling its resistance against her wings, might imagine that its flight would be freer still in empty space." But if we were to remove the air, the bird would plummet. Is faith—or a cosmic sense of mystery—like the air that allows some seekers to soar? Whatever mathematical or scientific advances humans make, we will always continue to swim in a sea of mystery.

Pascal's mystery. "There is a God-shaped vacuum in every heart" (Blaise Pascal, *Pensées*, 1670).

❓ Leaving mathematics and approaching God. What famous French mathematician and teenage prodigy finally decided that religion was more to his liking and joined his sister in her convent, where he gave up mathematics and social life? (See Answer 1.4.)

- - - - - -

❶ Ramanujan's gods. As mentioned in this book's introduction, the mathematician Srinivasa Ramanujan (1887–1920) was an ardent follower of several Hindu deities. After receiving visions from these gods in the form of blood droplets, Ramanujan saw scrolls that contained very complicated mathematics. When he woke from his dreams, he set down on paper only a fraction of what the gods showed him.

Throughout history, creative geniuses have been open to dreams as a source of inspiration. Paul McCartney said that the melody for the famous Beatles' song "Yesterday," one of the most popular songs ever written, came to him in a dream. Apparently, the tune seemed so beautiful and haunting that for a while he was not certain it was original. The Danish physicist Niels Bohr conceived the model of an atom from a dream. Elias Howe received in a dream the image of the kind of needle design required for a lock-stitch sewing machine. René Descartes was able to advance his geometrical methods after flashes of insight that came in dreams. The dreams of Dmitry Mendeleyev, Friedrich August Kekulé, and Otto Loewi inspired scientific breakthroughs. It is not an exaggeration to suggest that many scientific and mathematical advances arose from the stuff of dreams.

❶ Blaise Pascal (1623–1662), a Frenchman, was a geometer, a probabilist, a physicist, a philosopher, and a combinatorist. He was also deeply spiritual and a leader of the Jansenist sect, a Calvinistic quasi-Protestant group within the Catholic Church. He believed that it made sense to become a Christian. If the person dies, and there is no God, the person loses nothing. If there is a God, then the person has gained heaven, while skeptics lose everything in hell.

Legend has it that Pascal in his early childhood sought to prove the existence of God. Because Pascal could not simply command God to show Himself, he tried to prove the existence of a devil so that he could then infer the existence of God. He drew a pentagram on the ground, but the exercise scared him, and he ran away. Pascal said that this experience made him certain of God's existence.

One evening in 1654, he had a two-hour mystical vision that he called a "night of fire," in which he experienced fire and "the God of Abraham, Isaac, and Jacob . . . and of Jesus Christ." Pascal recorded his vision in his work "Memorial." A scrap of paper containing the "Memorial" was found in the lining of his coat after his death, for he carried this reminder about with him always. The three lines of "Memorial" are

Complete submission to Jesus Christ and to my director.
Eternally in joy for a day's exercise on the earth.
May I not forget your words. Amen.

🌀 **Transcendence.** "Much of the history of science, like the history of religion, is a history of struggles driven by power and money. And yet, this is not the whole story. Genuine saints occasionally play an important role, both in religion and science. For many scientists, the reward for being a scientist is not the power and the money but the chance of catching a glimpse of the transcendent beauty of nature" (Freeman Dyson, in the introduction to *Nature's Imagination*).

– – – – – –

❓ **The value of eccentricity.** "That so few now dare to be eccentric, marks the chief danger of our time" (John Stuart Mill, nineteenth-century English philosopher).

– – – – – –

❓ **Counting and the mind.** I quickly toss a number of marbles onto a pillow. You may stare at them for an instant to determine how many marbles are on the pillow. Obviously, if I were to toss just two marbles, you could easily determine that two marbles sit on the pillow. What is the largest number of marbles you can quantify, at a glance, without having to individually count them? (See Answer 1.5.)

– – – – – –

❓ **Circles.** Why are there 360 degrees in a circle? (See Answer 1.6.)

– – – – – –

🌀 **The mystery of Ramanujan.** After years of working through Ramanujan's notebooks, the mathematician Bruce Berndt said, "I still don't understand it all. I may be able to prove it, but I don't know where it comes from and where it fits into the rest of mathematics. The enigma of Ramanujan's creative process is still covered by a curtain that has barely been drawn" (Robert Kanigel, *The Man Who Knew Infinity*, 1991).

❓ **Calculating π.** Which nineteenth-century British boarding school supervisor spent a significant portion of his life calculating π to 707 places and died a happy man, despite a sad error that was later found in his calculations? (See Answer 1.8.)

– – – – – –

💡 **The special number 7.** In ancient days, the number 7 was thought of as just another way to signify "many." Even in recent times, there have been tribes that used no numbers higher than 7.

In the 1880s, the German ethnologist Karl von Steinen described how certain South American Indian tribes had very few words for numbers. As a test, he repeatedly asked them to count ten grains of corn. They counted "slowly

❓ **The world's most forgettable license plate?** Today, mathematics affects society in the funniest of ways. I once read an article about someone who claimed to have devised the most forgettable license plate, but the article did not divulge the secret sequence. What *is* the most forgettable license plate? Is it a random sequence of eight letters and numbers—for example, 6AZL4QO9 (the maximum allowed in New York)? Or perhaps a set of visually confusing numbers or letters—for example, MWNNMWWM? Or maybe a binary number like 01001100. What do you think? What would a *mathematician* think? (See Answer 1.7.)

but correctly to six, but when it came to the seventh grain and the eighth, they grew tense and uneasy, at first yawning and complaining of a headache, then finally avoided the question altogether or simply walked off." Perhaps *seven* means "many" in such common phrases as "seven seas" and "seven deadly sins." (These interesting facts come from Adrian Room, *The Guinness Book of Numbers*, 1989.)

- - - - - -

Carl Friedrich Gauss (1777–1855), a German, was a mathematician, an astronomer, and a physicist with a wide range of contributions. Like Ramanujan, after Gauss proved a theorem, he sometimes said that the insight did not come from "painful effort but, so to speak, by the grace of God." He also once wrote, "There are problems to whose solution I would attach an infinitely greater importance than to those of mathematics, for example, touching ethics, or our relation to God, or concerning our destiny and our future; but their solution lies wholly beyond us and completely outside the province of science."

Genius and eccentricity. "The amount of eccentricity in a society has been proportional to the amount of genius, material vigor and moral courage which it contains" (John Stuart Mill, *On Liberty*, 1869).

- - - - - -

Mathematics and God. "The Christians know that the mathematical principles, according to which the corporeal world was to be created, are co-eternal with God. Geometry has supplied God with the models for the creation of the world. Within the image of God it has passed into man, and was certainly not received within through the eyes" (Johannes Kepler, *The Harmony of the World*, 1619).

Isaac Newton (1642–1727), an Englishman, was a mathematician, a physicist, an astronomer, a coinventor of calculus, and famous for his law of gravitation. He was also the author of many books on biblical subjects, especially prophecy.

Perhaps less well known is the fact that Newton was a creationist who wanted to be known as much for his theological writings as for his scientific and mathematical texts. Newton believed in a Christian unity, as opposed to a trinity. He developed calculus as a means of describing motion, and perhaps for understanding the nature of God through a clearer understanding of nature and reality. He respected the Bible and accepted its account of Creation.

James Hopwood Jeans (1877–1946) was an applied mathematician, a physicist, and an astronomer. He sometimes likened God to a mathematician and wrote in *The Mysterious Universe* (1930), "From the intrinsic evidence of his creation, the Great Architect of the Universe now begins to appear as a pure mathematician." He has also written, "Physics tries to discover the pattern of events which controls the phenomena we observe. But we can never know what this pattern means or how it originates; and even if some superior intelligence were to tell us, we should find the explanation unintelligible" (*Physics and Philosophy*, 1942).

- - - - - -

🛑 **Leonhard Euler** (1707–1783) was a prolific Swiss mathematician and the son of a vicar. Legends tell of Leonhard Euler's distress at being unable to mathematically prove the existence of God. Many mathematicians of his time considered mathematics a tool to decipher God's design and codes. Although he was a devout Christian all his life, he could not find the enthusiasm for the study of theology, compared to that of mathematics. He was completely blind for the last seventeen years of his life, during which time he produced roughly half of his total output.

Euler is responsible for our common, modern-day use of many famous mathematical notations—for example, $f(x)$ for a function, e for the base of natural logs, i for the square root of -1, π for pi, Σ for summation. He tested Pierre de Fermat's conjecture that numbers of the form $2^n + 1$ were always prime if n is a power of 2. Euler verified this for $n = 1, 2, 4, 8,$ and 16, and showed that the next case $2^{32} + 1 = 4,294,967,297 = 641 \times 6,700,417$, and so is not prime.

found wide applications in the design of computers.

- - - - - -

❓ **The value of puzzles.** "It is a wholesome plan, in thinking about logic, to stock the mind with as many puzzles as possible, since these serve much the same purpose as is served by experiments in physical science" (Bertrand Russell, *Mind*, 1905).

- - - - - -

❓ **A mathematical nomad.** What legendary mathematician, and one of the most prolific mathematicians in history, was so devoted to math that he lived as a nomad with no home and no job? Sexual contact revolted him; even an accidental touch by anyone made him feel uncomfortable. (See Answer 1.9.)

- - - - - -

🛑 **George Boole** (1815–1864), an Englishman, was a logician and an algebraist. Like Ramanujan and other mystical mathematicians, Boole had "mystical" experiences. David Noble, in his book *The Religion of Technology*, notes, "The thought flashed upon him suddenly as he was walking across a field that his ambition in life was to explain the logic of human thought and to delve analytically into the spiritual aspects of man's nature [through] the expression of logical relations in symbolic or algebraic form. . . . It is impossible to separate Boole's religious beliefs from his mathematics."

Boole often spoke of his almost photographic memory, describing it as "an arrangement of the mind for every fact and idea, which I can find at once, as if it were in a well-ordered set of drawers."

Boole died at age forty-nine, after his wife mistakenly thought that tossing buckets of water on him and his bed would cure his flu. Today, Boolean algebra has

🛑 **Marin Mersenne** (1588–1648) was another mathematician who was deeply religious. Mersenne, a Frenchman, was a theologian, a philosopher, a number theorist, a priest, and a monk. He argued that God's majesty would not be diminished had

❓ **Mirror phobia.** What brilliant, handsome mathematician so hated mirrors that he covered them wherever he went? (See Answer 1.10.)

- -

🌀 **What is a mathematician?** "A mathematician is a blind man in a dark room looking for a black cat which isn't there" (Charles Darwin).

- -

❓ **Animal math.** Can animals count? (See Answer 1.11.)

He created just one world, instead of many, because the one world would be infinite in every part. His first publications were theological studies against atheism and skepticism.

Mersenne was fascinated by prime numbers (numbers like 7 that were divisible only by themselves and 1), and he tried to find a formula that he could use to find all primes. Although he did not find such a formula, his work on "Mersenne numbers" of the form $2^p - 1$, where p is a prime number, continues to interest us today. Mersenne numbers are the easiest type of number to prove prime, so they are usually the largest primes of which humanity is aware.

Mersenne himself found several prime numbers of the form $2^p - 1$, but he underestimated the future of computing power by stating that all eternity would not be sufficient to decide if a 15- or 20-digit number were prime. Unfortunately, the prime number values for p that make $2^p - 1$ a prime number seem to form no regular sequence. For example, the Mersenne number is prime when $p = 2, 3, 5, 7, 13, 17, 19, \ldots$ Notice that when p is equal to the prime number 11, $M_{11} = 2,047$, which is not prime because $2,047 = 23 \times 89$.

The fortieth Mersenne prime was discovered in 2003, and it contained 6,320,430 digits! In particular, the Michigan State University graduate student Michael Shafer discovered that $2^{20,996,011} - 1$ is prime. The number is so large that it would require about fifteen hundred pages to write on paper using an ordinary font. Shafer, age twenty-six, helped find the number as a volunteer on a project called the Great Internet Mersenne Prime Search. Tens of thousands of people volunteer the use of their personal computers in a worldwide project that harnesses the power of hundreds of thousands of computers, in effect creating a supercomputer capable of performing trillions of calculations per second. Shafer used an ordinary Dell computer in his office for nineteen days. What would Mersenne have thought of this large beast?

In 2005, the German eye surgeon Martin Nowak, also part of the Great Internet Mersenne Prime Search, discovered the forty-second Mersenne prime number, $2^{25,964,951} - 1$, which has over seven million digits. Nowak's 2.4-GHz Pentium-4 computer spent roughly fifty days analyzing the number before reporting the find. The Electronic Frontier Foundation, a U.S. Internet campaign group, has promised to give $100,000 to whoever finds the first ten-million-digit prime number.

Mathematics and God. "Before creation, God did just pure mathematics. Then He thought it would be a pleasant change to do some applied" (John Edensor Littlewood, *A Mathematician's Miscellany*, 1953).

The division symbol. The division symbol ÷ first appeared in print in Johann Heinrich Rahn's *Teutsche Algebra* (1659).

Donald Knuth (1938–) is a computer scientist and a mathematician. He is also a fine example of a mathematician who is interested in religion. For example, he has been an active Lutheran and a Sunday school teacher. His attractive book titled *3:16* consists entirely of commentary on chapter 3, verse 16, of each of the books in the Bible. Knuth also includes calligraphic renderings of the verses. Knuth himself has said, "It's tragic that scientific advances have caused many people to imagine that they know it all, and that God is irrelevant or nonexistent. The fact is that everything we learn reveals more things that we do not understand. . . . Reverence for God comes naturally if we are honest about how little we know."

Mathematics and sex. "A well-known mathematician once told me that the great thing about liking both math and sex was that he could do either one while thinking about the other" (Steven E. Landsburg, in a 1993 post to the newsgroup sci.math).

Mystery mathematician. Around A.D. 500, the Greek philosopher Metrodorus gave us the following puzzle that describes the life of a famous mathematician:

A certain man's boyhood lasted $\frac{1}{6}$ of his life; he married after $\frac{1}{7}$ more; his beard grew after $\frac{1}{12}$ more, and his son was born 5 years later; the son lived to half his father's final age, and the father died 4 years after the son.

Tell me the mystery man's name or his age at death. (See Answer 1.12.)

Mathematician starves. What famous mathematician deliberately starved himself to death in 1978? (Hint: He was perhaps the most brilliant logician since Aristotle.) (See Answer 1.13.)

Understanding brilliance. "Maybe the brilliance of the brilliant can be understood only by the nearly brilliant" (Anthony Smith, *The Mind*, 1984).

Brain limitation. "Our brains have evolved to get us out of the rain, find where the berries are, and keep us from getting killed. Our brains did not evolve to help us grasp really large numbers or to look at things in a hundred thousand dimensions" (Ronald Graham, a prior director of Information Sciences Research at AT&T Research, quoted in Paul Hoffman's "The Man Who Loves Only Numbers," *Atlantic Monthly*, 1987).

🕐 **Georg Friedrich Bernhard Riemann** (1826–1866) was a German mathematician who made important contributions to geometry, number theory, topology, mathematical physics, and the theory of complex variables. He also attempted to write a mathematical proof of the truth of the Book of Genesis, was a student of theology and biblical Hebrew, and was the son of a Lutheran minister.

The Riemann hypothesis, published by Riemann in 1859, deals with the zeros of a very wiggly function, and the hypothesis still resists modern mathematicians' attempts to prove it. Chapter 3 describes the hypothesis further.

🕐 **Calculating prodigy has plastic brain.** Rüdiger Gamm is shocking the world with his calculating powers and is changing the way we think about the human brain. He did poorly at mathematics in school but is now a world-famous human calculator, able to access regions of his brain that are off limits to most of us. He is not autistic but has been able to train his brain to perform lightning calculations. For example, he can calculate 53 to the ninth power in his head. He can divide prime numbers and calculate the answer to 60 decimal points and more. He can calculate fifth roots.

Amazing calculating powers such as these were previously thought to be possible only by "autistic savants." (Autistic savants often have severe developmental disabilities but, at the same time, have special skills and an incredible memory.) Gamm's talent has attracted the curiosity of European researchers, who have imaged his brain with PET scans while he performed math problems. These breathtaking studies reveal that Gamm is now able to use areas of his brain that ordinary humans can use for *other* purposes. In particular, he can make use of the areas of his brain that are normally responsible for long-term memory, in order to perform his rapid calculations. Scientists hypothesize that Gamm temporarily uses these areas to "hold" digits in so-called "working memory," the brain's temporary holding area. Gamm is essentially doing what computers do when they extend their capabilities by using swap space on the hard drive to increase their capabilities. Scientists are not sure how Gamm acquired this ability, considering that he became interested in mathematical calculation only when he was in his twenties. (You can learn more in Steve Silberman's "The Keys to Genius," *Wired*, no. 11.12, December 2003.)

🌀 **A dislike for mathematics.** "I'm sorry to say that the subject I most disliked was mathematics. I have thought about it. I think the reason was that mathematics leaves no room for argument. If you made a mistake, that was all there was to it" (Malcolm X, *The Autobiography of Malcolm X*, 1965).

🌀 **Mathematics and humanity.** "Anyone who cannot cope with mathematics is not fully human. At best he is a tolerable subhuman who has learned to wear shoes, bathe, and not make messes in the house" (Robert A. Heinlein, *Time Enough for Love*, 1973).

🔵 **Kurt Gödel** (1906–1978) is an example of a mathematical genius obsessed with God and the afterlife. As discussed in Answer 1.13 about the mathematician who starved himself to death, Gödel was a logician, a mathematician, and a philosopher who was famous for having shown that in any axiomatic system for mathematics, there are propositions that cannot be proved or disproved within the axioms of the system.

Gödel thought it was possible to show the logical necessity for life after death and the existence of God. In four long letters to his mother, Gödel gave reasons for believing in a next world.

- - - - - -

❓ **Math and madness.** "Cantor's work, though brilliant, seemed to move in half-steps. The closer he came to the answers he sought, the further away they seemed. Eventually, it drove him mad, as it had mathematicians before him" (Amir D. Aczel, *The Mystery of the Aleph: Mathematics, the Kabbalah, and the Search for Infinity*, 2000).

🔵 **Gottfried Wilhelm von Leibniz** (1646–1716), a German, was an analyst, a combinatorist, a logician, and the coinventor of calculus who also passionately argued for the existence of God. According to Leibniz, God chooses to actualize this world out of an infinite number of possible worlds. In other words, limited only by contradiction, God first conceives of every possible world, and then God simply chooses which of them to create.

Leibniz is also famous for the principle of "preestablished harmony," which states that God constructed the universe in such a way that corresponding mental and physical events occur simultaneously. His "monad theory" states that the universe consists of an infinite number of substances called monads, each of which has its own individual identity but is an expression of the whole universe from a particular unique viewpoint.

- -

🔵 **Greater-than symbol.** The greater-than and less-than symbols (> and <) were introduced by the British mathematician Thomas Harriot in his *Artis Analyticae Praxis*, published in 1631.

- -

🔵 **Greater-than or equal-to symbol.** The symbol ≥ (greater than or equal to) was first introduced by the French scientist Pierre Bouguer in 1734.

❓ **Mathematician murdered.** Why was the first woman mathematician murdered? (See Answer 1.14.)

- - - - - -

❓ **Going to the movies.** What was the largest number ever used in the title of an American movie? Name the movie! (See Answer 1.15.) What is the largest number less than a billion ever used in a major, full-length movie title? (Hint: The song was popular in the late 1920s and the early 1930s.) (See Answer 1.16.)

- - - - - -

🌐 **Math and madness.** Many mathematicians were depressed and religious at the same time. Which famous mathematician invented the concept of "transfinite numbers" (essentially, different "levels" of infinity), believed that God revealed mathematical ideas to him, and was a frequent guest of sanitariums? (See Answer 1.17.)

- - - - - -

🌐 **Mathematics and diapering.** "A human being should be able to change a diaper, plan an invasion, butcher a hog, conn a ship, design a building, write a sonnet, balance accounts, build a wall, set a bone, comfort the dying, take orders, give orders, cooperate, act alone, solve equations, analyze a new problem, pitch manure, program a computer, cook a tasty meal, fight efficiently, die gallantly. Specialization is for insects" (Robert A. Heinlein, *Time Enough for Love*, 1973).

- - - - - -

🌐 **Mathematicians and God.** "Mathematicians, astronomers, and physicists are often religious, even mystical; biologists much less often; econ- omists and psychologists very seldom indeed. It is as their subject matter comes nearer to man himself that their antireligious bias hardens" (C. S. Lewis, *The Grand Miracle: And Other Selected Essays on Theology and Ethics from God in the Dock*, 1983).

- - - - - -

🌐 **The Number Pope.** As I write this book, I realize that a thousand years ago, the last Pope-mathematician died. Gerbert of Aurillac (c. 946– 1003) was fascinated by mathematics and was elected to be Pope Sylvester II in 999. His advanced knowledge of mathematics convinced some of his enemies that he was an evil magician.

In Reims, he transformed the floor of the cathedral into a giant abacus. That must have been a sight to see! The "Number Pope" was also important because he adopted Arabic numerals (1, 2, 3, 4, 5, 6, 7, 8, 9) as a replacement for Roman numerals. He contributed to the invention of the pendulum clock, invented devices that tracked planetary orbits, and wrote on geometry. When he realized that he lacked knowledge of formal logic, he studied under German logicians. He said, "The just man lives by faith; but it is good that he should combine science with his faith."

🌐 **Hardy's six wishes.** In the 1920s, the British mathematician G. H. Hardy wrote a postcard to his friend, listing six New Year's wishes:

1. prove the Riemann hypothesis

2. score well at the end of an important game of cricket

3. find an argument for the nonexistence of God that convinces the general public

4. be the first man at the top of Mount Everest

5. be the first president of the USSR, Great Britain, and Germany

6. murder Mussolini

(*The London Mathematical Society Newsletter*, 1994)

Charles Babbage (1792–1871), an Englishman, was an analyst, a statistician, and an inventor who was also interested in religious miracles. He once wrote, "Miracles are not a breach of established laws, but . . . indicate the existence of far higher laws." Babbage argued that miracles could occur in a mechanistic world. Just as Babbage could program strange behavior on his calculating machines, God could program similar irregularities in nature. While investigating biblical miracles, he assumed that the chance of a man rising from the dead is one in 10^{12}.

Babbage is famous for conceiving an enormous hand-cranked mechanical calculator, an early progenitor of our modern computers. Babbage thought the device would be most useful in producing mathematical tables, but he worried about mistakes that would be made by humans who transcribed the results from its thirty-one metal output wheels. Today, we realize that Babbage was a hundred years ahead of his time and that the politics and the technology of his era were inadequate for his lofty dreams.

Tinkertoy computer. In the early 1980s, the computer geniuses Danny Hillis, Brian Silverman, and friends built a Tinkertoy computer that played tic-tac-toe. The device was made from 10,000 Tinkertoy pieces.

- - - - - -

Fantasy meeting of Pythagoras, Cantor, and Gödel. I often fantasize about the outcome of placing mathematicians from different eras in the same room. For example, I would be intrigued to gather Pythagoras, Cantor, and Gödel in a small room with a single blackboard to debate their various ideas on mathematics and God. What profound knowledge might we gain if we had the power to bring together great thinkers of various ages for a conference on mathematics? Would a roundtable discussion with Pythagoras, Cantor, and Gödel produce less interesting ideas than one with Newton and Einstein?

Could ancient mathematicians contribute *any* useful ideas to modern mathematicians? Would a meeting of time-traveling *mathematicians* offer more to humanity than a meeting of other scientists—for example, biologists or sociologists? These are all fascinating questions to which I don't yet have answers.

- - - - - -

Mathematicians as God's messengers. "Cantor felt a duty to keep on, in the face of adversity, to bring the insights he had been given as God's messenger to mathematicians everywhere" (Joseph Dauben, *Georg Cantor*, 1990).

- - - - - -

Power notation. In 1637, the philosopher René Descartes was the first person to use the superscript notation for raising numbers and variables to powers—for example, as in x^2.

- - - - - -

Numerical religion. What ancient mathematician established a numerical religion whose main tenets included the transmigration of souls and the sinfulness of eating beans? (See Answer 1.18.)

Modern mathematical murderer. Which modern mathematician murdered and maimed the most people from a distance? (See Answer 1.19.)

The ∞ symbol. Most high school students are familiar with the mathematical symbol for infinity (∞). Do you think this symbol was used a hundred years ago? Who first used this odd symbol? (See Answer 1.20.)

Mathematics of tic-tac-toe. In how many ways can you place Xs and Os on a standard tic-tac-toe board? (See Answer 1.21.)

The square root symbol. The Austrian mathematician Christoff Rudolff was the first to use the square root symbol $\sqrt{}$ in print; it was published in 1525 in *Die Coss*.

Mathematics and poetry. "It is impossible to be a mathematician without being a poet in soul" (Sofia Kovalevskaya, quoted in *Agnesi to Zeno* by Sanderson Smith, 1996).

Chickens and tic-tac-toe. The mathematics of tic-tac-toe have been discussed for decades, but can a chicken actually learn to play well? In 2001, an Atlantic City casino offered its patrons a tic-tac-toe "chicken challenge" and offered cash prizes of up to $10,000. The chicken gets the first entry, usually by pecking at X or O on a video display inside a special henhouse set up in the casino's main concourse. Gamblers standing outside the booth then get to make the next move by pressing buttons on a separate panel. There is no prize for a tie. A typical game lasts for about a minute, and the chicken seems to be trained to peck at an X or an O, depending on the human's moves.

Supposedly, the tic-tac-toe-playing chickens work in shifts of one to two hours to avoid stressing the animals. Various animal-rights advocates have protested the use of chickens in tic-tac-toe games. Can a chicken actually learn to play tic-tac-toe? (See Answer 1.22.)

Mathematics and God. "God exists since mathematics is consistent, and the devil exists since we cannot prove the consistency" (Morris Kline, *Mathematical Thought from Ancient to Modern Times*, 1990).

Lunatic scribbles and mathematics. "If a lunatic scribbles a jumble of mathematical symbols it does not follow that the writing means anything merely because to the inexpert eye it is indistinguishable from higher mathematics" (Eric Temple Bell, quoted in J. R. Newman's *The World of Mathematics*, 1956).

God's perspective. "When mathematicians think about algorithms, it is usually from the God's-eye perspective. They are interested in proving, for instance, that *there is* some algorithm with some interesting property, or that *there is no* such algorithm, and in order to prove such things you needn't actually locate the algorithm you are talking about . . . " (Daniel Dennett, *Darwin's Dangerous Idea: Evolution and the Meaning of Life*, 1996).

☺ Erdös contemplates death. Once, while pondering his own death, the mathematician Paul Erdös (1913–1996) remarked, "My mother said, 'Even you, Paul, can be in only one place at one time.' Maybe soon I will be relieved of this disadvantage. Maybe, once I've left, I'll be able to be in many places at the same time. Maybe then I'll be able to collaborate with Archimedes and Euclid."

- - - - - -

☺ Mathematics and the infinite. "Mathematics is the only infinite human activity. It is conceivable that humanity could eventually learn everything in physics or biology. But humanity certainly won't ever be able to find out everything in mathematics, because the subject is infinite. Numbers themselves are infinite" (Paul Erdös, quoted in Paul Hoffman's *The Man Who Loved Only Numbers*, 1998).

- - - - - -

❓ First female doctorate. Who was the first woman to receive a doctorate in mathematics, and in what century do you think she received it? (See Answer 1.23.)

☺ Creativity and madness. "Creativity and genius feed off mental turmoil. The ancient Greeks, for instance, believed in divine forms of madness that inspired mortals' extraordinary creative acts" (Bruce Bower, *Science News*, 1995).

- - - - - -

☺ Science, Einstein, and God. "The scientist's religious feeling takes the form of a rapturous amazement at the harmony of natural law, which reveals an intelligence of such superiority that, compared with it, all the systematic thinking and acting of human beings is an utterly insignificant reflection. This feeling is the guiding principle of his life and work. . . . It is beyond question closely akin to that which has possessed the religious geniuses of all ages" (Albert Einstein, *Mein Weltbild*, 1934).

- - - - - -

☺ Math in the movies. In the movie *A Beautiful Mind*, Russell Crowe scrolls the following formulas on the blackboard in his MIT class:

$$V = \{F : \mathbb{R}^3 - X \to \mathbb{R}^3 smooth$$
$$W = \{F = \nabla g\}$$
$$\dim(V/W) = ?$$

Movie directors were told by advisers that this set of formulas was subtle enough to be out of reach for most undergraduates, but accessible enough so that Jennifer Connelly's character might be able to dream up a possible solution.

- -

❓ Mathematics and homosexuality. Which brilliant mathematician was forced to become a human guinea pig and was subjected to drug experiments to reverse his homosexuality? (Hint: He was a 1950s computer theorist whose mandatory drug therapy made him impotent and caused his breasts to enlarge. He also helped to break the codes of the German Engima code machines during World War II.) (See Answer 1.24.)

🛈 **A famous female mathematician.** Maria Agnesi (1718–1799) is one of the most famous female mathematicians of the last few centuries and is noted for her work in differential calculus. When she was seven years old, she mastered the Latin, the Greek, and the Hebrew languages, and at age nine she published a Latin discourse defending higher education for women. As an adult, her clearly written textbooks condensed the diverse research writings and methods of a number of mathematicians. They also contained many of her own original contributions to the field, including a discussion of the cubic curve that is now known as the "Witch of Agnesi." However, after the death of her father, she stopped doing scientific work altogether and devoted the last forty-seven years of her life to caring for sick and dying women.

🌐 **Einstein's God.** "It was, of course, a lie what you read about my religious convictions, a lie which is being systematically repeated. I do not believe in a personal God and I have never denied this but have expressed it clearly. If something is in me which can be called religious then it is the unbounded admiration for the structure of the world so far as our science can reveal it" (Albert Einstein, personal letter to an atheist, 1954).

- - - - - -

❓ **Mathematician cooks.** What eighteenth-century French mathematician cooked himself to death? (See Answer 1.25.)

🛈 **Women and math.** Despite horrible prejudice in earlier times, several women have fought against the establishment and persevered in mathematics. Emmy Amalie Noether (1882–1935) was described by Albert Einstein as "the most significant creative mathematical genius thus far produced since the higher education of women began." She is best known for her contributions to abstract algebra and, in particular, for her study of "chain conditions on ideals of rings." In 1933, her mathematical achievements counted for nothing when the Nazis caused her dismissal from the University of Göttingen because she was Jewish.

🌐 **Science and religion.** "A contemporary has said, not unjustly, that in this materialistic age of ours the serious scientific workers are the only profoundly religious people" (Albert Einstein, *New York Times Magazine*, 1930).

- - - - - -

❓ **Mathematics and money.** What effect would doubling the salary of every mathematics teacher have on education and the world at large? (See Answer 1.26.)

- - - - - -

🛈 **A famous female mathematician.** Sophie Germain (1776–1831) made major contributions to number theory, acoustics, and elasticity. At age thirteen, Sophie read an account of the death of Archimedes at the hands of a Roman soldier. She was so moved by this story that she decided to become a mathematician. Sadly, her parents felt that her interest in mathematics was inappropriate, so at night she secretly studied the works of Isaac Newton and the mathematician Leonhard Euler.

- - - - - -

❓ **Mad mom tortures mathematician daughter.** What brilliant, famous, and beautiful woman mathematician died in incredible pain because her mother withdrew all pain medication? (Hint: The woman is recognized for her contributions to computer programming. The mother wanted her daughter to die painfully so that her daughter's soul would be cleansed.) (See Answer 1.27.)

- - - - - - - - - - - - - - - - - -

🎲 **Mathematics and relationships.** "'No one really understood music unless he was a scientist,' her father had declared, and not just a scientist, either, oh, no, only the real ones, the theoreticians, whose language is mathematics. She had not understood mathematics until he had explained to her that it was the symbolic language of relationships. 'And relationships,' he had told her, 'contained the essential meaning of life'" (Pearl S. Buck, *The Goddess Abides*, 1972).

❓ **Mathematician pretends.** What important eleventh-century mathematician pretended he was insane so that he would not be put to death? (Hint: He was born in Iraq and made contributions to mathematical optics.) (See Answer 1.28.)

- - - - - -

🎲 **Mathematical greatness.** "Each generation has its few great mathematicians, and mathematics would not even notice the absence of the others. They are useful as teachers, and their research harms no one, but it is of no importance at all. A mathematician is great or he is nothing" (Alfred Adler, "Reflections: Mathematics and Creativity," *The New Yorker*, 1972).

- - - - - -

🎲 **Mathematics, mind, universe.** "If we wish to understand the nature of the Universe we have an inner hidden advantage: we are ourselves little portions of the universe and so carry the answer within us" (Jacques Boivin, *The Single Heart Field Theory*, 1981).

🎲 **Christianity and mathematics.** "The good Christian should beware of mathematicians, and all those who make empty prophesies. The danger already exists that the mathematicians have made a covenant with the devil to darken the spirit and to confine man in the bonds of Hell" (St. Augustine, *De Genesi Ad Litteram*, Book II, c. 400).

- - - - - -

❓ **Mathematician believes in angels.** What famous English mathematician had not the slightest interest in sex and was also a biblical fundamentalist, believing in the reality of angels, demons, and Satan? (Hint: According to most scholars, he is the most influential scientist and mathematician to have ever lived.) (See Answer 1.29.)

- - - - - -

❓ **History's most prolific mathematician.** Who was the most prolific mathematician in history? (If you are unable to answer this, can you guess in what century he lived?) (See Answer 1.30.)

- - - - - -

❓ **Suicidal mathematician.** What mathematician accepted a duel, knowing that he would die? (Hint: He spent the night before the duel feverishly writing down his mathematical ideas, which have since had a great impact on mathematics.) (See Answer 1.31.)

- - - - - -

❓ **Marry a mathematician?** Would you rather marry the best mathematician in the world or the best chess player? (See Answer 1.32.)

- - - - - -

🌀 **Mathematics and truth.** "We who are heirs to three recent centuries of scientific development can hardly imagine a state of mind in which many mathematical objects were regarded as symbols of spiri-tual Truth" (Philip Davis and Reuben Hersh, *The Mathematical Experience*, 1981).

- - - - - -

🌀 **Mathematics and lust.** "I tell them that if they will occupy themselves with the study of mathematics they will find in it the best remedy against the lusts of the flesh" (Thomas Mann, *The Magic Mountain*, 1924).

- - - - - -

🌀 **Newton's magic.** "Had Newton not been steeped in alchemical and other magical learning, he would never have proposed forces of attraction and repulsion between bodies as the major feature of his physical system" (John Henry, "Newton, Matter, and Magic," in John Fauvel's *Let Newton Be!*, 1988).

❗ **Earliest known symbols.** The Egyptian Rhind Papyrus (c. 1650 B.C.) contains the earliest known symbols for mathematical operations. "Plus" is denoted by a pair of legs walking toward the number to be added.

- - - - - -

🌀 **Mathematics and the divine.** "Mathematical inquiry lifts the human mind into closer proximity with the divine than is attainable through any other medium" (Hermann Weyl, quoted in Philip Davis and Reuben Hersh, *The Mathematical Experience*, 1981).

- - - - - -

❗ **π and the law.** In 1896, an Indiana physician promoted a legislative bill that made π equal to 3.2, exactly. The Indiana House of Representatives approved the bill unanimously, 67 to 0. The Senate, however, deferred debate about the bill "until a later date."

- - - - - -

🌀 **The mathematical life.** "The mathematical life of a mathematician is short. Work rarely improves after the age of twenty-five or thirty. If little has been accomplished by

❓ **Mathematical corpse.** You come home and see a corpse on your foyer floor. Would you be more frightened if (1) scrawled on the floor is the Pythagorean theorem: $a^2 + b^2 = c^2$, or (2) scrawled on the floor is the following complicated formula:

$$\frac{1}{\pi} = \frac{2\sqrt{2}}{9801} \sum_{k=0}^{\infty} \frac{(4k!)(1103 + 26390k)}{(k!)^4 396^{4k}}$$

(See Answer 1.33.)

then, little will ever be accomplished" (Alfred Adler, "Mathematics and Creativity," *The New Yorker*, 1972).

The genesis of $x^0 = 1$. Ibn Yahya al-Maghribi Al-Samawal in 1175 was the first to publish

$$x^0 = 1$$

In other words, he realized and published the idea that any number raised to the power of 0 is 1. Al-Samawal's book was titled *The Dazzling*. His father was a Jewish scholar of religion and literature from Baghdad.

Euler's one-step proof of God's existence. In mathematical books too numerous to mention, we have heard the story of the mathematician Leonhard Euler's encounter with the French encyclopedist Denis Diderot. Diderot was a devout atheist, and he challenged the religious Euler to mathematically prove the existence of God. Euler replied, "Sir $(a + b^n)/n = x$; hence, God exists. Please reply!"

Supposedly, Euler said this in a public debate in St.

Blind date. You are single and going on a blind date. The date knocks on your door and is extremely attractive. However, you note that the person has the following formulas tattooed on the right arm:

$$\frac{dR}{dt} = -k_B B(t), R(0) = R_0$$

$$\frac{dB}{dt} = -k_R R(t), B(0) = B_0$$

From this little information, do you think you would enjoy the evening with this person? (See Answer 1.34.)

Fundamental Anagram of Calculus. Probably most of you have never heard of Newton's "Fundamental Anagram of Calculus":

6accdae13eff7i3l9n4o4qrr4s8t12ux

Can you think of any possible reason Newton would want to code aspects of his calculus discoveries? (See Answer 1.35.)

Petersburg and embarrassed the freethinking Diderot with this simple algebraic proof of God's existence. Diderot was shocked and fled. Was Euler deliberately demonstrating how lame these kinds of arguments can be?

Today we know that there is little evidence that the encounter ever took place. Dirk J. Struik, in his book *A Concise History of Mathematics*, third revised edition (New York: Dover, 1967, p. 129), says that Diderot was mathematically well versed and wouldn't have been shocked by the formula. Moreover, Euler wasn't the type of person to make such a zany comment. While people of the time did seek simple mathematical proofs of God, the "Euler versus Diderot" story was probably fabricated by the English mathematician De Morgan (1806–1871).

Science and religion.
"I have always thought it curious that, while most scientists claim to eschew religion, it actually dominates their thoughts more than it does the clergy" (Fred Hoyle, astrophysicist, "The Universe: Past and Present Reflections," *Annual Review of Astronomy and Astrophysics*, 1982).

- - - - - -

Parallel universes and mathematics. In theory, it is possible to list or "enumerate" all rational numbers. How has this mathematical fact helped certain cosmologists to "prove" that there is an infinite number of universes alongside our own? (As you will learn in the next chapter, "rational numbers" are numbers like $\frac{1}{2}$, which can be expressed as fractions.) (See Answer 1.36.)

- - - - - -

The mysterious 0^0.
Students are taught that any number to the zero power is 1, and zero to any power is 0. But serious mathematicians often consider 0^0 undefined. If you try to make a graph of x^y, you'll see it has a discontinuity at the point (0,0). The

π savants. In 1844, Johann Dase (a.k.a., Zacharias Dahse) computed π to 200 decimal places in less than two months. He was said to be a calculating prodigy (or an "idiot savant"), hired for the task by the Hamburg Academy of Sciences on Gauss's recommendation. To compute

$$\pi = 3.14159\ 26535\ 89793\ 23846\ 26433\ 83279$$
$$50288\ 41971\ 69399\ 37510\ 58209\ 74944\ 59230$$
$$78164\ 06286\ 20899\ 86280\ 34825\ 34211\ 70679$$
$$82148\ 08651\ 32823\ 06647\ 09384\ 46095\ 50582$$
$$23172\ 53594\ 08128\ 48111\ 74502\ 84102\ 70193$$
$$85211\ 05559\ 64462\ 29489\ 54930\ 38196$$

Dase supposedly used $\pi/4 = \arctan(1/2) + \arctan(1/5) + \arctan(1/8)$ ... with a series expansion for each arctangent. Dase ran the arctangent job in his brain for nearly sixty days.

Not everyone believes the legend of Dase. For example, Arthur C. Clarke recently wrote to me that he simply doesn't believe the story of Dase calculating pi to 200 places in his head. Clarke says, "Even though I've seen fairly well authenticated reports of other incredible feats of mental calculation, I think this is totally beyond credibility."

I would be interested in hearing from readers who can confirm or deny this story.

discussion of the value of 0^0 is very old, and controversy raged throughout the nineteenth century.

- - - - - -

One-page proof of God's existence. Which famous German mathematician "proved" God's existence in a proof that fit on just one page of paper? (See Answer 1.37.)

- - - - - -

Greek numerals. Did you know that the ancient Greeks had two systems of numerals? The earlier of these was based on the initial letters of the names of numbers: the number 5 was indicated by the letter pi; 10 by the letter delta; 100 by the antique form of the letter H; 1,000 by the letter chi; and 10,000 by the letter mu.

Our mathematical perceptions. "The three of you stare at the school of fish and watch them move in synchrony, despite their lack of eyes. The resulting patterns are hypnotic, like the reflections from a hundred pieces of broken glass. You imagine that the senses place a filter on how much humans can perceive of the mathematical fabric of the universe. If the universe is a mathematical carpet, then all creatures are looking at it through imperfect glasses. How might humanity perfect those glasses? Through drugs, surgery, or electrical stimulation of the brain? Probably our best chance is through the use of computers" (Cliff Pickover, *The Loom of God*, 1997).

Pierre de Fermat. In the early 1600s, Pierre de Fermat, a French lawyer, made brilliant discoveries in number theory. Although he was an "amateur" mathematician, he created mathematical challenges such as "Fermat's Last Theorem," which was not solved until 1994. Fermat's Last Theorem states that $x^n + y^n = z^n$ has no nonzero integer solutions for x, y, and z when $n > 2$.

Fermat was no ordinary lawyer indeed. He is considered, along with Blaise Pascal, a founder of probability theory. As the coinventor of analytic geometry, he is considered, along with René Descartes, one of the first modern mathematicians.

Ancient number notation lets humans "think big." The earliest forms of number notation, which used straight lines for grouping 1s, were inconvenient when dealing with large numbers. By 3400 B.C. in Egypt, and 3000 B.C. in Mesopotamia, a special symbol was adopted for the number 10. The addition of this second number symbol made it possible to express the number 11 with 2 symbols instead of 11, and the number 99 with 18 symbols instead of 99.

- - - - - -

Caterpillar vehicle. Many mathematicians were creative inventors, although not all of their inventions were practical. For example, Polish-born Josef Hoëné-Wronski (1778–1853), an analyst, a philosopher, a combinatorialist, and a physicist, developed a fantastical design for caterpillar-like vehicles that he intended to replace railroad transportation. He also attempted to build a perpetual motion machine and to build a machine to predict the future (which he called the *prognometre*).

- - - - - -

The Beal reward. In the mid-1990s, the Texas banker Andrew Beal posed a perplexing mathematical problem and offered $5,000 for the solution of this problem. In particular, Beal was curious about the equation $A^x + B^y = C^z$. The six letters represent integers, with x, y, and z greater than 2. (Fermat's Last Theorem involves the special case in which the exponents x, y, and z are the same.) Oddly enough, Beal noticed that for any solution of this general equation he could find, A, B, and C have a common factor. For example, in the equation $3^6 + 18^3 = 3^8$, the numbers 3, 18, and 3 all have the factor 3. Using computers at his bank, Beal checked equations with exponents up to 100 but could not discover a solution that didn't involve a common factor.

Progress in mathematics.

"In most sciences, one generation tears down what another has built and what one has established another undoes. In mathematics alone, each generation adds a new story to the old structure" (Hermann Hankel, 1839–1873, who contributed to the theory of functions, complex numbers, and the history of mathematics, quoted in Desmond MacHale, *Comic Sections*, 1993).

Hilbert's problems.

In 1900, the mathematician David Hilbert submitted twenty-three important mathematical problems to be targeted for solution in the twentieth century. These twenty-three problems extend over all fields of mathematics. Because of Hilbert's prestige, mathematicians spent a great deal of time tackling the problems, and many of the problems have been solved. Some, however, have been solved only very recently, and still others continue to daunt us. Hilbert's twenty-three wonderful problems were designed to lead to the furthering of various disciplines in mathematics.

God's math book.

"God has a transfinite book with all the theorems and their best proofs. You don't really have to believe in God as long as you believe in the book" (Paul Erdös, quoted in Bruce Schechter, *My Brain Is Open: The Mathematical Journeys of Paul Erdös*, 1998).

Song lyrics.

What is the largest number ever used in the lyrics to a popular song? (See Answer 1.38.)

The magnificent Erdös.

It is commonly agreed that Paul Erdös is the second-most prolific mathematician of all times, being surpassed only by Leonhard Euler, the great eighteenth-century mathematician whose name is spoken with awe in mathematical circles. In addition to Erdös's roughly 1,500 published papers, many are yet to be published after his death. Erdös was still publishing a paper a week in his seventies. Erdös undoubtedly had the greatest number of coauthors (around 500) among mathematicians of all times.

In 2004, an eBay auction offered buyers an opportunity to link their names, through five degrees of separation, with Paul Erdös. In particular, the mathematician William Tozier presented bidders with the chance to collaborate on a research paper. Tozier was linked to Erdös through a string of coauthors. In particular, he had collaborated with someone who had collaborated with someone who had collaborated with someone who had collaborated with Paul Erdös. A mathematician who has published a paper with Erdös has an Erdös number of 1. A mathematician who has published a paper with someone who has published a paper with Erdös has an Erdös number of 2, and so on. Tozier has an Erdös number of 4, quite a respectable ranking in the mathematical community. This means that the person working with Tozier would have an Erdös number of 5. During the auction, Tozier heard from more than a hundred would-be researchers. (For more information, see Erica Klarreich, "Theorems for Sale: An Online Auctioneer Offers Math Amateurs a Backdoor to Prestige," *Science News* 165, no. 24 (2004): 376–77.)

Rope and lotus symbols. The Egyptian hieroglyphic system evolved special symbols (resembling ropes, lotus plants, etc.) for the numbers 10, 100, 1,000, and 10,000.

Strange math title. I love collecting math papers with strange titles. These papers are published in serious math journals. For example, in 1992, A. Granville published an article with the strange title "Zaphod Beeblebrox's Brain and the Fifty-Ninth Row of Pascal's Triangle," in the prestigious *The American Mathematical Monthly* (vol. 99, no. 4 [April]: 318–31).

Urantia religion and numbers. In the modern-day Urantia religion, numbers have an almost divine quality. According to the sect, headquartered in Chicago, we live on the 606th planet in a system called Satania, which includes 619 flawed but evolving worlds. Urantia's grand universe number is 5,342,482,337,666. Urantians believe that human minds are created at birth, but the soul does not develop until about age six. They also believe

A calculating prodigy and 365,365,365,365,365,365. When he was ten years old, the calculating prodigy Truman Henry Safford (1836–1901) of Royalton, Vermont, was once asked to square, in his head, the number 365,365,365,365,365,365. His church leader reports, "He flew around the room like a top, pulled his pantaloons over the tops of his boots, bit his hands, rolled his eyes in their sockets, sometimes smiling and talking, and then seeming to be in agony, until in not more than a minute said he, 133,491,850,208,566,925,016,658,299,941, 583,225!"

Incidentally, I had first mentioned this large number in my book *Wonders of Numbers* and had misprinted one of the digits. Bobby Jacobs, a ten-year-old math whiz from Virginia, wrote to me with the corrected version that you see here. He was the only person to have discovered my earlier typographical error.

that when we die, our souls survive. Incidentally, Jesus Christ is number 611,121 among more than 700,000 Creator Sons.

Mathematics and God. "Philosophers and great religious thinkers of the last century saw evidence of God in the symmetries and harmonies around them— in the beautiful equations of classical physics that describe such phenomena as electricity and magnetism. I don't see the simple patterns underlying nature's complexity as evidence of God. I believe *that* is God. To behold

[mathematical curves], spinning to their own music, is a wondrous, spiritual event" (Paul Rapp, in Kathleen McAuliffe, "Get Smart: Controlling Chaos," *Omni* [1989]).

The discovery of calculus. The English mathematician Isaac Newton (1642–1727) and the German mathematician Gottfried Wilhelm Leibniz (1646–1716) are generally credited with the invention of calculus, but various earlier mathematicians explored the concept of rates and limits, starting with the ancient Egyptians, who

Newton's giants. "If I have seen further than others, it is by standing upon the shoulders of giants" (Isaac Newton, personal letter to Robert Hooker, 1675 [see next quotation]).

- -

Abelson's giants. "If I have not seen as far as others, it is because giants were standing on my shoulders" (Hal Abelson, MIT professor).

- -

Mathematical marvel. "Even stranger things have happened; and perhaps the strangest of all is the marvel that mathematics should be possible to a race akin to the apes" (Eric T. Bell, *The Development of Mathematics*, 1945).

developed rules for calculating the volume of pyramids and approximating the areas of circles.

In the 1600s, both Newton and Leibniz puzzled over problems of tangents, rates of change, minima, maxima, and infinitesimals (unimaginably tiny quantities that are almost but not quite zero). Both men understood that differentiation (finding tangents to curves) and integration (finding areas under curves) are inverse processes. Newton's discovery (1665–1666) started with his interest in infinite sums; however, he was slow to publish his findings. Leibniz published his discovery of differential calculus in 1684 and of integral calculus in 1686. He said, "It is unworthy of excellent men, to lose hours like slaves in the labor of calculation. . . . My new calculus . . . offers truth by a kind of analysis and without any effort of imagination." Newton was outraged. Debates raged for many years on how to divide the credit for the discovery of calculus, and, as a result, progress in calculus was delayed.

- - - - - -

The notations of calculus. Today we use Leibniz's symbols in calculus, such as $\frac{df}{dx}$ for the derivative and the $\int$ symbol for integration. (This integral symbol was actually a long letter S for *summa*, the Latin word for "sum.") The mathematician Joseph Louis Lagrange (1736–1813) was the first person to use the notation $f'(x)$ for the first derivative and $f''(x)$ for the second derivative. In 1696, Guillaume François de L'Hôpital, a French mathematician, published the first textbook on calculus.

- - - - - -

Jews, π, the movie. The 1998 cult movie titled π stars a mathematical genius who is fascinated by numbers and their role in the cosmos, the stock market, and Jewish mysticism. According to the movie, what is God's number? (See Answer 1.39.)

- - - - - -

Mathematics and romance. What romantic comedy has the most complicated mathematics ever portrayed in a movie? (See Answer 1.40.)

- - - - - -

Greek death. Why did the ancient Greeks and other cultures believe 8 to be a symbol of death? (See Answer 1.41.)

- - - - - -

The mechanical Pascaline. One example of an early computing machine is Blaise Pascal's wheel computer called a Pascaline. In 1644, this French philosopher and mathematician built a calculating machine to help his father compute business accounts. Pascal was twenty years old at the time. The machine used a series of spinning numbered wheels to add large numbers.

The wonderful Pascaline was about the size of a shoebox. About fifty models were made.

- - - - - -

The Matrix. What number is on Agent Smith's license plate in the movie *The Matrix Reloaded*? Why? (See Answer 1.42.)

- - - - - -

At the movies. What famous book and movie title contains a number that is greater than 18,000 and less then 38,000? (See Answer 1.43.)

- - - - - -

The mathematical life. "The mathematician lives long and lives young; the wings of the soul do not early drop off, nor do its pores become clogged with the earthly particles blown from the dusty highways of vulgar life" (James Joseph Sylvester, 1814–1897, a professor of mathematics at Johns Hopkins University, 1869 address to the British Mathematical Association).

Mathematical progress. "More significant mathematical work has been done in the latter half of this century than in all previous centuries combined" (John Casti, *Five Golden Rules*, 1997).

- - - - - -

Simultaneity in science. The simultaneous discovery of calculus by Newton and Leibniz makes me wonder why so many discoveries in science were made at the same time by people working independently. For example, Charles Darwin (1809–1882) and Alfred Wallace (1823–1913) both independently developed the theory of evolution. In fact, in 1858, Darwin announced his theory in a paper presented at the same time as a paper by Wallace, a naturalist who had also developed the theory of natural selection.

As another example of simultaneity, the mathematicians János Bolyai (1802–1860) and Nikolai Lobachevsky (1793–1856) developed hyperbolic geometry independently and at the same time (both perhaps stimulated indirectly by Carl Friedrich Gauss). Most likely, such simultaneous discoveries have occurred because the time was "ripe" for such discoveries, given humanity's accumulated knowledge at the time the discoveries were made. On the other hand, mystics have suggested that there is a deeper meaning to such coincidences. The Austrian biologist Paul Kammerer (1880–1926) wrote, "We thus arrive at the image of a world-mosaic or cosmic kaleidoscope, which, in spite of constant shufflings and rearrangements, also takes care of bringing like and like together." He compared events in our world to the tops of ocean waves that seem isolated and unrelated. According to his controversial theory, we notice the tops of the waves, but beneath the surface there may be some kind of synchronistic mechanism that mysteriously connects events in our world and causes them to cluster.

First mathematician. What is the name of the first human who was identified as having made a contribution to mathematics? (See Answer 1.44.)

- - - - - -

Game show. Why was the 1950's TV game show called *The $64,000 Question*? Why not a rounder number like $50,000? (See Answer 1.45.)

- - - - - -

God and the infinite. "Such as say that things infinite are past God's knowledge may just as well leap headlong into this pit of impiety, and say that God knows not all numbers. . . . What madman would say so? . . . What are we mean wretches that dare presume to limit His knowledge?" (St. Augustine, *The City of God*, A.D. 412).

- - - - - -

Who was Pythagoras?
You can tell from some of the following ● factoids that I love trivia that relates to the famous ancient Greek mathematician Pythagoras. His ideas continue to thrive after three millennia of mathematical science. The philosopher Bertrand Russell once wrote that Pythagoras was intellectually one of the most important men who ever lived, both when he was wise and when he was unwise. Pythagoras was the most puzzling mathematician of history because he founded a numerical religion whose main tenets were the transmigration of souls and the sinfulness of eating beans, along with a host of other odd rules and regulations. To the Pythagoreans, mathematics was an ecstatic revelation.

The Pythagoreans, like modern day fractalists, were akin to musicians. They created pattern and beauty as they discovered mathematical truths. Mathematical and theological blending began with Pythagoras and eventually affected all religious philosophy in Greece, played a role in religion of the Middle Ages, and extended to Kant in modern times. Bertrand Russell felt that if it were not for Pythagoras, theologians would not have sought logical proofs of God and immortality.

If you want to read more about Pythagoras, see my book *The Loom of God* and Peter Gorman's *Pythagoras: A Life*.

Mathematical scope. "Mathematics is not a book confined within a cover and bound between brazen clasps, whose contents it needs only patience to ransack; it is not a mine, whose treasures may take long to reduce into possession, but which fill only a limited number of veins and lodes; it is not a soil, whose fertility can be exhausted by the yield of successive harvests; it is not a continent or an ocean, whose area can be mapped out and its contour defined: it is limitless as that space which it finds too narrow for its aspirations; its possibilities are as infinite as the worlds which are forever crowding in and multiplying upon the astronomer's gaze; it is as incapable of being restricted within assigned boundaries or being reduced to definitions of permanent validity, as the consciousness, the life, which seems to slumber in each monad, in every atom of matter, in each leaf and bud and cell, and is forever ready to burst forth into new forms of vegetable and animal existence" (James Joseph Sylvester, *The Collected Mathematical Papers of James Joseph Sylvester,* Volume III, address on Commemoration Day at Johns Hopkins University, February 22, 1877).

The secret life of numbers. To Pythagoras and his followers, numbers were like gods, pure and free from material change. The worship of numbers 1 through 10 was a kind of polytheism for the Pythagoreans.

Pythagoreans believed that numbers were alive, independent of humans, but with a telepathic form of consciousness. Humans could relinquish their three-dimensional lives and telepathize with these number beings by using various forms of meditation. Meditation upon numbers was communing with the gods, gods who desired nothing from humans but their sincere admiration and contemplation. Meditation upon numbers was a form of prayer that did not ask any favors from the gods.

These kinds of thoughts are not foreign to modern mathematicians, who often debate whether mathematics is a creation of the human mind or is out there in the universe, independent of human thought. Opinions vary. A few mathematicians believe that mathematics is a form of human logic that is not necessarily valid in all parts of the universe.

Anamnesis and the number 216. Was the ancient Greek mathematician Pythagoras once a plant? This is a seemingly bizarre question, but Pythagoras claimed that he had been both a plant and an animal in his past lives, and, like Saint Francis, he preached to animals. Pythagoras and his followers believed in *anamnesis*, the recollection of one's previous incarnations. During Pythagoras's time, most philosophers believed that only men could be happy. Pythagoras, on the other hand, believed in the happiness of plants, animals, and women.

In various ancient Greek writings, we are told the exact number of years between each of Pythagoras's incarnations: 216. Interestingly, Pythagoreans considered 216 to be a mystical number, because it is 6 cubed ($6 \times 6 \times 6$). Six was also considered a "circular number" because its powers always ended in 6. The fetus was considered to have been formed after 216 days.

The number 216 continues to pop up in the most unlikely of places in theological literature. In an obscure passage from *The Republic* (viii, 546 B–D), Plato notes that $216 = 6^3$. It is also associated with auspicious signs on the Buddha's footprint.

Shades of ghosts. Pythagoras believed that even rocks possess a psychic existence. Mountains rose from the earth because of growing pains of the earth, and Pythagoras told his followers that earthquakes were caused by the shades of ghosts of the dead, which created disturbances beneath the earth.

- - - - - -

Pythagorean sacrifice. Although some historians report that Pythagoras joyfully sacrificed a hecatomb of oxen (a hundred animals) when he discovered his famous theorem about the right-angled triangle, this would have been scandalously un-Pythagorean and is probably not true. Pythagoras refused to sacrifice animals. Instead, the Pythagoreans believed in the theurgic construction of agalmata—statues of gods consisting of herbs, incense, and metals to attract the cosmic forces.

- - - - - -

Pythagoras and aliens. UFOs and extraterrestrial life are hot topics today. But who would believe that these same ideas enthralled the Pythagoreans a few millennia ago? In fact, Pythagoreans believed that all the planets in the solar system are inhabited, and humans dwelling on Earth were less advanced than these other inhabitants were. (The idea of advanced extraterrestrial neighbors curiously has continued, for some, to this day.) According to later Pythagoreans, as one travels farther from Earth, the beings on other planets and in other solar systems become less flawed.

Pythagoras went as far as to suggest that disembodied intelligences existed in the universe. These mind-creatures had very tenuous physical bodies. Many of Pythagoras's followers believed that Pythagoras himself had once been a superior being who inhabited the Moon or the Sun.

- - - - - -

Gods and sets. "The null set is also a set; the absence of a god is also a god" (A. Moreira).

Famous epitaphs. Replace x and y with the names of two famous mathematicians in these two puzzles. In Puzzle 1, the following couplet of Alexander Pope is the engraved epitaph on the mystery person's sarcophagus in Westminster Abbey, in London:

Nature and Nature's laws lay hid in the night;
God said, "Let x be" and all was light.
—Alexander Pope (1688–1744)

In Puzzle 2, the epitaph "$S = k \ln W$" is engraved on y's tombstone. Who are x and y? (See Answer 1.46.)

Matrix prayers. Some of you will be interested in Underwood Dudley's *Mathematical Cranks* (Washington, D.C.: Mathematical Association of America, 1992). The book contains musings of slightly mad mathematicians. My favorite chapter is on the topic of "Matrix Prayers," designed by a priest of the Church of England. The priest regularly prayed to God in mathematical terms using matrices, and he taught the children in his church to pray and think of God in matrices. *Mathematical Cranks* goes into great detail regarding "revelation matrices," "Polite Request Operators," and the like. The priest finally derives a beautiful prayer that he succinctly writes as

$$P < R\{S\} \rightarrow \{U\}, r > 0$$

He says, "This prayer should be sufficiently concise to be acceptable to Christ, yet every single Christian inhabitant of Northern Ireland has been separately included." The chapter concludes with information regarding the geometry of heaven.

- - - - - -

Secret mathematician. Who really was the famous and secretive French mathematician "N. Bourbaki"? (See Answer 1.47.)

- - - - - -

How much? How much mathematics can we know? (See Answer 1.48.)

Mathematics and reality. From string theory to quantum theory, the deeper one goes in the study of physics, the closer one gets to pure mathematics. Mathematics is the fabric of reality. Some might even say that mathematics "runs" reality in the same way that Microsoft's Windows runs your computer and shapes your interactions with the vast network beyond. Schrödinger's wave equation—which describes basic reality and events in terms of wave functions and probabilities—is the evanescent substrate on which we all exist:

$$\left[-\frac{\hbar^2}{2m} \nabla^2 + V(\vec{r}) \right] \psi(\vec{r}, t) = i\hbar \frac{\partial \psi}{\partial t}(\vec{r}, t)$$

Freeman Dyson, in the introduction to *Nature's Imagination*, speaks highly of this formula: "Sometimes the understanding of a whole field of science is suddenly advanced by the discovery of a single basic equation. Thus it happened that the Schrödinger equation in 1926 and the Dirac equation in 1927 brought a miraculous order into the previously mysterious processes of atomic physics. Bewildering complexities of chemistry and physics were reduced to two lines of algebraic symbols."

Factorial symbol. In 1808 Christian Kramp (1760–1826) introduced the "!" as the factorial symbol as a convenience to the printer.

- - - - -

Nobel Prize. Why is there no Nobel Prize for mathematics? (See Answer 1.49.)

- - - - -

Roman numerals. Why don't we use Roman numerals anymore? (See Answer 1.50.)

Insights and analysis. "Einstein's fundamental insights of space/matter relations came out of philosophical musings about the nature of the universe, not from rational analysis of observational data—the logical analysis, prediction, and testing coming only after the formation of the creative hypotheses" (R. H. Davis, *The Skeptical Inquirer*, 1995).

- - - - -

Einstein on mathematics and reality. "At this point an enigma presents itself which in all ages has agitated inquiring minds. How can it be that mathematics, being after all a product of human thought which is independent of experience, is so admirably appropriate to the objects of reality? Is human reason, then, without experience, merely by taking thought, able to fathom the properties of real things? In my opinion the answer to this question is briefly this: As far as the laws of mathematics refer to reality, they are not certain; and as far as they are certain, they do not refer to reality" (Albert Einstein's address to the Prussian Academy of Science in Berlin, 1921).

- - - - -

Ramanujan's tongue. In the spring of 2003, the Aurora Theater Company of Berkeley performed Ira Hauptman's play *Partition*, which focuses on the collaboration of the mathematicians Ramanujan and Hardy. In the play, Namagiri, Ramanujan's personal deity and inspiration in real life, is seen literally writing equations on Ramanujan's tongue with her finger.

⑤ **Mathematics and reality.** "My complete answer to the late 19th century question 'what is electrodynamics trying to tell us' would simply be this: *Fields in empty space have physical reality; the medium that supports them does not*. Having thus removed the mystery from electrodynamics, let me immediately do the same for quantum mechanics: *Correlations have physical reality; that which they correlate does not*" (N. David Mermin, "What Is Quantum Mechanics Trying to Tell Us?" *American Journal of Physics* 66, [1998]: 753–67).

- - - - - -

⑤ **More mathematics and reality.** "[Much of frontier mathematics] confounds even mathematicians and physicists, as they use math to calculate the inconceivable, undetectable, nonexistent and impossible. So what does it mean when mainstream explanations of our physical reality are based on stuff that even scientists cannot comprehend? When nonscientists read about the strings and branes of the latest physics theories, or the Riemann surfaces and Galois fields of higher mathematics, how close are we to a real understanding?" (Susan Kruglinski, "When Even Mathematicians Don't Understand the Math," *New York Times*, May 25, 2004).

- - - - - -

⑤ **Dyson on the infinite reservoir of mathematics.** "Gödel proved that the world of pure mathematics is inexhaustible; no finite set of axioms and rules of inference can ever encompass the whole of mathematics; given any finite set of axioms, we can find meaningful mathematical questions which the axioms leave unanswered. I hope that an analogous situation exists in the physical world. If my view of the future is correct, it means that the world of physics and astronomy is also inexhaustible; no matter how far we go into the future, there will always be new things happening, new information coming in, new worlds to explore, a constantly expanding domain of life, consciousness, and memory" (Freeman Dyson, "Time without End: Physics and Biology in an Open Universe," *Reviews of Modern Physics*, 1979).

- - - - - -

⑤ **More mathematics and reality.** "It is difficult to explain what math is, let alone what it says. Math may be seen as the vigorous structure supporting the physical world or as a human idea in development. [Dr. John Casti says that] 'the criteria that mathematicians use for what constitutes good versus bad mathematics is much more close to that of a poet or a sculptor or a musician than it is to a chemist'" (Susan Kruglinski, "When Even Mathematicians Don't Understand the Math," *New York Times*, May 25, 2004).

⑤ **Creativity and travel.** "One way of goosing the brain is traveling, particularly internationally. It helps shake up perspective and offers new experiences. Interviews with 40 MacArthur 'genius' award winners found 10 lived overseas permanently or temporarily, three traveled at least a few months each year, and at least two have a 'horror of a home'" (Sharon McDonnell, "Innovation Electrified," *American Way*).

🌀 **Ramanujan redux.** How would mathematics have been advanced if Ramanujan had developed in a more nurturing early environment? Although he would have been a better-trained mathematician, would he have become such a unique thinker? Could he have discovered so many wonderful formulas if he had been taught the rules of mathematics early on and pushed to publish his results with rigorous proofs? Perhaps his relative isolation and poverty enhanced the greatness of his mathematical thought. For Ramanujan, equations were not just the means for proofs or calculations. The *beauty* of the equation was of paramount value for Ramanujan.

— — — — — —

🌀 **The secret life of formulas.** "We cannot help but think that mathematical formulae have a life of their own, that they know more than their discoverers do and that they return more to us than we have invested in them" (Heinrich Hertz, German physicist, quoted in Eric Bell, *Men of Mathematics*, 1937).

— — — — — —

🌀 *The Fractal Murders.* In 2002, Mark Cohen, a lawyer and a judge, published *The Fractal Murders*, a novel in which three mathematicians, all of whom are experts in fractals, have died. Two were murdered. The third was an apparent suicide. In the novel, the math professor Jayne Smyers hires a private eye, Pepper Keane, to look into the three deaths, which seem to be related only because each victim was researching fractals.

❓ **Mathematical universe.** Why does the universe seem to operate according to mathematical laws? (See Answer 1.51.)

— — — — — —

❓ **A universe of blind mathematicians.** Sighted mathematicians generally work by studying vast assemblages of numbers and symbols scribbled on paper. It would seem extremely difficult to do mathematics without being able to see, and to be forced to keep the information "all in one's head." Can a great mathematician be totally blind? (See Answer 1.52.)

— — — — — —

🌀 **Einstein on comprehensibility and reality.** "The very fact that the totality of our sense experiences is such that by means of thinking . . . it can be put in order, this fact is one which leaves us in awe, but which we shall never understand. One may say 'the eternal mystery of the world is its comprehensibility.' It is one of the great realizations of Immanuel Kant that this setting up of a real external world would be senseless without this comprehensibility" (Albert Einstein, "Physics and Reality," 1936).

— — — — — —

🌀 **Wigner on mathematics and reality.** "The miracle of appropriateness of the language of mathematics for the formulation of the laws of physics is a wonderful gift which we neither understand nor deserve. We should be grateful for it, and hope that it will remain valid for future research, and that it will extend, for better or for worse, to our pleasure even though perhaps also to our bafflement, to wide branches of learning" (Eugene Wigner, "The Unreasonable Effectiveness of Mathematics," 1960).

❓ Close to reality. We've talked quite a bit about mathematics and reality. Who do you think is in more direct contact with reality, a mathematician or a physicist? What do famous twentieth-century mathematicians say on this subject? (See Answer 1.53.)

- - - - - -

◔ All reality is mathematics. "The Gedemondan chuckles. 'We read probabilities. You see, we see—perceive is a better word—the math of the Well of Souls. We feel the energy flow, the ties and bands, in each and every particle of matter and energy. All reality is mathematics, all existence—past, present, and future—is equations'" (Jack Chalker, *Quest for the Well of Souls*, 1985).

- - - - -

◔ Pythagoras on mathematics and reality. "All things are numbers."

❓ Why learn mathematics? It has been estimated that much greater than 99.99 percent of all Americans will never use the quadratic formula or most of the other algebraic or geometrical relations they learn in school. Why teach or learn mathematics beyond the basic operations of addition, subtraction, multiplication, and division? (See Answer 1.54.)

Cool Numbers

I N WHICH WE ENCOUNTER FASCINATING NUMBERS AND STRANGE NUMBER
sequences. We'll explore transcendental numbers, octonions, surreal numbers, obstinate numbers, cyclic numbers, Vibonacci numbers, perfect numbers,
automorphic numbers, prime numbers, Wilson primes, palindromic primes,
Fibonacci primes, Sophie Germain primes, Baxter-Hickerson primes, star-
congruent primes, narcissistic numbers, amenable numbers, amicable
numbers, p-adic numbers, large palindromes, factorions, hyperfactorials,
primorials, palindions and hyperpalindions, exotic-looking formulas for π,
the Golay-Rudin-Shapiro sequence, the wonderful Pochhammer notation, and
famous and curious math constants (such as Liouville's constant, the
Copeland-Erdös constant, Brun's
constant, Champernowne's number,
Euler's gamma, Chaitin's constant,
the Landau-Ramanujan constant,
Mills's constant, the golden ratio,
Apéry's constant, and constants even
more bizarre).

Numbers percolate like bubbles in the
ocean of mathematics.
The mathematician's job is to transport
us to new seas, while deepening the
waters and lengthening horizons.

? Pi. Who discovered pi (π)? (See Answer 2.1.)

? An avalanche of digits. How do obsessed mathematicians calculate π to trillions of decimal digits? (See Answer 2.2.)

? Evenness. Is 0 an even number? (See Answer 2.3.)

? Billion. In America, a billion has 9 zeros (1,000,000,000). In England, a billion has 12 zeros (1,000,000,000,000). Why? (See Answer 2.4.)

? Pick an integer, any integer. If I asked you to select an integer number at random from all the integers, is this a possible task? (Integers are the numbers we're mostly familiar with and consist of the natural numbers [0, 1, 2, . . .] and their negatives [–1, –2, –3, . . .]). (See Answer 2.5.)

Definition of a transfinite number. We briefly mentioned transfinite numbers in chapter 1 when discussing the German mathematician Georg Cantor. A transfinite number is an infinite cardinal or ordinal number. (A cardinal number is a whole number, an integer, that is used to specify how many elements there are in a set. An ordinal number is considered a place in the ordered sequence of whole numbers. For example, it is used in counting as first, second, third, fourth, etc., to nth in a set of n elements.) The smallest transfinite number is called "aleph-nought," written as $\aleph_0$, which counts the number of integers. $\aleph$ is the first letter in the Hebrew alphabet. If the number of integers is infinite (with $\aleph_0$ members), are there yet higher levels of infinity? It turns out that even though there are an infinite number of integers, *rational numbers* (numbers that can be

Street-corner integers. David Chalmers, a professor of philosophy and the director of the Center for Consciousness Studies (www.u.arizona.edu/~chalmers/), once conducted an experiment in which he stood on a busy street corner in Oxford and asked passersby to "name a random number between zero and infinity." He wondered what this "random" distribution would look like.

The results, sorted in order of most popular, were 3, 7, 5, 12, 1, 4, 10, 77, 2, 47, infinity, 15, 17, 20, 27, 18, 23, 26, 30, 42, and 99. The remaining random numbers were given by only one person each: 6, 13, 14, 19, 21, 22, 25, thirteen more 2-digit numbers, twenty 3-digit numbers, twelve 4-digit numbers, one 5-digit number, one 6-digit number, four 7-digit numbers, one 8-digit number, one non-integer (328.39), and one huge number:

$$9.265 \times 10^{10^{10}}$$

Dr. Chalmers notes that "Of course, a uniform distribution is a priori impossible so I couldn't have expected that. Even a logarithmic distribution is impossible (it has an infinite integral). Interestingly enough, this distribution, taken coarsely, was quite close to logarithmic up to 1,000 or so. There were roughly the same number of 2-digit responses as 1-digit responses, and a few less 3-digit responses."

expressed as fractions), and *irrational numbers* (like $\sqrt{2} = 1.141421356\ldots$, which cannot be expressed as a fraction), the infinite number of irrationals is in some sense greater than the infinite number of rationals and integers. Similarly, there are more *real* numbers (which include rational and irrational numbers) than there are integers. To denote this difference, mathematicians refer to the infinity of rationals or integers as $\aleph_0$ and the infinite number of irrationals or real numbers as C, which stands for the cardinality of the real number "continuum." There is a simple relationship between C and $\aleph_0$. It is $C = 2^{\aleph_0}$. In other words, $2^{\aleph_0}$ is denoted by C and is also the cardinality of the set of real numbers, or *the continuum*, whence the name. (The real numbers are sometimes called *the continuum*.) Some of these cardinalities are shown in figure 2.1. This figure is complicated and will be referred to throughout the chapter, as you build your knowledge and see passing references to other kinds of numbers.

Mathematicians also think about greater infinities, symbolized by $\aleph_1$, $\aleph_2$, and so on. For example, the set theory symbol $\aleph_1$ stands for the smallest infinite set larger than $\aleph_0$. The "continuum hypothesis" states that $C = \aleph_1 = 2^{\aleph_0}$; however, the question of whether C truly equals $\aleph_1$ is considered undecidable in our present set theory. In other words, great mathematicians such as Kurt Gödel proved that the hypothesis was a consistent assumption in one branch of mathematics. However, another mathematician, Paul Cohen, proved that it was also consistent to assume that the continuum hypothesis is false!

Interestingly, the number of rational numbers is the same as the number of integers. The

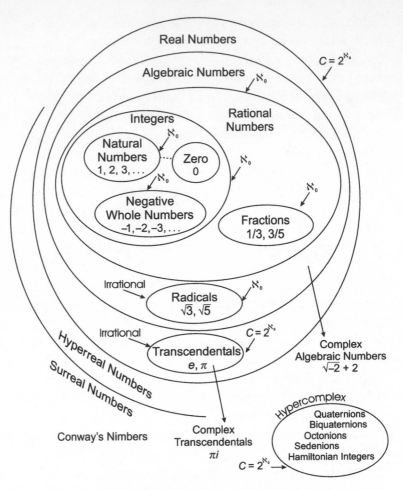

Figure 2.1   The universe of numbers.

number of irrationals is the same as the number of real numbers. (Mathematicians usually use the term *cardinality* when talking about the "number" of infinite numbers. For example, true mathematicians would say that the cardinality of the irrationals is known as the continuum.)

What do we do with the paradox of the continuum hypothesis? Cantor's colleague, Constantin Gutberlet, believed that God could resolve the problem of the continuum hypothesis. How many of the great mathematical paradoxes would melt away if humanity had a higher level of intelligence? How many would remain because they are somehow part of the mathematical tapestry underpinning our universe? These are questions not easily answered, at least not by *Homo sapiens*. Our minds have not sufficiently evolved to comprehend all the mysteries of "God" and mathematics.

- - - - - -

Definition of a Fibonacci number. Fibonacci numbers— 1, 1, 2, 3, 5, 8, 13 . . . —are named after the Italian merchant Leonardo Fibonacci of Pisa (c. 1200). Notice that except for the first two numbers, every successive number in the sequence equals the sum of the two previous. These numbers appear in an amazing number of places in various mathematical disciplines, and I've scattered just a few throughout this book.

- - - - - -

Definition of a prime number. A number larger than 1, such as 5 or 13, that is divisible only by itself or by 1. The number 14 is not prime because $14 = 7 \times 2$. Primeness is a property of the number itself. For example, 5 is prime whether it is written as 5 or as the binary form 101 or in any other system of numeration. The largest known prime as of 2005 has 7,816,230 digits.

- - - - - -

Definition of number theory. Number theory—the study of properties of the integers—is an ancient discipline. Much mysticism accompanied early treatises; for example, Pythagoreans based all events in the universe on whole numbers. Only a few hundred years ago, courses in numerology—the study of the mystical and religious properties of numbers—were required by all college students, and even today such numbers as 13, 7, and 666 conjure up emotional reactions in many people.

Today, integer arithmetic is important in a wide spectrum of human activities and has repeatedly played a crucial role in the evolution of the natural sciences. For a description of the use of number theory in communications, computer science, cryptography, physics, biology, and art, see Manfred Schroeder's *Number Theory in Science and Communication*. Primes and prime factorization are especially important in number theory.

Fibonacci and 998999. Here's a mathematical curiosity. The fraction 1/998999 contains a number of obvious instances of Fibonacci numbers, 1, 1, 2, 3, 5, 8 . . . , in which each successive number is the sum of the previous two. I've underlined the Fibonacci numbers to make them easy to find:

$1/998999 = 0.000001001002003005008013021034055089 \ldots$

Integer God. "Is God a mathematician? Certainly, the world, the universe, and nature can be reliably understood using mathematics. Nature *is* mathematics. The arrangement of seeds in a sunflower can be understood using Fibonacci numbers. Sunflower heads, like other flowers, contain two families of interlaced spirals—one winding clockwise, the other counter clockwise. The numbers of seeds and petals are almost always Fibonacci numbers" (Clifford Pickover, *The Loom of God*, 1997).

Gears and prime numbers. In gears whose purpose is to reduce speed, it is often helpful to use wheels whose numbers of teeth are prime numbers, in order to reduce uneven wear. On such gears, the same gear teeth will mesh only at long intervals. In old eggbeater hand drills, prime numbers of teeth were also used so that individual pairs of teeth in the mating gears did not revisit each other as often as would be the case for composite numbers.

Numbers and number representations. In our daily lives today, most of us use a number system based on powers of 10, so that 1492, for example, means $1 \times 10^3 + 4 \times 10^2 + 9 \times 10^1 + 2 \times 10^0$, noting that $10^0 = 1$. The popularity of a "base-10" system is due, in part, to the fact that humans are accustomed to counting on their ten fingers. If we had six fingers on each hand, we might use a base-12 number system, in which the value 147 (base 10) would be written 103 (base 12) because 103 (base 12) $= 1 \times 12^2 + 0 \times 12^1 + 3$. Humans have not always preferred a base-10 system. For example, the Yuki Pomo of Northern California used a base-8 system because they counted using the spaces between the fingers. The Sumerians used a base-60 system, and this is why we have 60 seconds in a minute and 60 minutes in an hour.

Some properties of numbers are independent of the number system in which they are represented. For example, 8, a cube number, is adjacent to 9, a square number, regardless of how we represent the values 8 and 9. As just discussed, the value 5 is a prime, and 6 is a composite, regardless of the representation. However, other properties of numbers depend on the number system that is used in their representation. For example, "a number is divisible by 9 if and only if the sum of its digits is divisible by 9" is true in base 10 but not necessarily in other number systems. The number 353,535 is an "undulating" or "oscillating" number when written in base 10. In base 12, it's a very ordinary-looking 150,713. You can tell from this book that I'm crazy about all kinds of number properties. Therefore, in this chapter, you'll see questions and facts about numbers that are independent of their representation, as well as observations that assume a base-10 system. Some of my colleagues, like Dr. Bob Ewell, care only about properties of numbers independent of their representation. I like observations that involve all kinds of properties. As you read, see if you can determine which kind of property is being discussed (number system independent or number system dependent) and decide which flavor you like the best.

Why care about integers? The brilliant mathematician Paul Erdös was fascinated by number theory and the notion that he could pose problems using integers that were often simple to state but notoriously difficult to solve. Erdös believed that if one can state a problem in mathematics that is unsolved and more than 100 years old, it is a problem in number theory. There is a harmony in the universe that can be expressed by whole numbers. Numerical patterns describe the arrangement of florets in a daisy, the reproduction of rabbits, the orbit of the planets, the harmonies of music, and the relationships between elements in the periodic table.

Leopold Kronecker (1823–1891), a German, was an algebraist and a number theorist who once said, "The integers came from God and all else was man-made." His implication was that the primary source of all mathematics is the integers. Since the time of Pythagoras, the role of integer ratios in musical scales has been widely appreciated. More important, integers have been crucial in the evolution of humanity's scientific understanding. For example, the English chemist John Dalton (1766–1844) discovered that chemical compounds are composed of fixed proportions of elements that correspond to the ratios of small integers. This was very strong evidence for the existence of atoms. In 1925, certain integer relations between the wavelengths of spectral lines emitted by excited atoms gave early clues to the structure of atoms. The near-integer ratios of atomic weights were evidence that the atomic nucleus is made up of an integer number of similar nucleons (protons and neutrons). The deviations from integer ratios led to the discovery of elemental isotopes (variants with nearly identical chemical behavior but with different radioactive properties). Small divergences in pure isotopes' atomic weights from exact integers confirmed Einstein's famous equation $E = mc^2$ and also the possibility of atomic bombs. Integers are everywhere in atomic physics.

Integer relations are fundamental strands in the mathematical weave—or, as the German mathematician Carl Friedrich Gauss said, "Mathematics is the queen of sciences—and number theory is the queen of mathematics."

Prime proof. The first person known to have proved that there are an infinite number of primes was Euclid (third century B.C.).

– – – – –

Definition of a composite number. A composite number is a positive integer, greater than 1, that is not prime. In other words, a composite number can be written as a product of two or more integers, each larger than 1. For example, 42 is composite because $42 = 7 \times 2 \times 3$. The first few composite numbers are 4, 6, 8, 9, 10, 12, 14, 15, 16, 18, . . .

– – – – –

Definition of set theory. Set theory is a branch of mathematics that involves sets and membership. A set may be considered any collection of objects, called the *members* (or elements) of the set. One mathematical example is the set of positive integers {1, 2, 3, 4, . . .}. There are a number of different versions of set theory, each with its rules and axioms.

– – – – –

Alien set theory. "Would intelligent beings evolving on a planet or environment less contiguous than our own ever come up with set theory? If beings exist in amorphous globules in a dynamic flowing ecosystem without the benefit of solid matter, and never witness two identical objects or actions, would such creatures be familiar with the number 1?" (Todd Redden, personal communication).

– – – – – –

Drowning in pi digits. "The digits of pi beyond the first few decimal places are of no practical or scientific value. Four decimal places are sufficient for the design of the finest engines; ten decimal places are sufficient to obtain the circumference of the earth within a fraction of an inch if the earth were a smooth sphere" (Petr Beckmann, *A History of Pi*, 1976).

– – – – – –

The loom of God. "The shape assumed by a delicate spider web suspended from fixed points, or the cross-section of sails bellying in the wind, is a catenary—a simple curve defined by a simple formula. Seashells, animal's horns, and the cochlea of the ear are logarithmic spirals which can be generated using a mathematical constant known as the golden ratio. Mountains and the branching patterns of blood vessels and plants are fractals, a class of shapes which exhibit similar structures at different magnifications. Einstein's $E = mc^2$ defines the fundamental relationship between energy and matter. And a few simple constants—the gravitational constant, Planck's constant, and the speed of light—control the destiny of the universe. I do not know if

Definition of Euler's number e. The constant e is the base of the natural logarithm. It is approximately equal to

$$e = 2.71828\ 18284\ 59045\ 23536\ 02874\ldots$$

Along with π, e is the most important constant in mathematics since it appears in countless mathematical contexts. Roughly 2 billion digits of e have been determined.

Euler's number can be defined as follows:

$$e = \sum_{k=0}^{\infty} \frac{1}{k!}$$

In other words, the number e can be defined as the sum of a series in which the series terms are the reciprocals of the factorial numbers: $e = 1/0! + 1/1! + 1/2! + \ldots = 2.7182818284\ 590\ldots$ (Recall that for a positive integer n, $n!$ is the product of all the positive integers less than or equal to n. $0!$ is equal to 1.) Here's another way to look at it. Euler's number, e, is the limit value of the expression $(1 + 1/n)$ raised to the nth power, when n increases indefinitely:

$$e = \lim_{n \to \infty} \left(1 + \frac{1}{n}\right)^n = 2.71828\ldots$$

The symbol e was first used by the Swiss mathematician Leonhard Euler (1707–1783). The symbol e also appeared in his 1736 *Mechanica*, perhaps inspired by the word *exponential*.

📖 Definition of a rational number. A rational number can be expressed as a ratio of two integers, where the denominator is non-zero. Example: $\frac{1}{3}$, $\frac{1}{2}$, 4. If you multiply, add, subtract, or divide a rational by a rational, another rational number is produced (see figure 2.1).

- - - - - - - - - - - - - - -

📖 Definition of complex numbers. Complex numbers are an extension of the real numbers; complex numbers contain a number i, called the *imaginary unit*, with $i = \sqrt{-1}$. Every complex number can be represented in the form $x + iy$, where x and y are real numbers.

God is a mathematician, but mathematics is the loom upon which God weaves the fabric of the universe" (Clifford Pickover, *The Loom of God*, 1997).

- - - - - -

❗ A cool pi formula. The following little-known formula yields the correct decimal digits of pi to an amazing 42 billion digits; however, it is not a perfect formula for pi! In other words, it's only a very high-precision approximation. The $\approx$ symbol signifies an approximation (Borwein and Borwein, "Strange Series and High Precision Fraud," 1992).

$$\left(\frac{1}{10^5} \sum_{n=-\infty}^{n=\infty} e^{-\left(n^2 / 10^{10} \right)} \right)^2 \approx \pi$$

📖 Definition of transcendental numbers. These numbers are so exotic that they were only "discovered" 150 years ago. They're so rare in common usage that you may be familiar with only one of them, π. These numbers cannot be expressed as the root of any algebraic equation with rational coefficients. This means that π could not exactly satisfy equations of the type: $x^2 = 10$ or $9x^4 - 240x^2 + 1{,}492 = 0$. These are equations involving simple integers with powers of π. The numbers π and e (Euler's number) can be expressed as an endless continued fraction or as the limit of the sum of an infinite series.

Proving that a number is transcendental is no easy task. Charles Hermite proved that e was transcendental in 1873 and Ferdinand von Lindemann proved that π was transcendental in 1882. In 1874, Cantor surprised most mathematicians by demonstrating that almost all real numbers are transcendental. Thus, if you could somehow put all the numbers in a big jar and pull one out, it would be virtually certain to be transcendental.

There are many more transcendental numbers than algebraic ones. In fact, if we could imagine the number line with just the algebraic numbers represented and the transcendental ones chopped out, such a line would have more holes than points. Yet despite the fact that transcendental numbers are "everywhere," only a few are known and named. There are lots of stars in the sky, but how many can you name? (See figure 2.1.)

All transcendental numbers are irrational numbers. However, a transcendental number (unlike $\sqrt{2}$) is not the root of any polynomial equation with integer coefficients.

- - - - - -

An irrational number cannot be expressed as a ratio of two integers. Examples: e, π, and $\sqrt{2}$. The irrational numbers are either *algebraic* irrational numbers like $\sqrt{2}$ and other radicals or surds, which are the roots of polynomial equations with rational coefficients (e.g., $\sqrt{2}$ is the root of $a^2 - 2 = 0$), or *transcendental* numbers like e and π. The decimal digits of irrational numbers go on and on, with no period or repetition. Sometimes mathematicians look for patterns in the endless string of digits to determine whether the arrangement of digits is similar to what would be expected of a completely random sequence. Writing an irrational number in decimal form, like $\sqrt{2} = 1.414213\ldots$, produces an endless sequence of decimal digits, in particular, a nonperiodic nonterminating decimal number (see figure 2.1).

- - - - - -

🌀 **Number zoo.** Figure 2.1 is a schematic illustration (or a visual definition) of various kinds of numbers. Natural numbers are often called "whole numbers" or "positive integers," for example, 1, 2, 3, . . . Sometimes, 0 is also included in the list of "natural" numbers (e.g., in books such as Bourbaki's *Elements of Mathematics: Theory of Sets* or Halmos's *Naive Set Theory*), and there seems to be no general agreement about whether to include zero.

In figure 2.1, I have focused on scalar numbers at the expense of complex numbers. Here I refer to "scalar" as a quantity that can be described by a single number. Scalar quantities have magnitude but not a direction. A complex number has the form $x + iy$, where i is $\sqrt{-1}$. If y is zero, the complex number is real. If x is zero, the complex number is imaginary.

Algebraic numbers and transcendental numbers can be complex, as hinted at by the downward arrows piercing the real number bubble and pointing to numbers like πi and $\sqrt{-2} + 1$. Of course, if I wanted to fully include complex numbers, I might also include hypercomplex numbers—for example, higher-dimensional complex numbers like the quaternions of the form $w + xi + yj + zk$, which have more than one imaginary component. A quaternion with $y = 0$ and $z = 0$ is a simple complex number. If x is also 0, the quaternion becomes a scalar, that is, a vector with one element. Would intelligent aliens on other worlds develop a similar scheme?

Several features of figure 2.1 are not perfectly accurate, due to the difficulty of making bubble diagrams of this sort. For example, my floating of the complex numbers outside of the algebraic numbers is a simplification in order to make the diagram easier to draw. For instance, $\sqrt{-2} + 2$ is algebraic and contains real and imaginary parts. Radicals are algebraic; thus there are real radicals, imaginary radicals, and complex radicals that are neither real nor imaginary but a mixture of both.

🌀 **Definition of algebraic numbers.** Most of the ordinary positive and negative numbers you know about are algebraic numbers. More specifically, these numbers are real or complex numbers that are solutions of polynomial expressions having rational coefficients. Here's an example. Consider the polynomial $ax^3 + bx^2 + cx + d$. If r is the root or the solution $ax^3 + bx^2 + cx + d = 0$, then r is an algebraic number. As we said in the definition of irrational numbers, the irrational numbers are considered to be of two kinds, the *algebraic* irrational numbers and *transcendental* numbers (see figure 2.1).

- - - - - -

🔮 **Superthin dart.** I randomly toss a superthin dart at a number line that represents all numbers from 0 to 1. Given what you have already learned in this chapter, is the dart more likely to land on an algebraic number, such as 0.6, or a transcendental one, such as π? If you feel that transcendentals are more likely, how much more likely? (See Answer 2.6.)

- - - - - -

📖 **Definition of hypercomplex numbers.** Hypercomplex numbers are higher-dimensional extensions of the complex numbers; they include such numbers as quaternions, octonions, and sedenions. Complex numbers, such as $2 + 3i$, can be viewed as points in a plane. Similarly, hypercomplex numbers can be viewed as points in some higher-dimensional Euclidean space (4 dimensions for the quater-

❗ **Transcendental mystery.** We know that π, Euler's number e, and e^π are transcendental. However, today, we don't know whether π^e, π^π, or e^e are transcendental. Perhaps we will never know.

- -

📖 **Definition of a Hamiltonian integer.** This is a linear combination of basis quaternions with integer coefficients.

nions, 8 for the octonions, 16 for the sedenions).

- - - - - -

📖 **Definition of quaternions.** A quaternion is an extension of the complex plane, discovered in 1843 by William Hamilton while he was attempting to define three-dimensional multiplications. Quaternions have since been used to describe the dynamics of motion in 3-space. The space shuttle's flight software uses quaternions in its computations for guidance, navigation, and flight control for reasons of compactness, speed, and avoidance of singularities. (In mathematics, a singularity is a point at which a given mathematical object is not defined or lacks some useful property, such as differentiability.)

Quaternions define a four-dimensional space that contains the complex plane. As

we mentioned, quaternions can be represented in four dimensions by $Q = a_0 + a_1 i + a_2 j + a_3 k$, where i, j, and k are (like the imaginary number i) unit vectors in three orthogonal directions, and they are perpendicular to the real axis. To add or multiply two quaternions, we treat them as polynomials in i, j, and k but use the following rules to deal with products:

$$i^2 = j^2 = k^2 = -1$$
$$ij = -ji = k$$
$$jk = -kj = i$$
$$ki = -ik = j$$

- - - - - -

🌀 **Quaternions and quadrupeds.** "It is as unfair to call a vector a quaternion as to call a man a quadruped" (Oliver Heaviside, 1892, quoted in Paul Nahin, *Oliver Heaviside*, 1988).

- - - - - -

Definition of a magic square. Some of the problems in this book relate to magic squares. A *magic square* is a square matrix drawn as a checkerboard that's filled with numbers or letters in particular arrangements. Mathematicians are most interested in *arithmetic* squares consisting of N^2 boxes, called *cells*, filled with integers that are all different. Such an array of numbers is called a magic square if the sums of the numbers in the horizontal rows, the vertical columns, and the main diagonals are all equal.

If the integers in a magic square are the consecutive numbers from 1 to N^2, the square is said to be of the Nth order, and the *magic number*, or the sum of each row, is a constant symbolized as S:

$$S = \frac{N(N^2 + 1)}{2}$$

(The magic number is sometimes referred to as the *magic sum* or the *magic constant*.)

A few examples will help to demystify these mathematical definitions. The simplest magic square possible is one of the third order, with 3×3 cells containing the integers 1 through 9, and with the magic sum 15 along the three rows, three columns, and two diagonals. In some sense, only one unique arrangement of digits, and its mirror image, is possible for a third-order square:

4	9	2
3	5	7
8	1	6

Third-Order Magic Square

2	9	4
7	5	3
6	1	8

Mirror Image

Here $N = 3$, because there are 3 rows and 3 columns, and the magic sum S is 15 because the numbers in the rows, the columns, and two diagonals sum to 15. For example, if you look at the square on the left, you'll see that the sum of the numbers in the first row is $4 + 9 + 2 = 15$. The sum of the numbers in the first column is $4 + 3 + 8 = 15$. One of the diagonal sums is $4 + 5 + 6 = 15$, and so forth. We can also use the magic sum formula to compute the magic sum: $3(3^2 + 1)/2 = 15$. Notice that the mirror image is also a magic square. By rotating the square four times by 90 degrees, you can produce eight third-order magic squares.

As far back as 1693, the 880 different fourth-order magic squares were published posthumously by the French mathematician Bernard Frénicle de Bessy (1602–1675). Frénicle, one of the leading magic square researchers of all time, was an eminent amateur mathematician working in Paris during the great period of French mathematics in the seventeenth century.

— — — — —

Definition of a sedenion. A sedenion is a hypercomplex number constructed from sixteen basic elements. A sedenion may be represented as an ordered pair of two octonions. The word *sedenion* is derived from *sexdecim*, meaning "sixteen."

📖 **Definition of an octonion.** An octonion is a hypercomplex number that can be written as a linear combination of eight basic elements. An octonion may also be represented and written as an ordered pair of two 4-dimensional quaternions.

- - - - - -

📖 **Definition of nimbers.** Nimbers are ordinal numbers that have been recently developed by John Conway and are fascinating because of their addition and multiplication operations. Two equal nimbers always add to 0. If the nimbers are different, and if the larger of the two nimbers is 1 or 2 or 4 or 8 or 16 or . . . , then you add them just as you would add the corresponding ordinary numbers. Multiplication of nimbers is even more exotic, and you can learn more at the Wikipedia online encyclopedia, en.wikipedia.org/wiki/Nimber, or in John Conway's book *On Numbers and Games*.

- - - - - -

📖 **Definition of a biquaternion.** A biquaternion is a quaternion with complex coefficients.

- - - - - -

🕹 **On counting.** "Waclaw Sierpinski, the great Polish mathematician, was very interested in infinite numbers. The story, presumably apocryphal, is that once when he was traveling, he was worried that he'd lost one piece of his luggage. 'No dear!' said his wife, 'All six pieces are here.' 'That can't be true,' said Sierpinski. 'I've counted them several times: zero, one, two, three, four, five'" (John Conway and Richard Guy, *The Book of Numbers*, 1996). As an aside, in 1944, the Nazis burned Sierpinski's house, destroying his library and personal letters. More than half of the mathematicians who lectured in Poland's academic schools were killed, and Nazis burned down the Warsaw University Library, which contained several thousand volumes, magazines, and mathematical books. Sierpinski was the author of 724 papers and 50 books.

- - - - - -

❓ **Bombs on magic squares.** I've dropped bombs on several of the numbers in this magic square, where consecutive numbers from 0 to 63 are used, and each row, each column, and two main diagonals have the same sum. Can you replace the bombs with the proper numbers? (See Answer 2.7.)

💣	💣	2	60	11	53	9	💣
💣	💣	13	51	4	58	6	56
16	46	18	44	27	37	25	39
31	33	29	💣	💣	42	22	40
52	10	54	💣	💣	1	61	3
59	5	57	7	48	14	50	12
36	26	38	24	47	17	💣	💣
💣	21	41	23	32	30	💣	💣

Liouville constant. In 1844, the French mathematician Joseph Liouville (1809–1882) considered the following interesting number

$$L = 0.110001000000000000000001000\ldots$$

known today as the Liouville constant. Can you guess its significance or what rule he used to create it? (See Answer 2.8.)

- -

Viete's formula for π. The first person to uncover an infinite product formula for pi was the French mathematician François Viete (1540–1603). This remarkable gem involves just two numbers—π and 2:

$$\pi = 2 \cdot \frac{2}{\sqrt{2}} \cdot \frac{2}{\sqrt{2+\sqrt{2}}} \cdot \frac{2}{\sqrt{2+\sqrt{2+\sqrt{2}}}} \ldots$$

Integer satisfaction. "What could be more beautiful than a deep, satisfying relation between whole numbers. How high they rank, in the realms of pure thought and aesthetics, above their lesser brethren: the real and complex numbers . . ." (Manfred Schroeder, *Number Theory in Science and Communication*, 1984).

- - - - - -

Math anagrams. Apart from the fact that they are equal, ONE PLUS TWELVE and TWO PLUS ELEVEN are anagrams of each other.

The paradox of pepperonis. Perhaps you've heard this sort of fallacy when you were a child. (A fallacy produces a wrong answer using explanations that sometimes appear to be very logical.) Before you are six pepperonis, three on each of two pizzas:

○○○　○○○

A child will now try to prove there are really seven pepperonis. Try this on friends. Start counting "1, 2, 3" on the first pizza, and then pause and continue counting, "4, 5, 6" on the second pizza. Now count *backward* while pointing to the second pizza, "6, 5, 4." You are now pointing to the first pepperoni in the second pizza and have said the word *four*. Next, you say, "*four*, and three more on the first pizza makes seven."

What is wrong with this argument? How would children of various ages respond? (If you have access to a child, please tell me how he or she responded.) Sure, this is crazy, but can you articulate precisely what's wrong with the child's argument? (See Answer 2.9.)

- - - - - -

Taxicab numbers. What do British taxicabs have to do with the number 1,729? Why is 1,729 such a special and famous number in the history of mathematics? (See Answer 2.10.)

- - - - - -

Transcendence. We know that π (3.1415 . . .) and *e* (2.7182 . . .) are the most famous transcendental numbers. Is (π + *e*) transcendental? Is (π × *e*) transcendental? (See Answer 2.11.)

❓ Special class of numbers. This puzzle is for extreme math lovers. The following numbers represent a special class of numbers that mathematicians have studied for years: 12, 18, 20, 24, 30, 36, 40, 42, 48, 54, 56, 60, 66, 70, 72, 78, 80, 84, 88, 90 . . . These numbers are so unique that mathematicians have a special name for them. Can you determine what mathematical property these numbers have in common, aside from the fact that the numbers in this limited list are all even numbers? (See Answer 2.12.)

- - - - - -

❶ Goldbach conjecture. In 1742, the mathematician Christian Goldbach conjectured that every integer greater than 5 can be written as the sum of three primes, like 21 = 11 + 7 + 3. As reexpressed by Euler, an equivalent conjecture (called the "strong" Goldbach conjecture) asserts that all positive even integers greater than 2 can be expressed as the sum of two primes. (Note that 1 is not considered a prime number.) The publishing giant Faber and Faber offered a $1,000,000 prize to anyone who proved Goldbach's conjecture between March 20, 2000, and March 20, 2002, but the prize went unclaimed, and the conjecture remains open. In 2003, the conjecture was verified up to 6×10^{16}.

- - - - - -

❶ Fermat numbers. In 1650, the French scholar and lawyer Pierre de Fermat rashly proposed that all numbers of the form $F_p = 2^{2^p} + 1$ were prime numbers. He knew that this was true for the first five such numbers corresponding to $p = 0, 1, 2, 3, 4$. Alas, Leonhard Euler in 1732 proved that the fifth number was not prime. In fact, today, many believe that all Fermat numbers greater than F_4 are composite.

- - - - - -

❶ Fermat numbers. In 2003, Thomas Koshy proved that every Fermat number (numbers of the form $F_p = 2^{2^p} + 1$) ends in the digits 17, 37, 57, or 97, where $p > 1$ (*Journal of Recreational Mathematics*).

- - - - - -

❓ Champernowne's number. Here is a very famous number called Champernowne's number:

0.1234567891011121314 . . .

Do you see how it is created? What makes it so interesting? (See Answer 2.13.)

- - - - - -

❶ Calculating π. The first computer calculation of π occurred in 1947, yielding 2,037 decimal digits in 70 hours. Today we know π to over a trillion digits.

- - - - - -

❶ π on a tombstone. The fencing master Ludolph van Ceulen (1540–1610) gave π to 20 decimal digits in 1596. Several digits of π were engraved on his tombstone.

❶ Wallis's equation for π. In 1655, the English mathematician John Wallis (1616–1703) devised this wonderful-looking infinite product involving only rational numbers to calculate pi:

$$\frac{\pi}{2} = \prod_{n=1}^{\infty} \left[\frac{(2n)^2}{(2n-1)(2n+1)} \right] = \frac{2 \cdot 2}{1 \cdot 3} \frac{4 \cdot 4}{3 \cdot 5} \frac{6 \cdot 6}{5 \cdot 7} \cdots$$

No train stops at π. "Pi is not the solution to any equation built from a less-than-infinite series of whole numbers. If equations are trains threading the landscape of numbers, then no train stops at pi." (Richard Preston, "The Mountains of Pi," *New Yorker*, 1992).

- - - - - -

Chung-chih formula for π. Astronomer Tsu Chung-chih (430–501) gave the following value of π, which is correct to six decimal digits.

$$\pi = \tfrac{355}{113} = 3.141592\ldots$$

This value was not improved upon in Europe until the sixteenth century, more than a thousand years later.

- - - - - -

Self-location and π. Humanity is aware of just a few "self-locating" strings within π. Defining the first digit after the decimal point as digit 1, Jeff Roulston has found the following numbers that can self-locate in the first 50 million decimal digits of π; 1; 16,470; and 44,899. For example, the digit 1 is found in position 1 in 3.1415. The

Copeland-Erdös constant. Here's a famous number that caused quite a stir in 1945:

$$0.23571113171923\ldots$$

Can you determine how this number was constructed? (See Answer 2.14.)

- -

Thue constant. This number is called the Thue constant:

$$0.1101101111101101111101101101101101111110110\ldots$$

Can you figure out how it was generated or determine its remarkable properties? (See Answer 2.15.)

string 16,470 starts in position 16,470, and so forth.

- - - - - -

Numerical axis of evil. The Russian computer specialist Timofei Shatrov has discovered a relationship between 13, 666, and 911, the date of the World Trade Center disaster: *911 = 13 + 666 + 2 × 116.* As you can see, 116 is 911, flipped upside down.

- - - - - -

Exclusionary squares. I have a particular penchant for an unusual class of numbers called "exclusionary squares," such as the very special number 639,172. It turns out that this is the largest integer with distinct digits whose square is made

up of digits not included in itself: $639,172^2 = 408,540,845,584$. Can you find the only other six-digit example? Do any exclusionary cubes or exclusionary numbers of higher orders exist? (See Answer 2.16.)

- - - - - -

The grand search for isoprimes. Note that 11 is an *isoprime*, a prime number with all digits the same. (A prime number is divisible only by itself and 1.) Do any other isoprimes exist in base 10?

Similarly, 101 is an *oscillating bit* prime (base 10). Do any others exist? For example, 10,101 is not prime. Neither is 1,010,101. (See Answer 2.17.)

❓ Special augmented primes. One of my favorite bizarre number quests involves a specific kind of prime number called a "special augmented prime." You can augment a prime simply by placing a 1 before and after the number. The augmented prime is "special" if it yields an integer when divided by the original prime number. For example, 137 is such a number because 137 is prime and because 11371/137 yields an integer— namely, 83. Are there other such numbers? How rare are special augmented primes? (See Answer 2.18.)

- - - - - -

❓ Unreasonable effectiveness of mathematics. "The enormous usefulness of mathematics in natural sciences is something bordering on the mysterious, and there is no rational explanation for it. It is not at all natural that 'laws of nature' exist, much less that man is able to discover them. The miracle of the appropriateness of the language of mathematics for the formulation of the laws of physics is a wonderful gift which we neither understand nor deserve" (Eugene P. Wigner, "The Unreasonable Effectiveness of Mathematics in the Natural Sciences," 1960).

- - - - - -

❓ Body weights. What would happen if everyone's body weight was quantized and came in multiples of π pounds? (See Answer 2.20.)

- - - - - -

❓ Jesus and negative numbers. Would Jesus of Nazareth ever have worked with a negative number, like –3? (See Answer 2.21.)

- - - - - -

❓ Jesus notation. What kinds of written numbers did Jesus of Nazareth, or a comparable figure of his era, use? Did these people use numbers that looked like the numbers we use today? (See Answer 2.22.)

- - - - - -

❓ Jesus and multiplication. Could Jesus of Nazareth multiply two numbers? (See Answer 2.23.)

❓ Triangle of the Gods. An angel descends to Earth and shows you the following simple progression of numbers:

$$1$$
$$12$$
$$123$$
$$1234$$
$$12345$$
$$123456$$
$$1234567$$
$$12345678$$
$$123456789$$
$$1234567890$$
$$12345678901$$
$$123456789012$$
$$1234567890123 \ldots$$

The angel will let you enter the afterlife if you can determine what is the smallest prime number of this kind. Can you do so? (See Answer 2.19.)

? The digits of π. Is it true that I can find consecutive digits, like 1, 2, 3, . . . 1,000,000, all neatly in a row in the decimal digits of π? (See Answer 2.24.)

- - - - - -

? Living in the π matrix. Is it true that we all live forever, coded in the endless digits of π? (See Answer 2.25.)

- - - - - -

? Numbers and sign language. In American sign language, a person puts up one finger for 1, two fingers for 2, and so forth. How would a deaf person sign "100" without growing one hundred fingers? (See Answer 2.26.)

- - - - - -

? Adding numbers. It would be a tough job to add all the numbers between 1 and 1,000. What formula would you use to do this quickly? (See Answer 2.27.)

- - - - - -

? The mystery of 0.33333. We all know that ⅓ = 0.3333 . . . repeating. Multiplying both sides of the equation by 3, we find that 1 = 0.9999 . . . How can this be? (See Answer 2.28.)

? The grand Internet undulating obstinate number search. In 1848, Alphonse Armand Charles Georges Marie, better known as the "Prince de Polignac," conjectured that every odd number is the sum of a power of 2 and a prime. (For example, $13 = 2^3 + 5$.) He claimed to have proved this to be true for all numbers up to three million, but de Polignac probably would have kicked himself if he had known that he missed 127, which leaves residuals of 125, 123, 119, 111, 95, and 63 (all composites—that is, nonprimes) when the possible powers of 2 are subtracted from it. There are another 16 of these odd numbers— which my colleague Andy Edwards calls *obstinate numbers*—that are less than 1,000. There is an infinity of obstinate numbers greater than 1,000.

Most obstinate numbers that we have discovered are prime themselves. The first composite obstinate number is 905. What is the largest obstinate number that humanity can compute? What is the smallest difference between adjacent obstinates?

Do any obstinates undulate? (Undulating numbers are of the form abababababab . . . For example, 171,717 is an undulating number, but it's not obstinate.) (See Answer 2.29.)

- - - - - -

? Mystery sequence. What is the missing number in the following sequence? No numbers may repeat in this sequence.

13, 24, 33, 40, 45, 48, ?

(See Answer 2.30.)

? Strange code. If ..--- + ...- equals -...., what does .---- + ..--- equal? (See Answer 2.31.)

- - - - - - - - - - - - - - - -

? Mystery sequence. What number comes next?

1, 9, 17, 3, 11, 19, 5, 13, 21, 7, 15, ?

(See Answer 2.32.)

Mystery sequence. Replace the 🚚 with the correct numbers in this interesting sequence:

1, 8, 15, 3, 🚚, 19, 9, 18, 10, 🚚,
14, 7, 5, 4, 🚚, 13, 0, 12, 16, 🚚

(See Answer 2.33.)

- -

Mystery sequence. What is the significance of the following sequence?

2357, 1113, 1719, 2329, 3137, 4143

(See Answer 2.34.)

Time-travel integer. Pete tells Penny, "I want to travel back in time to a year that has all even digits, like A.D. 246."

Penny replies, "How about the year 2000?"

"No," Pete says, "I want to travel to a year before 2000. Given this constraint, what's the most recent year that I can travel to that has all even digits, with each digit different? I want a recent year so that I'll have a chance of enjoying some modern amenities." (See Answer 2.35.)

- - - - - -

Mystery sequence. What is the next number in the sequence?

1, 2, 9, 64, 625, ?

(See Answer 2.36.)

Ostracism. Which of these numbers is the odd one out?

3 8 15 24 35 48 55 63

(See Answer 2.37.)

- - - - - -

Natives and mathematics. "In Samoa, when elementary schools were first established, the natives developed an absolute craze for arithmetical calculations. They laid aside their weapons and were to be seen going about armed with slate and pencil, setting sums and problems to one another and to European visitors. The Honourable Frederick Walpole declares that his visit to the beautiful island was positively embittered by ceaseless multiplication and division" (Robert Briffault

[1876–1948], quoted in John Barrow's *Pi in the Sky*, 1992).

- - - - - -

$a^2 + b^2 + 1$? In 1996, Daniel Schepler, a student at Washington University, St. Louis, Missouri, answered the following cool question: "For what pairs of integers a, b does ab exactly divide $a^2 + b^2 + 1$?" Schepler proved that the solution pairs (a, b) are always (F_{2n+1}, F_{2n-1}), where F_n is the nth Fibonacci number. (For further reading, see Richard Guy and Richard Nowakowski, "A Recurrence of Fibonacci," *The American Mathematical Monthly* 103, no. 10 [December 1996]: 854–69).

- - - - - -

Mystery sequence. What rule am I using to determine the numbers in this sequence?

18, 20, 24, 30, 32, 38, 42, . . .

(See Answer 2.38.)

- - - - - -

Mystery sequence. The great trumpet player Dizzy Lizzy plays a run of 767 notes, then a run of 294, then one of 72, then one of 14. How many notes does she play next? (See Answer 2.39.)

Cellular communication. Marvin is studying cells that live in a warm sea on another planet. These cells emit puffs of tiny bubbles when they communicate with each other. The bubbles come in two colors: red and blue, which we can symbolize as ∩ and ∪.

Today, Marvin is trying to determine whether these cells have any intelligence. The first cell emits a red bubble ∩, then its neighbor emits a red, followed by a blue bubble ∩∪. The sequence of bubbles grows rapidly:

∩

∩∪

∩∪∪∩

∩∪∪∩∪∩∩∪

∩∪∪∩∪∩∩∪∪∩∩∪∩∪∪∩

What rules are the cells using to determine bubble colors in this growing sequence? (See Answer 2.40.)

- - - - - -

Definition of an automorphic number. An automorphic number is a number with a power (such as a square or a cube) that ends in that number. For example, 6 is automorphic because $6^2 = 36$. Here's another: $625^2 = 390,625$. The number 40,081,787,109,376 is a magnificent example, because

$$40081787109376^2 = 1606549657881\underline{340081787109376}$$

- - - - - -

Automorphic acumen. Here is a 100-digit automorphic number from Mr. R. A. Fairbairn of Toronto:

6,046,992,680,891,830,197,061,490,109,937,833,490, 419,136,188,999,442,576,576,769,103,890,995,893, 380,022,607,743,740,081,787,109,376

(The square of this number ends with the digits of this number. The source for the Fairbairn number is Joseph S. Madachy's *Madachy's Mathematical Recreations* [New York: Dover, 1979]).

Mystery sequence. There is a logical pattern to the following sequence of numbers. What is the next number in the sequence?

1, 5, 12, 22, 35, 51, 70, ?

(See Answer 2.41.)

- - - - - -

Mystery sequence. You and your friend are enjoying desserts. Your friend, who is eating apple pie with whipped cream, asks you to supply the missing number in this very difficult sequence, which he has written using vanilla ice cream:

1, 41, 592, 6535, ?

Can you determine the missing number? Your reward is the cake or the pie of your choice, which your friend will deliver to you personally. Hurry, the ice-cream numbers are melting! (See Answer 2.42.)

- - - - - -

Ostracism. Which product does not belong in this list?

$$21 \times 60 = 1,260$$
$$15 \times 93 = 1,395$$
$$15 \times 87 = 1,305$$
$$30 \times 51 = 1,530$$
$$21 \times 87 = 1,827$$
$$80 \times 86 = 6,880$$

(See Answer 2.43.)

? The amazing 5. Can you list four amazing mathematical facts about the number 5? (See Answer 2.44.)

? Mystery sequence. What is the value of the missing digit in this sequence?

6 2 5 5 4 5 6 3 ?

(I have never known anyone who was able to solve this puzzle.) (See Answer 2.45.)

? Mystery sequence. Supply the missing number in this very difficult sequence:

2, 71, 828, ?

(See Answer 2.46.)

📖 Definition of an untouchable number. An untouchable number is a number that is never the sum of the factors of any other number. In particular, an untouchable number is an integer that is not the sum of the proper divisors of any other number. (A proper divisor of a number is any divisor of the number, except itself. Example: the proper divisors of 12 are 1, 2, 3, 4, and 6.) The first few untouchables are 2, 5, 52, 88, 96, 120, 124,

? Mystery sequence. A violinist plays a seemingly random riff of short and long notes over and over again, which can be represented as a string of 0s (long notes) and 1s (short notes):

01101010001010001010001000001010000010001010...

What rule is the violinist using to produce this sequence? (See Answer 2.47.)

? Mystery sequence. What rule is used to generate this very difficult sequence?

3, 4, 5, 7, 11, 13, 17, 23, 29, 43, 47, 83, 131, 137, 359, 431, 433, 449, 509, 569, 571, 2971, 4723, 5387, . . .

(See Answer 2.48.)

146, . . . Paul Erdös has proved that there are infinitely many untouchables.

? The Lego sequence. What rule is used to generate the Lego sequence?

1, 3, 7, 19, 53, 149, 419, . . .

(See Answer 2.49.)

? Vampire numbers. Vampire numbers are the products of two progenitor numbers that when multiplied survive, scrambled together, in the vampire number. Consider one such case: $27 \times 81 = 2{,}187$. Another vampire number is 1,435, which is the

product of 35 and 41. Can you find any others? (See Answer 2.50.)

? Jewel thief. A jewel thief with long, spindly fingers has a burlap bag containing 5 sets of emeralds, 4 sets of diamonds, and 3 sets of rubies. A "set" consists of a large, a medium, and a small version of each of these gems. The electricity is out, and it is dark. How many gems must he withdraw from his bag to be sure that he has a complete set of one of the gems? How many gems must he withdraw to ensure that he has removed all of the large gems? (See Answer 2.51.)

The amazing 1/89. Although not widely known, the decimal expansion of 1/89 (0.01123 . . .) relates to the Fibonacci series when certain digits are added together in a specific way. Examine the following sequence of decimal fractions, arranged so that the right-most digit of the nth Fibonacci number is in the $n + 1$th decimal place:

n	
1	.01
2	.001
3	.0002
4	.00003
5	.000005
6	.0000008
7	.00000013
	.0112359 . . .

Unbelievably,

$$1/89 =$$
$$0.01123595505617977528089887640449438202247191 . . .$$

Fantastic! Why should this be so? Why on Earth is 89 so special?

Palinpoints of arithmetical functions. Dr. Joseph Pe and others have studied arithmetical functions having domain and range that are sets of integers. Let $rev(n)$ denote the digit reversal of the integer n; for example, $rev(123) = 321$ and $rev(-29) = -92$. The palinpoints ("palindronomic points") of an arithmetical function f are arguments n at which f commutes with $rev(n)$, that is, $f(rev(n)) = rev(f(n))$. Ignore leading 0s.

For example, let $f(n) = n^2$. Because $311^2 = 96,721$ and $113^2 = 12,769$. And $f(rev(311)) = rev(f(311))$, so that 311 is a palinpoint of f. Can you find any multidigit palinpoints of $f(n) = Prime(n)$, that is, palinpoints for prime numbers? (See Answer 2.52.)

Dr. Brain's Mystery sequence. Dr. Brain asks you to consider this number sequence:

$$1, 3, 6, 10, 15, 21, . . .$$

What number comes next? (See Answer 2.53.)

- - - - - -

Poseidon's sequence. Poseidon, the Greek god of the oceans, extended his pitchfork and drew on a seashell:

$$727, 98, 72, 14, 4$$

He turned back to his class. "Can anyone tell me how the following sequence arises?" Jessica instantly stuck up her hand. "Sir, in '727, 98, 72, 14, 4,' each term is the product of the digits of the previous one."

"Jessica, you are amazing. Now let me tell you about 727's persistence. The *persistence* of a number is the number of steps (4 in our example) before the number collapses to a single digit. Now, consider my mighty difficult question: What is the smallest number with persistence 4? I will take you on a tour of all of the Earth's oceans if you can solve this problem." (See Answer 2.54.)

Mystery sequence. Each day a zookeeper chooses several additional animals for the zoo. Row 1 shows his choices for day 1. Row 2 are his choices for day 2, and so forth. What rule is he using?

(See Answer 2.55.)

- -

Constructions with 1, 2, and 3. Today, we will construct integers using just 1s, 2s, and 3s, and any number of +, −, and × signs. You are also allowed exponentiation, but you cannot concatenate numbers to form multidigit numbers. As an example, let's first consider the problem where only the digit 1 is allowed. The number 40 could be written:

$$40 = (1 + 1 + 1 + 1 + 1) \times (1 + 1 + 1 + 1) \times (1 + 1)$$

If we let $f(n)$ be the least number of digits that can be used to represent n, then we see that $f(40) \leq 11$. A contest that allows only 1s for forming small numbers turns out not to be very interesting. However, once the digits 2 and 3 are also allowed, the problem becomes fascinating. Here is an example: $121 = (2^{(2+1)} + 3)^2$ Here $f(121) \leq 5$. Is this the best you can do?

Here is my challenge to you for this contest! Your goal is to represent the numbers 40, 61, 263, and 500 with as few digits as possible. (See Answer 2.56.)

Mystery sequence. What number comes next?

$$1;\ 8;\ 81;\ 1{,}024;\ ?$$

(See Answer 2.57.)

- - - - - -

Very rare square pyramidal numbers that are also square. In the answer to the 69,696 question (see page 68), we defined a square number as an integer of the form $y = x^2$. Square numbers count the number of balls in the following figure: 1, 4, 9, 16, . . . :

O OO OOO OOOO
OO OOO OOOO
OOO OOOO
OOOO

Square pyramidal numbers count the number of balls in a growing pyramid of balls formed by square layers and grow according to the sequence 1, 5, 14, 30, . . . For years, people had wondered how many numbers existed that were both square and square pyramidal. Finally, in 1918, G. N. Watson proved that only one multidigit square pyramidal number exists that is also square: 4,900.

Blue liquid. A mad scientist gazes into a vial of blue liquid. "This substance is still a liquid at minus 40 degrees," he tells his assistant, Boris.

Boris replies, "Is that Centigrade or Fahrenheit?"

With his dark eyes, the scientist looks at Boris and says, "It doesn't matter." Why did the scientist say that? (See Answer 2.58.)

Grasshopper sequences. Consider the numbers generated by the following expressions, where x is an integer, and we start with $x = 1$.

$$x \to 2x + 2$$
$$x \to 6x + 6$$

These mappings generate two branches of a "binary tree" (figure 2.2).

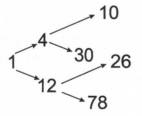

Figure 2.2 Grasshopper sequences.

Pretend that each generation of children requires a day to compute. For example, after one day, we have 4 and 12 as "children" of the "parent" 1. The next day produces 10,

Mystery consecutive integers. Consider the sequence of integers consisting of all square numbers and cube numbers, starting at 4:

$$4, 8, 9, 16, 25, 27, 36, 49, 64, 81, 100, \ldots$$

Notice how 8 and 9 are consecutive integers. Are there any other consecutive numbers in this sequence? For years, mathematicians and computer scientists have searched for other examples, but, besides 8 and 9, no consecutive powers were ever observed. (For further reading, see Paulo Ribenboim, "Catalan's Conjecture," *The American Mathematical Monthly* 103, no. 7 [August–September 1996]: 529-32.) The conjecture that 8 and 9 are the only such consecutive integers was finally proved by Preda Mihailescu of the University of Paderborn in Germany and published in "A Class Number Free Criterion for Catalan's Conjecture," *Journal of Number Theory* 99, no. 2 (2003): 225–31.

30, 26, and 78. All the numbers that have appeared so far, when arranged in numerical order, are 1, 4, 10, 12, 26, 30, 78, . . . No number seems to appear twice in a row—for example, 1, 4, 10, 10, . . .

Does a number ever appear twice? Maybe we don't see a repetition now, but would we see one after thousands of years? (See Answer 2.59.)

The loneliness of the factorions. Factorions are numbers that are the sum of the factorial values for each of their digits. (For a positive integer n, the product of all the positive integers less than or equal to n is called "n factorial," usually denoted as $n!$. For example, $3! = 3 \times 2 \times 1$.)

The number 145 is a factorion because it can be expressed as

$$145 = 1! + 4! + 5!$$

How many others can you find? (See Answer 2.60.)

69,696. What is special about the number 69,696? (See Answer 2.61.)

The beauty of 153. What's so special about the number 153? (See Answer 2.62.)

Definition of a googol. A "googol" (not to be confused with the Internet search engine Google) is a large number equal to 10^{100} (1 followed by 100 zeroes). The word *googol* was coined by the American mathematician Edward Kasner (1878–1955) in 1938. According to the story, Kasner asked his young nephew what name he would give to an incredibly large number, and "googol" was the nephew's response. An equivalent name for googol is "ten dotrigintillion."

- - - - - -

Christians and 153. St. Augustine, the famous Christian theologian, thought that 153 was a mystical number and that 153 saints would rise from the dead in the eschaton. St. Augustine was fascinated by a New Testament event (John 21:11) where the Apostles caught 153 fish from the Sea of Tiberias. Seven disciples hauled in the fish, using nets. St. Augustine reasoned that these seven were saints. Why, you ask? Because there are seven gifts from the Holy Ghost that enable men to obey the Ten Commandments, Augustine thought that the disciples must therefore be saints. Moreover, $10 + 7 = 17$, and if we add together the numbers 1 through 17, we get a total of 153. The hidden meaning of all this is that 153 saints will rise from the dead after the world has come to an end.

The majesty of 153. My favorite integer is 153. Why? First, as we just mentioned, $153 = 1^3 + 5^3 + 3^3$. But, more than that, $153 = 1! + 2! + 3! + 4! + 5!$, where the "!" symbols mean factorial. Also, when the cubes of the digits of any three-digit number that is a multiple of 3 are added, and the digits of the resulting number are cubed and added, and the process continued, the final result is 153. For instance, start with 369, and you get the sequence 369, 972, 1,080, 513, 153. Want more reasons? Because 153 is also the 17th triangular number, and I just told you why St. Augustine thought that 17 was important. It's the sum of 10 (for the Ten Commandments of the Old Testament) and 7 (for the Gifts of the Spirit in the New Testament).

- - - - - -

Erdös equation. I know of only three different solutions to $n! + 1 = m^2$, where n and m are integers. Can you find at least one of these solutions? (See Answer 2.64.)

- - - - - -

Ten silver boxes. In each of ten shiny, silver boxes in Row 2 in the following diagram, write a one-digit number. The digit in the first box indicates the total number of zeros in Row 2. The box beneath "1" in Row 1 indicates the total number of 1s in Row 2. The box marked "2" indicates the total number of 2s in Row 2, and so on.

0	1	2	3	4	5	6	7	8	9	Row 1
										Row 2

Is there a solution to this problem? Are there many solutions to this problem? (See Answer 2.63.)

String theory and ancient math. Ancient math can find obscure applications centuries later and can even describe the very fabric of reality. For example, in 1968, Gabriele Veneziano, a researcher at CERN (a European particle accelerator lab), observed that many properties of the strong nuclear force are perfectly described by the Euler beta-function, an obscure formula devised for purely mathematical reasons two hundred years earlier by Leonhard Euler. In 1970, three physicists, Nambu, Nielsen, and Susskind, published their theory behind the beta-function; this eventually led to modern string theory, which says that all of the fundamental particles of the universe consist of tiny strings of energy.

– – – – – –

Square numbers. Can you tell me every positive integer whose square contains the same decimal digits as its double? (See Answer 2.66.)

– – – – – –

$\sqrt{2}$ is irrational. Can you prove, in just a few simple steps, that $\sqrt{2}$ is irrational? Do this in such a way that a high-school student could understand it and agree with you. (See Answer 2.67.)

Perrin sequence. Consider a simple sequence defined by addition:

$$u(n + 3) = u(n) + u(n + 1).$$

We can "seed" the sequence with three terms: $u(0) = 3$, $u(1) = 0$, and $u(2) = 2$. Then, the first few terms are

n	0	1	2	3	4	5	6	7	8	9	10	11	12	13	14
$u(n)$	3	0	2	3	2	5	5	7	10	12	17	22	29	39	51
n prime?			p	p		p		p				p		p	

Notice that it is easy to locate prime numbers n, because they seem to occur whenever $u(n) = n$ or is a multiple of n. How can this be? Is this always true? If not, what is the first example to the contrary? (See Answer 2.65.)

Reversed numbers and palindromes. If you select an integer, reverse its digits, add the two numbers together, and continue to reverse and add, the result often becomes a palindrome—that is, the number reads the same in both directions (see the example on page 71). With some numbers, this happens in a single step. For example, $18 + 81 = 99$, which is a palindrome. Other numbers may require more steps. For example, $19 + 91 = 110$. And $110 + 011 = 121$. Of all the numbers under 10,000, only 249 fail to form palindromes in 100 steps or less.

In 1984, Gruenberg noted that the smallest number that *seems* never to become palindromic by this process is 196. (It has been tested through 50,000 steps.) I have tested the starting number 879 for 19,000 steps, producing a 7,841-digit number—with no palindrome resulting. Statistical tests indicate an approximately equal percentage of occurrence of digits 0 through 9 for this large number. Similarly, I have tested 1,997 for 8,000 steps, with no palindrome occurring.

Fibonacci snakes. You go to a neighborhood pet store and buy a pair of small snakes and breed them. The pair produces one pair of young after one year, and a second pair after the second year. Then they stop breeding. Each new pair also produces two more pairs in the same way, and then stops breeding. How many new pairs of snakes would you have each year? (See Answer 2.68.)

— — — — — —

Pascal's triangle. One of the most famous integer patterns in the history of mathematics is Pascal's triangle. Blaise Pascal was the first to write a treatise about this progression in 1653, although the pattern had been known by Omar Khayyam as far back as A.D. 1100. The first seven rows of Pascal's triangle can be represented as

```
          1
         1 1
        1 2 1
       1 3 3 1
      1 4 6 4 1
    1 5 10 10 5 1
  1 6 15 20 15 6 1
```

Catalan numbers. Catalan numbers are defined by the following rules. The first two Catalan numbers are 1, which we can write as $C(0) = 1$ and $C(1) = 1$. The nth Catalan number is defined as

$$C_n = \sum_{i=0}^{n-1} C_i C_{n-i-1}$$

Can you write a program to print the first twenty Catalan numbers? The first ten Catalan numbers are 1; 1; 2; 5; 14; 42; 132; 429; 1,430; and 4,862. The Catalan numbers answer the question "In how many ways can a regular n-gon be divided into n-2 triangles if different orientations are counted separately?"

The nth Catalan number can also be computed from

$$C_n = \frac{(2n)!}{n!(n+1)!}$$

The nth Catalan number can be approximated by

$$C_n \approx \frac{4n}{\sqrt{\pi} n^{3/2}}$$

Each number in the triangle is the sum of the two numbers above it. The role that Pascal's triangle plays in probability theory, in the expansion of binomials of the form $(x + y)^n$, and in various number theory applications has been discussed in numerous references. If similar expansions are made for $(x + y + z)^n$ for successive powers of n, Pascal's Pyramid can be generated.

— — — — — —

Palindromes on parade. Palindromic numbers are positive integers that "read" the same backward or forward. For example, 12,321; 11; 261,162; and 454 are all palindromic numbers. How many palindromic numbers are there that are less than or equal to the integer P? We call this quantity $W(P)$. For example, $W(3) = 3$, since 1, 2, and 3 are palindromes. (See Answer 2.69.)

Alpha and omega. Dr. Brain writes two unknown integers, alpha and omega, on a scrap of paper and places the scrap in a black box. Each integer is between 2 and 99 inclusive. Dr. Brain tells his wife the product of alpha and omega. She looks at the product and says, "Simply by looking at the product, I can't tell what alpha and omega are."

From this little information, can we exclude any values for alpha and omega to make a more informed guess at the possible values of alpha and omega? (See Answer 2.70.)

- - - - - -

"A mathematician is a machine for turning coffee into theorems" (Paul Erdös, quoted in Paul Hoffman, *The Man Who Loved Only Numbers*, 1999).

- - - - - -

Does pi contain pi? Pi is an infinite string of digits and contains an infinite number of finite strings of digits. In fact, it probably includes every possible string of finite digits. Does pi contain itself? (See Answer 2.71.)

Word frequency. What is your estimate of the relative occurrences of the English words *one*, *two*, *three*, *four*, . . . , *twenty* on the Internet? Do you think that the number of occurrences smoothly decreases, starting from *one* to *twenty*? (See Answer 2.72.)

- - - - - -

Bad luck. Why is the number 13 considered unlucky? (See Answer 2.73.)

The height of irrationality. A wise man once challenged, "Show, by a simple example, that an irrational number raised to an irrational power need not be irrational." Can you do it? (See Answer 2.74.)

- - - - - -

A common fear. What is paraskevidekatriaphobia? (See Answer 2.75.)

- - - - - -

Too many 3s. One unusual paper in a prestigious chemical journal reported that many products and quotients of fundamental constants have values very close to the number 3 multiplied by a power of 10:

Function	Numerical Value
c	2.9976×10^{10}
$(\varepsilon/m_0)^{1/2}$	2.9995×10^{1}
$m_0/(2\pi\varepsilon)^{1/2}$	3.0009×10^{9}
$\sqrt{m_0}$	2.9990×10^{-14}
$2\pi\varepsilon$	2.9971×10^{-9}
$3Gc/2$	2.9967×10^{3}

The author, J. E. Mills, claims that the high occurrence of values extremely close to the number 3 is "amazing." (Do you think it is amazing if the results depend on the units in which the constants are expressed?) In the table, m_0 is the mass of an electron, c is the speed of light, G is the gravitational constant, and ε the electronic charge. (J. E. Mills, "Relations between Fundamental Physical Constants," *Journal of Physical Chemistry* 36 (1932): 1089–1107.)

Apocalyptic Fibonacci numbers. The 3,184th Fibonacci number is apocalyptic, having 666 digits. For numerologist readers, the apocalyptic number is

1167243740814955412334357645792141840689747174434394372363312
8273626208245238531296068232721031227888076824497987607345597
1975198631224699392309001139062569109651074019651076081705393
2060237984793918970003774751244713440254679507687069905503229
7133437094009365444241181520685790404104340056856808119437950
3001967669356633792347218656896136583990327918167352721163581
6503595776865522931027088272242471094763821154275682688200402
5850498611340877333322087361645911672649719869891579135588343
1385556958002121928147052087175206748936366171253380422058802
6552914033581456195146042794653576446729028117115407601267725
6157286715574607026067859229791790424885389235886177163

Definition of an emirp. An emirp is a prime number that turns into another prime number when the digits are reversed. *Emirp* is *prime* spelled backward. Example: 37 and 73. When you reverse the digit of most primes, you get a composite (43 becomes 34). Here are the first few emirps: 13, 17, 31, 37, 71, 73, 79, 97, 107, 113. . . .

The number 1,597 turns out to be an emirp because 7,951 is also prime.

- - - - - -

Definition of a perfect number. A perfect number, like 6, is the sum of its proper divisors ($6 = 3 \times 2 \times 1 = 3 + 2 + 1$). The ancient mathematician Nicomachus (A.D. 60–120) knew of the first four: 6, 28, 496, and 8,128. The next two are 33,550,336 and 8,589,869,056. All perfect numbers discovered thus far are even.

- - - - - -

Judaism and perfect numbers. The ancient Jews tried to use numbers to prove to atheists that the Old Testament was part of God's revelation. The famous Jewish philosopher and theologian Philo of Alexandria (20 B.C.–A.D. 40) justified the story of Genesis by first validating the assertion that God created the world in six days. He claimed that the six-day creation must be correct because 6 was the first perfect number (since 6 is the sum of its divisors,

$6 = 1 + 2 + 3$). The perfection of the number 6 led to it becoming the symbol of creation and reinforcing the notion of God's existence.

So important were perfect numbers to the Jews in their search for God that Rabbi Josef ben Jehuda Ankin, in the twelfth century, recommended the study of perfect numbers in his book *Healing of Souls*. Why did the ancients have such a fascination with numbers? Could it be that in difficult times, numbers were the only constant thing in a chaotic world?

- - - - - -

Ten and the Jews. Ten appears often in Judaism. There are 10 commandments. The *Zohar*, the central text of the Kabbalah, says that the world was created in 10 words, because in Genesis 1, the phrase "And God spoke" is repeated not less than 10 times. There are 10 generations between Adam and Noah. There are 10 plagues in Egypt. On 10 Tishri, the Jewish Day of Atonement, the confession of sins is repeated 10 times. On Rosh Hashanah, the Jewish New Year, 10 biblical verses are read in groups of 10.

Christianity, Islam, and perfect numbers. Christians in the twelfth century were fascinated by the second perfect number, 28. For example, since the lunar cycle is 28 days, and because 28 is perfect, Albertus Magnus (A.D. 1200–1280), a philosopher and a theologian, expressed the idea that the mystical body of Christ in the Eucharist appears in 28 phases.

The perfect number 28 also plays an important role in Islam, because religious Muslims connect the 28 letters of the alphabet in which the Koran is written with the 28 "lunar mansions." For example, the famous medieval mathematician al-Biruni (1048) suggested that this connection proves the close relation between the cosmos and the word of God. Note also that the Koran names 28 prophets before Mohammed.

- - - - - -

Number 7. There is an incredibly large number of occurrences of 7 in all religions. In the Old Testament, Lamech, the father of Noah and the son of the famous long-lived Methuselah, is born 7 generations after Adam. Lamech lives for 777

Sequence of primes. Carlos Rivera (primepuzzles.net) found the largest known sequence of primes $p_1, p_2, p_3, \cdots p_k$ such that $p_{i+1} = 4p_i^2 + 1$. The largest sequence has five members, starting with the prime 197,072,563:

- 197072563 (9 digits) →
- 155350380349555877 (18) →
- 96534962699006707074223324580956517 (35) →
- 37275996093194465195683744682486635943427798514276247679661962579085157 (71) →
- 55579995389593961295924052448331802040613737326783974746942903264272487529696010558819635123974590676633087823266633903412411149427482308585 97 (142)

years. Another Lamech should be avenged 77-fold (Genesis 4:24). Zechariah, a major biblical prophet, speaks of the 7 eyes of the Lord. The idea of 7 divine eyes occurs in Sufism in connection with 7 important saints who are the eyes of God. God is praised by creatures with 70,000 heads, each of which has 70,000 faces. There are 7 points in the body upon which mystics concentrate their spiritual power. Seven is important for Kabbalists. In fact, Trachtenberg, in his *Jewish Magic and Superstition*, mentions the following cure for tertian (malarial) fever: "Take 7 pickles from 7 palmtrees, 7 chips from 7 beams, 7 nails from 7 bridges, 7 ashes from

7 ovens, 7 scoops of earth from 7 door sockets, 7 pieces of pitch from 7 ships, 7 handfuls of cumin, and 7 hairs from the beard of an old dog, and tie them to the neck-hole of the shirt with a white twisted cord."

- - - - - -

The number 666 and society. On July 10, 1991, Procter & Gamble announced that it was redesigning its moon-and-stars company logo. The company said that it eliminated the curly hairs in the man-in-the-moon's beard that to some looked like 6s. The fall 1991 issue of the *Skeptical Inquirer* notes that "the number 666 is linked to Satan in the Book of Revelation, and this helped fuel the false rumors fostered by fundamentalists." A federal judge in Topeka, Kansas, has approved settlements in the last of a dozen lawsuits filed by Procter & Gamble Co. to halt rumors associating the company with Satanism.

⚠ **More on 666.** President Ronald Reagan altered his California address to avoid the number 666. His name becomes 666 if you count the letters in his name.

⚠ **Even more on 666.** Do you find it interesting that when you add up the Roman numerals I = 1, V = 5, X = 10, L = 50, C = 100, and D = 500, you get 1 + 5 + 10 + 50 + 100 + 500 = 666?

⚠ **Too much more on 666.** On May 1, 1991, the British vehicle licensing office stopped issuing license plates bearing the number 666. The winter 1992 issue of the *Skeptical Inquirer* reports two reasons given for the decision: cars with "666" plates were involved in too many accidents, and there were "complaints from the public."

⚠ **Continued fraction.** Here's a nice-looking continued fraction for you to ponder late at night:

$$\sqrt{2} = 1 + \cfrac{1}{2 + \cfrac{1}{2 + \cfrac{1}{2 + \cfrac{1}{\cdots}}}}$$

Here's another:

$$\sqrt{3} = 1 + \cfrac{1}{1 + \cfrac{1}{2 + \cfrac{1}{1 + \cfrac{1}{2 + \cfrac{1}{1 + \cdots}}}}}$$

📖 **Definition of coprime or relatively prime.** Number theorists call two numbers A and B that have no common factors (except 1) "relatively prime" or "coprime." The probability that two numbers selected at random are coprime is $6/\pi^2$. Pi crops up in all areas of mathematics.

✍ **Picasso on computers.** "Computers are useless. They can only give you answers" (Pablo Picasso).

⚠ **The incredible $\pi^2/6$.** The value $\pi^2/6$, denoted by λ, is everywhere in mathematics. For example, it appears in the sum of the reciprocals of the squares of the positive integers:

$$\lambda = \frac{\pi^2}{6} = \sum_{n=1}^{\infty} \frac{1}{n^2}$$

The probability that a randomly chosen integer is square free (not divisible by a square) is $1/\lambda$. The hypervolume of a 4-dimensional hypersphere = $3\lambda r^4$. The integral from 0 to infinity of $x/(e^x - 1)dx$ is λ. We also have:

$$\frac{\pi^2}{6} = -2e^3 \sum_{n=1}^{\infty} \frac{1}{n^2} \cos\left(\frac{9}{n\pi + \sqrt{n^2 \pi^2 - 9}} \right)$$

$$\frac{\pi^2}{6} = \pi - 1 + \sum_{n=1}^{\infty} \frac{\cos(2n)}{n^2}$$

Yes, there are countless examples of $\pi^2/6$ in the realm of mathematics. In fact, Clive Tooth is so excited about the fantastic occurrences of $\pi^2/6$ in mathematics and beyond that he has devoted and dedicated a web page to this topic: www.pisquaredoversix.force9.co.uk/.

Ramanujan's most "beautiful" formula. Ramanujan's formula, as follows, draws a shocking connection between an infinite series (at left) and a continued fraction (middle). It is wonderful that both the series and the continued fraction can be expressed through the famous numerical constants π and e, and their sum mysteriously equals $\sqrt{\pi e/2}$.

$$1+\frac{1}{1\cdot3}+\frac{1}{1\cdot3\cdot5}+\frac{1}{1\cdot3\cdot5\cdot7}+\frac{1}{1\cdot3\cdot5\cdot7\cdot9}+...=$$

$$\cfrac{1}{1+\cfrac{1}{1+\cfrac{2}{1+\cfrac{3}{1+\cfrac{4}{1+...}}}}}=\sqrt{\frac{\pi e}{2}}$$

Calculating pi. In 1998, seventeen-year-old Colin Percival calculated the five trillionth binary digit of pi. His accomplishment is significant not only because it was a record-breaker, but because, for the first time ever, the calculations were distributed among twenty-five computers around the world. In all, the project, dubbed PiHex, took five months of real time to complete and one and a half years of computer time.

The temple of mathematics. "Like many great temples of some religions, mathematics may be viewed only from the outside by those uninitiated into its mysteries. . . . Understanding [its] methods is reserved for those who devote years to the study of mathematics" (Andrew Gleason, "Evolution of an Active Mathematical Theory," *Science*, 1964).

Amateurs and e. In 1998, the self-taught inventor Harlan Brothers and the meteorologist John Knox developed an improved way of calculating a fundamental constant e (often rounded to 2.718). Studies of exponential growth—from bacterial colonies to interest rates—rely on e, which can't be expressed as a fraction and can only be approximated using computers. Knox demonstrated that amateurs continue to make strides in mathematics and can help to find more accurate ways of calculating fundamental mathematical constants.

The vast application of π. As an example of how far π has drifted from its simple geometrical interpretation, consider the book *A Budget of Paradoxes*, where Augustus De Morgan explains an equation to an insurance salesman. The formula, which gives the chances that a particular group of people would be alive after a certain number of days, involves the number π. The insurance salesman interrupts and exclaims, "My dear friend, that must be a delusion. What can a circle have to do with the number of people alive at the end of a given time?"

Even more recently, π has turned up in equations that describe subatomic particles, light, and other quantities that have no obvious connection to circles.

The ubiquity of π. "Every now and again one comes across an astounding result that closely relates two foreign objects which seem to have nothing in common. Who would suspect, for example, that on the average, the number of ways of expressing a positive integer n as a sum of two integral squares, $x^2 + y^2 = n$, is $π$" (Ross Honsberger, *Mathematical Gems III*, 1997).

Definition of Chaitin's constant. Chaitin's constant (Ω) is an irrational number that gives the probability that a "Universal Turing Machine" (for any set of instructions) will halt. The digits in Ω are random and cannot be computed prior to the machine halting. (A Turing Machine is a theoretical computing machine that consists of an infinitely long magnetic tape on which instructions can be written and erased, a single-bit register of memory, and a processor capable of carrying out certain simple instructions. The machine keeps processing instructions until it reaches a particular state, causing it to halt.) Chaitin's constant has implications for the development of human

and natural languages and gives insight into the ultimate potential of machines. You can learn more about this constant here: "Chaitin's Constant," en2.wikipedia.org/wiki/Chaitin's_constant and "Chaitin's Constant at Mathworld," mathworld.wolfram.com/ChaitinsConstant.html.

Even prime number. The numeral 2 is the only even prime number. In the words of Richard Guy, this makes it the "oddest prime of all." Notice that $2 + 2 = 2 \times 2$, which gives it a unique arithmetic property among the positive integers.

Definition of Euler's gamma. Euler's gamma ($γ$) has a numerical value of 0.5772157 . . . This number links the exponentials and logs to number theory, and it is defined by the limit of $(1 + 1/2 + 1/3 + . . . + 1/n - \log n)$ as n approaches infinity. The reach of $γ$ is far and wide, playing roles in such diverse areas as infinite series, products, probability, and definite integral representations.

Calculating $γ$ has not attracted the same public interest as calculating $π$, but $γ$ has still inspired many ardent devotees. While we presently know $π$ to trillions of decimal places, only several thousand places of $γ$ are known. The evaluation of $γ$ is considerably more difficult than $π$. Here are the first few digits:

0.57721 56649 01532 86060 65120 90082 40243
10421 59335 93992 35988 05767 23488 48677
26777 66467 09369 47063 29174 67495 . . .

Today, $γ$ is often known as the *Euler-Mascheroni constant*; however, the term *Euler's constant* is also used because in 1781, Euler symbolized this constant by $γ$ and calculated it to sixteen digits. No one knows whether $γ$ is rational, algebraic, or transcendental. We do know that if $γ$ is rational, its denominator is huge. I talk more about $γ$ in chapter 3 when discussing the harmonic series.

John Conway and Richard Guy write in *The Book of Numbers*, "Nobody has shown that $γ$ cannot be rational. We're prepared to bet that it is transcendental, but we don't expect to see a proof *during our lifetime*."

Poussin proof. In 1838, Dirichlet proved that the average number of divisors of all the numbers from 1 to n is very close to $\ln n + 2\gamma - 1$. In 1898, de la Vallee Poussin proved that if a large number n is divided by *all* the primes up to n, then the average fraction by which the quotient falls short of the next whole number is γ. For example, if we divide 29 by 2, we get 14.5, which falls short of 15 by 0.5.

Beautiful equation. Here's a beautiful and wondrous expression involving a limit that connects not only π and e, but also radicals, factorials, and infinite limits. Surely, this little-known beauty makes the gods weep for joy:

$$\lim_{n \to \infty} \frac{e^n n!}{n^n \cdot \sqrt{n}} = \sqrt{2\pi}$$

Ubiquitous 2. Powers of 2 appear more frequently in mathematics and physics than do those of any other number.

Cool equation. The following equation is certainly eye-catching. Notice that the sums on each side of the "=" sign total 365—the number of days in a year.

$$10^2 + 11^2 + 12^2 = 13^2 + 14^2$$

Mathematics and passion. "The union of the mathematician with the poet, fervor with measure, passion with correctness, this surely is the ideal" (William James, *Collected Essays*, 1892).

Calculating π. In 1996, the researchers David Bailey, Peter Borwein, and Simon Plouffe found a novel formula to calculate any digit of π without having to know any of the preceding digits, an accomplishment that was assumed for centuries to be impossible.

$$\pi = \sum_{k=0}^{\infty} \frac{1}{16^k} \left[\frac{4}{8k+1} - \frac{2}{8k+4} - \frac{1}{8k+5} - \frac{1}{8k+6} \right]$$

(This formula permits one to compute the nth binary or hexadecimal digit of π by means of a simple scheme that requires very little memory and no multiple precision software.)

Large triangular numbers. We can construct arbitrarily large triangular numbers by adding zeros to 55, such as in 55; 5,050; 500,500; and 50,005,000. These are all triangular! Therefore, one large triangular number is

50000000000000000000000000000000 \
5000000000000000000000000000000

(The "\" indicates continuation of the number on the next line.) We defined triangular numbers earlier, in the puzzle titled "Dr. Brain's mystery sequence."

Prime triangle. In the seventeenth century, mathematicians showed that the following numbers are all prime:

31
331
3331
33331
333331
3333331
33333331

At the time, some mathematicians were tempted to assume that all numbers of this form were prime; however, the next number in the pattern—333,333,331—turned out not to be prime because $333,333,331 = 17 \times 19,607,843$.

- - - - - -

73,939,133. Amazingly, this is the largest number known such that all of its digits produce prime numbers as they are stripped away from the right!

73939133
7393913
739391
73939
7393
739
73
7

The Golay-Rudin-Shapiro sequence. This is a famous and unusual number sequence in the history of mathematics. We can examine the terms in the GRS sequence $\{a(n)\}$, which can be defined recursively by the equations

$$a(2n) = a(n),$$

$$a(2n + 1) = (-1)^n a(n), n \geq 0,$$

$$a(0) = 1$$

We know from past work completed in the 1990s that the solution to this recurrence may be expressed as

$$a(n) = (-1)^{e_0 e_1 + \ldots + e_{k-1} e_k}, \text{ where } n = \sum_{r=0}^{k} e_r \, 2^r, e_r = 0 \text{ or } 1$$

Thus, the GRS sequence is a sequence of ± 1s. The first few terms of the sequence, starting with $n = 0$, are 1, 1, 1, –1, 1, 1, –1, 1, and the big question is, Does the number of +1s exceed the number of –1s as we examine more terms? In other words, as we add up more terms, do successive sums remain positive? Mathematically, the question can be stated, Is $s(n) > 0$ for $n \geq 0$, where $s(n)$ is defined as

$$s(n) = \sum_{k=0}^{n} a(k), n \geq 0.$$

The values of $s(n)$ start out as 1, 2, 3, 2, 3, 4, 3, 4, 5, 6, 7, 6, 5, 4, 5, 4, . . . I'd be curious to hear from readers who have solved this mystery. (For further reading, see John Brillhart and Patrick Morton, "A Case Study in Mathematical Research: The Golay-Rudin-Shapiro Sequence," *The American Mathematical Monthly* 103, no. 10 [December 1996]: 854–69.)

Prime number theorem. In the nineteenth century, it was shown that the number of primes less than or equal to n approaches $n/(\ln n)$ as n gets very large. This is often called the prime number theorem and written as

$$\pi(n) \approx \frac{n}{\ln n}$$

A rough estimate for the nth prime is $n \times \ln n$.

Pi-prime. The number 31,415,926,535,897,932,384, 626,433,832,795,028,841 contains the first 38 decimal digits of pi and is also a prime number. Mark Ganson has searched for other pi-primes and has found no others with more than 38 digits, even after he scanned successive digits to create increasingly large numbers up to the first 3,000 digits of pi.

- - - - - -

God's formula. This formula is one of the most profound and enigmatic formulas known to humans:

$$1 + e^{i\pi} = 0$$

Some people believe that this compact formula is surely proof of a Creator. Others have actually called $1 + e^{i\pi} = 0$ "God's formula." Edward Kasner and James Newman, in *Mathematics and the Imagination*, note, "We can only reproduce the equation and not stop to inquire into its implications. It appeals equally to the mystic, the scientist, the mathematician." This formula of Leonhard Euler (1707–1783) unites the five most important symbols of mathematics: 1, 0, π, e, and i (the square root of −1). This

Undulating primes. Undulating prime numbers are primes with oscillating digits. Here's a nice one for you:

$$7 + 720 \times (100^{49} - 1)/99 =$$

7272727272727272727272727272727272727272727272727

2727272727272727

27272727272727272727272727272727

Mark Ganson discovered the following 515-digit example in 2004:

9292929292929292929292929292929292929292929292929292
9292929292929292929292929292929292929292929292929292
9292929292929292929292929292929292929292929292929292
9292929292929292929292929292929292929292929292929292
9292929292929292929292929292929292929292929292929292
9292929292929292929292929292929292929292929292929292
9292929292929292929292929292929292929292929292929292
9292929292929292929292929292929292929292929292929292
9292929292929292929292929292929292929292929292929292
9292929292929292929292929292929292929292929292929292
929292929292929

union was regarded as a *mystic union*, containing representatives from each branch of the mathematical tree: arithmetic is represented by 0 and 1, algebra by the symbol i, geometry by π, and analysis by the transcendental number e. The Harvard mathematician Benjamin Pierce said about the formula, "That is surely true, it is absolutely paradoxical; we cannot understand it, and we don't know what it means, but we have proved it,

and therefore we know it must be the truth."

- - - - - -

Golden ratio. The golden ratio, symbolized by ϕ, appears with amazing frequency in mathematics and nature. I can explain the proportion most easily by dividing a line into two segments so that the ratio of the whole segment to the longer part is the same as the ratio of the longer part to the shorter part: $AB/AC = AC/CB = 1.61803 \ldots$ (See figure 2.3a).

Figure 2.3a   Golden ratio.

If the lengths of the sides of a rectangle are in the golden ratio, then the rectangle is a golden rectangle. It's possible to divide a golden rectangle into a square and a golden rectangle. Then we can cut the smaller golden rectangle into a smaller square and a golden rectangle. We could continue this process indefinitely, producing smaller and smaller golden rectangles.

If you draw a diagonal from the top left of the original rectangle to the bottom right, and then from the bottom left of the baby golden rectangle to the top right, the intersection point shows the point to which all the baby golden rectangles converge (see figure 2.3b). And the diagonals' lengths are in golden ratio to one another. In honor of the various "divine" properties attributed to the golden ratio over the centuries, I call the point to which all the golden rectangles converge "The Eye of

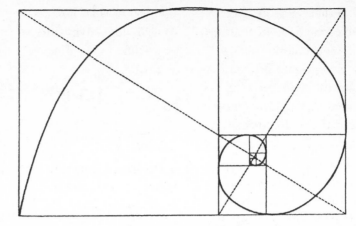

Figure 2.3b   Eye of God.

God." We can keep magnifying the figure but can never get to the Eye using finite magnifications.

The golden rectangle is the *only* rectangle from which a square can be cut so that the remaining rectangle will always be similar to the original rectangle. If we connect the vertices, we form a logarithmic spiral that "envelops" the Eye of God. Logarithmic spirals are everywhere— seashells, horns, the cochlea of the ear—anywhere that nature needs to fill space economically and regularly. A spiral is strong and uses a minimum of materials. While expanding, it alters its size but never its shape.

Mathematical cranks. According to Underwood Dudley's *Mathematical Cranks,* one eccentric mathematician put forth this seemingly elementary equation that, alas, incorrectly relates the golden ratio ϕ and π in a compact formula:

$$\frac{6}{5}\phi^2 = \pi$$

Another favorite but incorrect formula relates the value of pi to various physical constants, such as the speed of light, c, and Planck's constant, h:

$$\pi = \sqrt{\frac{E}{1/2\,mc^2}}\, J_\lambda \cdot \lambda^5 \,(e^{hc/k_\lambda T} - 1)$$

Golden ratio. The golden ratio $\phi = (1 + \sqrt{5})/2$ pops up in the most unlikely of places. In figure 2.4, the radius of the large circle divided by the diameter of one of the small circles is the golden ratio.

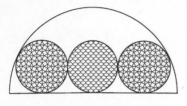

Figure 2.4   Golden circles.

Approximation to e. Ed Pegg Jr. derived the following approximation for Euler's number:

$$e \approx 3 - \sqrt{\frac{5}{63}}$$

Approximation for π. Borwein, Borwein, and Bailey once showed that

$$\left(1 + \frac{1}{\pi}\right)^{1+\pi} \approx \pi$$

Cool formula for π. I like this formula (Calvin Clawson, *Mathematical Mysteries*, 1996, p. 138):

$$\frac{4}{\pi} = 1 + \cfrac{1}{3 + \cfrac{4}{5 + \cfrac{9}{7 + \cfrac{16}{9 + \cfrac{25}{11 + \cfrac{36}{13+\ldots}}}}}}$$

Cool formula for π. I also like this formula:

$$\frac{4}{\pi} = \cfrac{1}{1 + \cfrac{1^2}{2 + \cfrac{3^2}{2 + \cfrac{5^2}{2 + \cfrac{7^2}{\ldots}}}}}$$

Cool formula for π. In 1666, Isaac Newton found π to 16 places using 22 terms of this series:

$$\pi = \frac{3\sqrt{3}}{4} + 24\left(\frac{1}{12} - \frac{1}{5 \cdot 2^5} - \frac{1}{28 \cdot 2^7} - \frac{1}{72 \cdot 2^9} - \ldots\right)$$

Regarding this queer formula, he wrote, "I am ashamed to tell you to how many figures I carried these computations, having no other business at the time."

Ramanujan π formula. Srinivasa Ramanujan (1887–1920) proposed this formula for computing pi:

$$\frac{1}{\pi} = \frac{2\sqrt{2}}{9801} \sum_{k=0}^{\infty} \frac{(4k)!(1103 + 26390k)}{(k!)^4 \, 396^{4k}}$$

Ramanujan's notebooks contain about 4,000 theorems, with about two-thirds being new to mathematics. The other one-third represented independent rediscoveries of other mathematicians' work (Robert Kanigel, *The Man Who Knew Infinity*, 1991, p. 203).

Physics and π. In quantum physics, the Heisenberg Uncertainty Principle involves π. Heisenberg's principle states that one cannot simultaneously know both the position and the momentum of a given object to arbitrary precision. One can measure the uncertainty in position Δx and the uncertainty of the momentum Δp and get

$$\Delta x \Delta p \geq \frac{h}{4\pi}$$

where h is Planck's constant.

Isaac Asimov and π. Isaac Asimov once proposed the mnemonic technique to memorize a sentence in which the number of letters in each word in turn is equal to each corresponding digit of pi, 3.1415 . . . : *How I want a drink, alcoholic, of course, after the heavy lectures involving quantum mechanics!*

- - - - - -

Cool formula for π. I love this pretty-looking formula for pi:

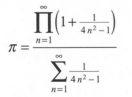

(From J. Sondow, "Problem 88," *Math Horizons* [September 1997], pp. 32 and 34).

- - - - - -

Mathematics and God. "One merit of mathematics few will deny: it says more in fewer words than any other science. The formula, $e^{i\pi} = -1$ expressed a world of thought, of truth, of poetry, and of the religious spirit. 'God eternally geometrizes'" (David Eugene Smith, in N. Rose, *Mathematical Maxims and Minims*, 1988.)

Chudnovsky π formula. Near the end of the twentieth century, the Chudnovsky brothers proposed this formula for computing π:

$$\frac{1}{\pi} = 12 \sum_{k=0}^{\infty} \frac{(-1)^k (6k)!(13591409 + 545140134k)}{(3k)!(k!)^3 640320^{3k+3/2}}$$

— — — — — — — — — —

Pi. Here is π to 1,000 decimal digits:

3.14159265358979323846264338327950288419716939937510582097494
45923078164062862089986280348253421170679821480865132823066470
93844609550582231725359408128481117450284102701938521105559644
62294895493038196442881097566593344612847564823378678316527120
19091456485669234603486104543266482133936072602491412737245870
06606315588174881520920962829254091715364367892590360011330530
54882046652138414695194151160943305727036575959195309218611738
19326117931051185480744623799627495673518857527248912279381830
11949129833673362440656643086021394946395224737190702179860943
70277053921717629317675238467481846766940513200056812714526356
08277857713427577896091736371787214684409012249534301465495853
71050792279689258923542019956112129021960864034418159813629774
77713099605187072113499999983729780499510597317328160963185950
24459455346908302642522308253344685035261931188171010003137838
75288658753320838142061717766914730359825349042875546873115956
28638823537875937519577818577805321712268066130019278766111959
092164201989

Euler and primes. Leonhard Euler expressed a pleasing and remarkable infinite product formula that relates pi to the nth prime number:

$$\pi = \frac{2}{\displaystyle\prod_{n=2}^{\infty}\left[1 + \frac{(-1)^{(p_n - 1)/2}}{p_n}\right]}$$

On the value of Euler's number e: "Dead bodies lose heat exponentially, and therefore e can be used in an appropriate equation to determine how long individuals have been dead" (Calvin Clawson, *Mathematical Mysteries*, 1996).

Cool π equation. Beginning with any positive integer n, your job is to round it up to the nearest multiple of $(n-1)$, then up to the nearest multiple of $(n-2)$, and so on, up to the nearest multiple of 1. Let $f(n)$ denote the result. Then the ratio turns out to be, amazingly,

$$\lim_{n \to \infty} \frac{n^2}{f(n)} = \pi$$

For example, $f(10) = 34$, because the procedure yields: $10 \to 18 \to 24 \to 28 \to 30 \to 30 \to 32 \to 33 \to 34$. It's possible to prove that as n tends to infinity, the ratio $n^2/f(n)$ tends to π.

Note that π is the limit of many infinite sums and infinite products. However, this is one of the most remarkable procedures that I've ever seen. If we let $g(n) = n^2/f(n)$, then, for example, $g(22) = 22/7$, a useful approximation. The most accurate approximation with $n < 10{,}000$ is $g(5{,}076) \sim 3.141592357$, which has the first six decimal places correct. Figure 2.5 is a graph of $g(n)$ versus n, with n in the range of 0 to 200. A horizontal line is at $y = \pi$. Notice that $g(n)$ does exceed π occasionally.

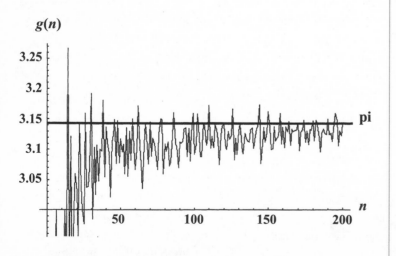

Figure 2.5 $g(n)$ is a close approximation to π as n increases.

Cool π formula. James Gregory (1638–1675) and Gottfried Leibniz (1646–1716) found this eye candy for pi:

$$\pi = 4\left(1 - \tfrac{1}{3} + \tfrac{1}{5} - \tfrac{1}{7} \ldots\right)$$

Alas, this sum converges so slowly that 300 terms are not sufficient to calculate π correctly to two decimal places.

– – – – – – – –

Rivera palindromic π:

$$\pi \approx \frac{666}{212}$$

– – – – – – – –

Mathematical laws. "Numbers written on restaurant bills within the confines of restaurants do not follow the same mathematical laws as numbers written on any other pieces of paper in any other parts of the Universe. This single statement took the scientific world by storm. It completely revolutionized it. So many mathematical conferences got held in such good restaurants that many of the finest minds of a generation died of obesity and heart failure and the science of math was put back by years" (Douglas Adams [1952–2001], *Life, the Universe and Everything*, 1995).

"It can be of no practical use to know that π is irrational, but if we can know, it surely would be intolerable not to know" (E. C. Titchmarsh, in N. Rose, *Mathematical Maxims and Minims*, 1988).

❶ Cool π formula. Leonhard Euler found this cute-looking formula for π:

$$\pi = \sqrt{6 \cdot \left(1 + \frac{1}{2^2} + \frac{1}{3^2} + \frac{1}{4^2} + \ldots\right)}$$

I presented this formula in a slightly different format a few pages earlier when I talked about how the value $\pi^2/6$ is everywhere in mathematics. Of this and related formulas for π, Calvin Clawson writes in *Mathematical Mysteries*: "How can this be? How can the sum of an infinite series be connected to the ratio of the circumference of a circle to its diameter? This demonstrates one of the most startling characteristics of mathematics—the interconnectedness of, seemingly, unrelated ideas. . . . It is as if there existed some great landscape of meta-mathematics, and we are only seeing the peaks of mathematical mountains above the valley fog."

Could some intelligent aliens have such superior brains that the fog is lifted from their eyes and all the interconnections become apparent?

- - - - - - - -

❶ Cool π formula:

$$\frac{\pi}{2} = 1 + \frac{1}{3}\left(1 + \frac{2}{5}\left(1 + \frac{3}{7}\left(1 + \frac{4}{9}\left(1 + \ldots\right)\right)\right)\right)$$

- - - - - - - -

❶ Cool Ramanujan π formula. Ramanujan discovered this gem for computing π—a method that involves two attractive-looking double factorials.

$$\frac{1}{\pi} = \sqrt{8} \sum_{n=0}^{\infty} \frac{(1103 + 26390n)(2n-1)!!(4n-1)!!}{99^{4n+2}\, 32^n\, (n!)^3}$$

❶ Gosper equation. William Gosper found this nice-looking relationship:

$$\lim_{n \to \infty} \prod_{i=n}^{2n} \frac{\pi}{2 \tan^{-1} i} = 4^{1/\pi}$$

- - - - - - - -

❶ Cool π equation. An approximation for π, from Schroeppel and Gosper's "Hakmem":

$$\sum_{n=0}^{\infty} \frac{n!\,n!}{(2n)!} = \frac{4}{3} + \frac{2\pi}{9\sqrt{3}}$$

- - - - - - - -

❶ Cool π equation. The value of $(\pi + 20)^i$ is extremely close to the integer −1. In particular, $(\pi + 20)^i \approx -0.9999999992 - 0.0000388927i$.

- - - - - - - -

⊘ **Mathematical microscope.** "If the entire Mandelbrot set were placed on an ordinary sheet of paper, the tiny sections of boundary we examine would not fill the width of a hydrogen atom. Physicists *think about* such tiny objects; only mathematicians have microscopes fine enough to actually observe them" (John Ewing, "Can We

See the Mandelbrot Set?" *The College Mathematics Journal*, 1995).

🌀 **1/137.** The fine-structure constant in physics is a dimensionless number that is very nearly equal to 1/137. It is also called the "electromagnetic coupling constant" and measures the strength of the electromagnetic force that governs how electrically charged elementary particles, such as electrons, interact with photons. The German physicist Arnold Sommerfeld introduced the quantity in 1916. Because the constant is nearly equal to 1/137, some mathematicians and mystics have wondered whether it has a more cosmic significance or a mathematical significance like other fundamental constants, such as pi or *e*. The physicist Arthur Eddington at one time argued that the number somehow related to the number of electrons in the universe—a theory that most physicists did not accept. Just this year, I even saw a paper by a creative Russian researcher that relates the fine structure constant α to π and the golden ratio ϕ:

$$\alpha^{20} = \sqrt[13]{\pi \phi^{14}} \cdot 10^{-43}$$

At his "World of Physics" Web site (scienceworld.wolfram.com/physics/), Eric W. Weisstein notes that the fine structure constant continues to fascinate numerologists, who have claimed that connections exist between α, the Cheops pyramid, and Stonehenge! Weisstein includes such curious very close approximations given by

$$\alpha^{-1} \approx 44\pi - \cos^{-1}(e^{-1}) = 137.03600\ldots$$

and

$$\alpha^{-1} \approx 96(e^{1/2} + 2^{1/3})^{1/3} = 137.03599\ldots$$

🌀 **Prime numbers and *e*.** The mathematician Joseph Pe enjoys trying to write *e* (2.7182 . . .) as a concatenation of prime numbers. Here is the beginning of his research:

2,

7,

182818284590452353602874713526624977572470936999595749669676277240766303535475945713821785251664274274663919320030599218174135966290435729003342952605956307381323286279434907632338298807531952510190115738341879307021540891499348841675092447614606680822648001684774118537423454424371075390777449920695517027618386062613313845830007520449338265602976067371132007093287091274437470472306969772093101416928368190255151086574637721112523897844250569536967707854499699679468644549059879316368892300987931277361782154249992295763514822082698951936680331825288693984964651058209392398294887933203625094431173012381970684161403970198376793206832823764648042

9 (649 digits),

5,

3,

11,

· · ·

To find these numbers, Pe continues to scan by adding a digit at a time until he finds a prime. For example, scanning "1719" from left to right, we see first the prime, 17; then, continuing with the leftover digits, we see the prime 19.

Feynman and 1/137. "There is a most profound and beautiful question associated with the observed coupling constant, the amplitude for a real electron to emit or absorb a real photon. [A simple number 1/137] has been a mystery ever since it was discovered more than fifty years ago, and all good theoretical physicists put this number up on their wall and worry about it. Immediately you would like to know where this number for a coupling comes from: is it related to pi or perhaps to the base of natural logarithms? Nobody knows. It's one of the greatest damn mysteries of physics: a magic number that comes to us with no understanding by man. You might say the 'hand of God' wrote that number, and 'we don't know how He pushed His pencil.' We know what kind of a dance to do experimentally to measure this number very accurately, but we don't know what kind of dance to do on the computer to make this number come out, without putting it in secretly!" (Richard Feynman, *QED*, 1988).

- - - - - -

Cool π equation. Note that $e^\pi - \pi$ is almost an integer. It equals 19.9990999 . . .

Mandelbrot set and π. In 1991, David Boll discovered a strange connection between π and the classic Mandelbrot set (figure 2.6), which can be visualized as a bushy object that describes the behavior of $z = z^2 + c$, where z and c are complex numbers.

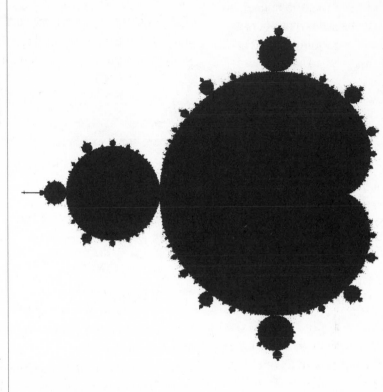

Figure 2.6   Mandelbrot set.

Boll studied points lying on a vertical line through the point $z = (-0.75, 0)$ and discovered that the number of iterations needed for these points to escape a circle of radius 2 (centered at the origin) is related to π. In particular, consider points $z = -0.75 \pm \varepsilon i$ and the n iterations needed for orbits of these points to escape the circle. As ε goes to 0, $n\varepsilon$ approaches π.

Prime contest. Researchers sometimes hold zany but curious contests in which they search for prime numbers within the decimal digit expansion of $1/137 = 0.0072992700729927700 \ldots$ For example, the mathematician Jason Earls has discovered that the following is prime:

72992700729927007299270072992
70072992700729927007299270072
99270072992700729927007299270
07299270072992700729927007299
27007299270072992700729927007
29927007299270072992700729927
00729927007299270072992700729
92700729927007299270072992700
72992700729927007299270072992
70072992700729927007299270072
99270072992700729927007299270
07299270072992700729927007299
27007299270072992700729927007

This number includes the first 371 digits of $1/137$, with the first two zeros omitted.

Hermann Schubert, Sirius, microbes, and pi. In 1889, the Hamburg mathematics professor Hermann Schubert described how there is no practical or scientific value to knowing pi to more than a few decimal places: "Conceive a sphere constructed with the earth at its center, and imagine its surface to pass through Sirius, which is 8.8 light years distant from the earth [8.8 years × 186,000 miles per second]. Then imagine this enormous sphere to be so packed with microbes that in every cubic millimeter millions of millions of these diminutive animalcula are present. Now conceive these microbes to be unpacked and so distributed singly along a straight line that every two microbes are as far distant from each other as Sirius from us, 8.8 light years. Conceive the long line thus fixed by all the microbes as the diameter of a circle, and imagine its circumference to be calculated by multiplying its diameter by π to 100 decimal places. Then, in the case of a circle of this enormous magnitude even, the circumference so calculated would not vary from the real circumference by a millionth part of a millimeter. This example will suffice to show that the calculation of π to 100 or 500 decimal places is wholly useless."

Prime test. Is it possible for a person of average intelligence to determine whether this large number is prime or composite (not prime) within 19 seconds?

$$5{,}230{,}096{,}303{,}003{,}196{,}309{,}630{,}967$$

(See Answer 2.76.)

Definition of an amenable number. An amenable number is an integer that can be constructed from integers $m_1, m_2, m_3, \ldots, m_k$ by both multiplication and addition such that

$$\sum_{i=1}^{k} m_i = \prod_{i=1}^{k} m_i$$

Here are some examples:

$$2 + 2 = 2 \times 2 = 4$$
$$1 + 1 + 2 + 4 = 1 \times 1 \times 2 \times 4 = 8$$

For more information, see Harry Tamvakis, "Problem 10454," *The American Mathematical Monthly* 102 (1995): 463.

π^3 and antennas. I have stumbled upon a practical example of the term π^3 occurring in physics. In particular, π^3 occurs in equations that relate to gravitational wave antennas and energy absorption rates, which have terms that look like

$$2\pi^3 ML^3 QG/3C^2\lambda S$$

where M is the reduced mass, L is the length, Q is the quality factor, G is Newton's gravitational constant, C is the speed of light in a vacuum, λ is the wavelength, and S is the length occupied by a single plane quadrupole. (See Melvin A. Lewis, "Gravitational Wave vs. Electromagnetic Wave Antennas," June 1995 issue of IEEE *Antennas and Propagation Magazine*.) Can you find any other examples of π^3 in physics or geometry?

- - - - - -

Never prime? Does this formula always produce a number that is never prime: $a^2 + b^2 + c^2 + d^2$? Variables a, b, c, and d are different positive integers. (See Answer 2.77.)

- - - - - -

Definition of a father prime. According to the retired math teacher Terry Trotter, a "father" prime is a prime number for which the sum of the squares of its digits is also a prime. The sum is therefore the "child" prime. For example, 23 is a "father" prime because $2^2 + 3^2 = 4 + 9 = 13$. That is, 23 is the "father" of a prime child, 13. Here is an example of an "ancestral line" of fathers and children:

191 → 83 → 3. Until recently, the longest known ancestral lines consisted of 5 generations, such as 1,499 → 179 → 131 → 11 → 2. My colleague Mark Ganson discovered a series of related ancestral lines consisting of 6 generations each. The line with the smallest starting prime is 28,999,999,999 → 797 → 179 → 131 → 11 → 2.

- - - - - -

Definition of a cyclic number. A cyclic number, C, is an integer that—when multiplied by any number from 1 to the number of digits of C—always contains the same digits as C. Also, these digits will appear in the same order but begin at a different point. An example will clarify this. 142,857 is cyclic because

$$1 \times 142{,}857 = 142{,}857$$
$$2 \times 142{,}857 = 285{,}714$$
$$3 \times 142{,}857 = 428{,}571$$
$$4 \times 142{,}857 = 571{,}428$$
$$5 \times 142{,}857 = 714{,}285$$
$$6 \times 142{,}857 = 857{,}142$$

Here are two more cyclic numbers: 588,235,294,117,647 and 52,631,578,947,368,421. Notice that these numbers can be constructed from certain primes in the following way. For example, $1/7 = 0.142857\ldots$, $1/17 = 0.0588235294117647\ldots$, and $1/19 = 0.052631578947368421\ldots$ It has been conjectured, but not yet proved, that an infinite number of cyclic numbers exist.

🌑 **Coincidental constant.** This number is pretty cool:

2.5061841455887692562929409223778472717713960521332128301 4 . . .

It is the solution to $x^x = 10$, and, as Robert Munafo says, it is curious that this is close to $\sqrt{2\pi}$, which is 2.5066 . . .

- -

🌑 **Almost an integer.** $62^{1/2} \times (127/1000) = 0.999998999999499999499999374 \ldots$ (Angel Garcia, sci.math).

- -

🌑 **666 equation.** $666 = 1^3 + 2^3 + 3^3 + 4^3 + 5^3 + 6^3 + 5^3 + 4^3 + 3^3 + 2^3 + 1^3$.

🌀 **Universe.** "It is a mathematical fact that the casting of this pebble from my hand alters the center of gravity of the universe" (Thomas Carlyle [1795–1881], *Sartor Resartus III*, 1831).

- - - - - -

🌑 **0.065988 . . . .** I'm in love with the constant $0.065988 \ldots = e^{-e} = (1/e)^e$. It is the lowest value of the function $y = x^{(1/x)}$. Moreover, if you consider this beautiful infinite power tower,

$$f(x) = x^{x^{x^{x^{x \cdots}}}}$$

then 0.065988 . . . is the lowest value of x for which the function converges to a finite value.

🌑 **0.692200 . . . = $(1/e)^{(1/e)}$.** This cool constant is the lowest point in the function $y = x^x$.

- - - - - -

🌑 **1.202056 . . .** is Apéry's constant. Please select three positive integers at random. The odds of them having no common divisor are 1 in 1.202056 . . .

- - - - - -

🌑 **1.506592 . . .** is the area of the famous fractal shape known as the Mandelbrot set. Cyril Soler, a French researcher at the Institut National de Recherche en Informatique et Automatique, conjectures that the value is

precisely $\sqrt{6\pi - 1} - e$, or 1.5065916514855 . . . (from Robert P. Munafo's "Mandelbrot Set Glossary and Encyclopedia," www.mrob.com/pub/muency.html).

- - - - - -

🌀 **Asimov universe.** "I believe that scientific knowledge has fractal properties, that no matter how much we learn, whatever is left, however small it may seem, is just as infinitely complex as the whole was to start with. That, I think, is the secret of the Universe" (Isaac Asimov, *I. Asimov: A Memoir*, 1994).

- - - - - -

🌑 **666 equation.** $666 = 3^6 - 2^6 + 1^6$.

- - - - - -

❓ **Integer hoax.** Is $e^{\pi\sqrt{163}}$ an integer? Why would I ask such a strange question? (See Answer 2.78.)

- - - - - -

🌑 **Impressive π formula.** $\pi \approx \ln(640320^3 + 744/\sqrt{163}$, which is correct to 30 digits after the decimal point!

- - - - - -

🌑 **Impressive π formula.** $(\pi^4 + \pi^5)/(e^6) \sim 1$. (To be precise, it equals 0.999999956 . . .)

Impressive π formula.
$\pi \approx (2143/22)^{1/4} =$
$3.14159265258264 \ldots$
(Reichenbacher, 1900).

- - - - - -

Smallest cube. Note that
8,000 is the smallest cube
that can be expressed as the
sum of 4 consecutive cubes
$(8{,}000 = 20^3 = 11^3 + 12^3 +$
$13^3 + 14^3)$.

- - - - - -

Cubes. In 1939, L. E.
Dickson proved that all positive integers can be represented as the sum of at most
9 positive cubes. Interestingly, only two numbers exist
that require all 9 cubes: 23
and 239 (L. E. Dickson,
*History of the Theory of
Numbers, Vol. 2: Diophantine
Analysis* [New York: Chelsea,
1966]).

- - - - - -

666 equation. Let $\phi(n)$ be
the number of integers smaller
than n and relatively prime to
n. Amazingly, we find that

$$\phi(666) = 6 \cdot 6 \cdot 6.$$

(Number theorists call two
numbers A and B that have no
common factors "relatively
prime" or "coprime.")

Triangles and 666. As you can see in figure 2.7, 666 is a
triangular number. Because it stands on a base of 36 marks,
it is designated the 36th triangular number. (We defined triangular numbers in the puzzle titled "Dr. Brain's mystery
sequence.")

Note that 36 is also triangular (the 8th triangular number).
The 666th triangular number contains only 2s and 1s and is
222,111.

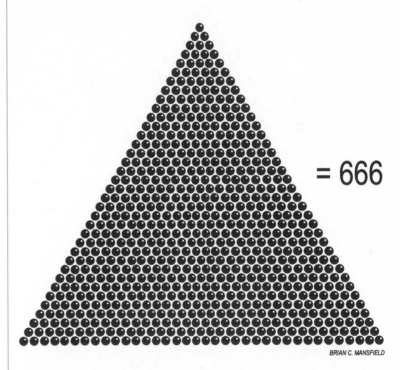

= 666

BRIAN C. MANSFIELD

Figure 2.7 A visual proof that 666 is a triangular number.

666 equation. The
number 666 is the sum of
the squares of the first 7
primes: $2^2 + 3^2 + 5^2 + 7^2 +$
$11^2 + 13^2 + 17^2$.

- - - - - -

666 equation. Examine
the prime factors of
$666 = 2 \times 3 \times 3 \times 37$ and
add their digits. You get
$18 = 6 + 6 + 6$.

- - - - - -

Tolstoy and the number 28. The great Russian novelist Leo Tolstoy believed 28 to be a special number for him. He was born on August 28, 1828. His son Sergei was born on June 28. When Tolstoy was asked to choose poems for one of his essays, he deliberately chose poems from page 28 of different poetry books.

During Tolstoy's youth, his older brother Nikolai said he'd written the secret of happiness on a green stick and hid it by a road in the Zakaz forest. Tolstoy later asked to be buried where the mystical green stick was thought to have been hidden.

Definition of primorial. Primorials are the product of the first N primes. For example, 210 is a primorial because $210 = 2 \times 3 \times 5 \times 7$. Here are the first few primorials: 2; 6; 30; 210; 2,310; 30,030; 510,510; 9,699,690; 223,092,870; 6,469,693,230; 200,560,490,130; 7,420,738,134,810; and 304,250,263,527,210.

Cube square puzzle. Consider an integer, N, and its cube, C, and its square, S. Together, C and S use all digits from 0 to 9 once. What is N? (See Answer 2.79.)

1,001 and primes. The number 1,001 is a product of three consecutive primes: $1,001 = 7 \times 11 \times 13$.

An approximation for π. $3.14159267204 \ldots = [e(\phi + 6)]/5 - 1$ is a close approximation to π.

Digits of e and π. I wonder why we find such a weird correspondence (the digits 314 and 926) in the following:

$$23.1\underline{40692}6327779269005 \ldots = e^{\pi}$$

$$3.14\underline{1592}653589793238 \ldots = \pi$$

Is this pure coincidence, or does it signify something much deeper?

------

666 formula. Perhaps I'm obsessed with 666 equations, but here is yet another:

$$666 = 6 + 6 + 6 + 6^3 + 6^3 + 6^3$$

Definition of hyperfactorial. Hyperfactorials are of the form $1^1 \times 2^2 \times 3^3 \times 4^4 \ldots$ The first few hyperfactorials are 1; 4; 108; 27,648; 86,400,000; 4,031,078,400,000; 331,976,639,877,120,000; and 55,696,437,941,726,556,979,200,000.

Palindromic triangular number. The 2,662nd triangular number is 3,544,453, so both the number and its index, 2,662, are palindromic.

Cube numbers. Consider an integer, N, that is the smallest cube number that is the sum of three cube numbers. What is N? (A cube number is simply a number created by cubing another. For example, 125 is a cube number because $125 = 5^3$.) (See Answer 2.80.)

Definition of amicable numbers. Two numbers are considered friendly or amicable to each other if the sums of their proper divisors equal each other. For example, 284 and 220 are amicable. To see why, let's list all the numbers by which 220 is evenly divisible: 1, 2, 4, 5, 10, 11, 20, 22, 44, 55, and 110 all go into 220. Now add up all those divisors: $1 + 2 + 4 + 5 + 10 + 11 + 20 + 22 + 44 + 55 = 284$. Now let's try the same approach with 284. Its perfect divisors are 1, 2, 4, 71, and 142. Now, add them up to get 220. Therefore 220 and 284 are amicable numbers. The sums of their divisors are equal to each other.

- - - - - -

Os and Is. Here are some strange equivalences:

$$011 = 8$$
$$1110 = ?$$
$$0011 = 12$$
$$10001 = 13$$
$$00111 = 23$$
$$010000001 = 26$$

Can you determine what logic I am using to equate these numbers and then can you solve the missing equivalence? (See Answer 2.81.)

Brun's constant and Barbra Streisand. Brun's constant, 1.90216 . . . , is the number obtained by adding the reciprocals of the odd twin primes

$$\left(\tfrac{1}{3}+\tfrac{1}{5}\right)+\left(\tfrac{1}{5}+\tfrac{1}{7}\right)+\left(\tfrac{1}{11}+\tfrac{1}{13}\right)+\left(\tfrac{1}{17}+\tfrac{1}{19}\right)+\ldots$$

Twin primes are pairs of primes that differ by two. The first twin primes are {3, 5}, {5, 7}, {11, 13,} and {17, 19}. It has been conjectured that there are infinitely many twin primes. No one knows for sure.

The conjecture was even mentioned in the 1996 movie *The Mirror Has Two Faces*, which starred Barbra Streisand. It is fascinating that the series converges to a definite, finite number, even if there are an infinite number of twin primes. This fact suggests the "scarcity" of twin primes relative to primes, because the sum of the reciprocals of all prime numbers diverges. We do not know very much about Brun's constant. For example, we don't even know if it is rational, algebraic, or transcendental.

Cube mystery. These numbers are very special: 1, 8, 17, 18, 26, and 27. Can you guess why? (Hint: The solution has to do with the cubes of these numbers.) (See Answer 2.82.)

- - - - - -

Surreal world. "Just as the *real* numbers fill in the gap between the integers, the *surreal* numbers fill in the gaps between Cantor's ordinal numbers" (John Conway and Richard Guy, *The Book of Numbers*, 1996).

- - - - - -

Hard to say. "In John Conway's surreal number system, it's possible to talk about whether ω is odd or even, to add 1 to infinity, to divide infinity in half, to take its square root or logarithm, and so on. Equally accessible and amenable to manipulation are the infinitesimals—the numbers generated by the reciprocals of these infinities (for example, 1/ω). What can you do with such numbers? It's still hard to say because very little research has been done on them" (Ivars Peterson, "Computing in a Surreal Realm," 1996).

Definition of surreal numbers. Surreal numbers are a superset of the real numbers, invented by John Conway for the analysis of games, although the name was coined by Donald Knuth in his popular novelette *Surreal Numbers*, perhaps one of the few times that a major mathematical discovery was published first in a work of fiction. Surreal numbers have all kinds of bizarre properties.

Surreal numbers include the real numbers plus much more. Knuth explains in a postscript that his aim was not so much to teach Conway's theory as

to teach how one might go about developing such a theory. . . . Therefore, as the two characters in this book gradually explore and build up Conway's number system, I have recorded their false starts and frustrations as well as their good ideas. I wanted to give a reasonably faithful portrayal of the important principles, techniques, joys, passions, and philosophy of mathematics, so I wrote the story as I was actually doing the research myself.

Martin Gardner has written in *Mathematical Magic Show*,

Surreal numbers are an astonishing feat of legerdemain. An empty hat rests on a table made of a few axioms of standard set theory. Conway waves two simple rules in the air, then reaches into almost nothing and pulls out an infinitely rich tapestry of numbers that form a real and closed field. Every real number is surrounded by a host of new numbers that lie closer to it than any other "real" value does. The system is truly "surreal."

A surreal number is a pair of sets $\{X_L, X_R\}$ where indices indicate the relative position (left and right) of the sets in the pair. Surreal numbers are fascinating because they are built upon an extremely small and simple foundation. In fact, according to Conway and Knuth, surreal numbers follow two rules: (1) every number corresponds to two sets of previously created numbers, such that no member of the left set is greater than or equal to any member of the right set; and (2) one number is less than or equal to another number if, and only if, no member of the first number's left set is greater than or equal to the second number, and no member of the second number's right set is less than or equal to the first number.

Surreal numbers include infinity and infinitesimals, numbers smaller than any imaginable real numbers. For more information, see John Horton Conway and Richard K. Guy, *The Book of Numbers* (New York: Copernicus/Springer, 1996). See also *Surreal Numbers: How Two Ex-Students Turned On to Pure Mathematics and Found Total Happiness* by Donald E. Knuth (Reading, Mass.: Addison-Wesley, 1974). "The Story of How Surreal Numbers Came to Be Written" is told in *Mathematical People* by Donald J. Albers and G. L. Alexanderson (Boston: Birkhauser, 1985). John Conway's presentation of the theory appears in *On Numbers and Games* (Natick, Mass.: A. K. Peters, Ltd., 2001).

– – – – – –

📖 **Definition of hyperreal numbers.** Hyperreal numbers are an extension of the real numbers that adds infinitely large, as well as infinitesimal, numbers to the real numbers. By "infinite numbers," I mean numbers whose absolute value is greater than any positive real number. By "infinitesimal numbers," I mean numbers whose absolute value is less than any positive real number. For more information, see "Hyperreal number," en2.wikipedia.org/wiki/Hyperreal_number.

⟳ **Fibonacci formula.** As we explore larger and larger Fibonacci numbers, F_n, we find that the ratio of a Fibonacci number to the one before approaches the golden mean $\phi = 1.61803 \ldots$

$$\lim_{n \to \infty} \frac{F_n}{F_{n-1}} = \phi$$

- - - - - -

⟳ **Fibonacci formula.** We can compute the nth Fibonacci number, F_n, directly from

$$F_n = \frac{\phi^n - (\phi')^n}{\sqrt{5}}$$

where $\phi = 1.61803 \ldots$ is the golden ratio and $\phi' = -0.618034 \ldots$ is the negative inverse of ϕ; in other words, $\phi' = -1/\phi$.

- - - - - -

🌀 **P-adic numbers.** "Almost your entire mathematical life has been spent on the real line and in real space working with real numbers. Some have dipped into complex numbers, which are just the real numbers after you throw in i. Are these the only numbers that can be built from the rationals? The answer is no. There are entire parallel universes of numbers that are totally unrelated to the real and complex numbers. Welcome to the world of p-adic analysis—where arithmetic replaces the tape measure and numbers take on a whole new look" (Professor Edward Burger, from a Web page description of the Williams College undergraduate mini-course "Exploring p-adic Numbers" [www.math.ksu.edu/main/course_info/courses/crs-des/burger.htm] [p-adic numbers can be used to help determine whether some equations have solutions in which the variables are all integers.]).

⟳ **Fibonacci formula.** The sum of the first n Fibonacci numbers can be calculated from

$$\sum_{i=1}^{n} F_i = F_1 + F_2 + \ldots F_n = F_{n+2} - 1$$

For example, consider the Fibonacci numbers: 1, 1, 2, 3, 5, 8, 13, . . . The sum of the first five terms is equal to $13 - 1 = 12$. We can use this handy formula to compute the sum of all Fibonacci numbers up to any term.

– –

⟳ **Fibonacci formula.** The sum of the consecutive Fibonacci numbers squared is

$$\sum_{i=1}^{n} F_i^2 = F_1^2 + F_2^2 + \ldots F_n^2 = F_n F_{n+1}$$

Consecutive prime numbers. The difference between consecutive prime numbers is always even, except for two particular prime numbers. What are they? (See Answer 2.83.)

Composite number sequence. If you were challenged to write down a sequence of consecutive composite numbers that is N numbers long, you could simply choose any N and create your list as follows: $(N+1)! + 2$, $(N+1)! + 3$, $(N+1)! + 4$, $(N+1)! + 5$ $\ldots$, $(N+1)! + (N+1)$. For example, try this for $N = 4$. The results are guaranteed to be composites.

Gaps between primes. In the 1980s, the largest gap between primes of which I was aware was a run of 803 composite numbers that exists between the primes 90,874,329,411,493 and 90,874,328,412,297 (J. Young and A. Potler, "First Occurrence Prime Gaps," *Mathematics of Computation* 52 [1989]: 221–24). A little later, Baugh and O'Hara discovered a prime gap of length 4,247 following $10^{314} - 1{,}929$. (D. Baugh and F. O'Hara, "Large Prime Gaps," *Journal of Recreational Mathematics* 24 [1992]: 186–87).

Wilson's theorem: p is a prime number if and only if $(p-1)! + 1$ is divisible by p.

Prime mysteries. "Mathematicians have tried in vain to this day to discover some order in the sequence of prime numbers, and we have reason to believe that it is a mystery into which the human mind will never penetrate" (Leonhard Euler, in George Simmons, *Calculus Gems*, 1992).

Strange radical formulas. Integers can be produced by certain nested radicals; for example,

$$2 = \sqrt{2 + \sqrt{2 + \sqrt{2 + \sqrt{2 + \ldots}}}}$$

or

$$6 = \sqrt{30 + \sqrt{30 + \sqrt{30 + \sqrt{30 + \ldots}}}}$$

In general, you can impress your friends with an infinite number of these identities of the form

$$K = \sqrt{m + \sqrt{m + \sqrt{m + \sqrt{m + \ldots}}}}$$

where $m = K^2 - K$ for $m > 0$. What do you think happens if $m = 0$ and $K = 1$? (For more information, see Calvin Clawson's *Mathematical Mysteries*, p. 141.)

Square root of zero. Although the square root of zero is zero, look at this amazingly odd beast in which we examine the limit of the following nested radicals as n approaches zero.

$$1 = \lim_{n \to 0} \sqrt{n + \sqrt{n + \sqrt{n + \sqrt{n + \ldots}}}}$$

📖 **Definition of a Wilson prime.** A prime number, p, is a Wilson prime when $(p-1)! + 1$ is evenly divisible by p^2. Today, humanity knows of only three Wilson primes: 5, 13, and 563. It is conjectured that the number of Wilson primes is infinite. (For example, 5 is a Wilson prime because 25 divides $4! + 1 = 25$.)

- - - - - -

💡 **A near isoprime.** In 1991, Harvey Dubner discovered a prime number with a total of 6,400 digits that is composed of all 9s except one 8. The precise value is $10^{6400} - 10^{6352} - 1$.

- - - - - -

🌀 **Beautiful rules.** "The mathematical rules of the universe are visible to men in the form of beauty" (John Michel).

- - - - - -

💡 **Palindromic primes.** The smallest palindromic prime containing all 10 digits is 1,023,456,987,896,543,201. One of the largest known palindromic primes was discovered by Harvey Dubner, a retired electrical engineer in New Jersey. The number consists of just 1s and 0s and has 30,803 digits. (Notice that 30,803 is also a palindrome.)

💡 **Exotic prime probe.** The function $F(j)$ is fascinating because it is equal to 1 when j is prime and 0 whenever j is a composite number:

$$F(j) = \left[\cos^2 \pi \frac{(j-1)!+1}{j} \right]$$

The quantity in brackets is truncated to an integer; for example, 4.5 would be truncated to 4, and 0.7 is truncated to 0. Calvin Clawson remarks, "This is truly an amazing function. How does it know when it's dealing with a prime or a composite?" (Calvin Clawson, *Mathematical Mysteries*, 1996).

- - - - - - - - - - - - - - - - -

💡 **Exotic prime number generator.** To calculate the nth prime number, we may use the following formula:

$$p_n = 1 + \sum_{m=1}^{2^n} \left[\left(\frac{n}{\sum_{j=1}^{m} F(j)} \right)^{\frac{1}{n}} \right]$$

where $F(j)$ is defined in the previous factoid. Square brackets indicate truncation of the number to an integer (Calvin Clawson, *Mathematical Mysteries*, 1996).

💡 **Exotic prime number generator.** The following formula yields only primes:

$$g(n) = \left[2^{2^{\cdots 2^{\beta}}} \right]$$

where $\beta = 1.92878$, and the number of exponents to the first 2 is equal to n. The brackets indicate truncation to an integer. For example, for $n = 2$, we get the prime number 13. Alas, because β is only an approximation and the precise value has not been calculated, at some point this formula fails humanity (Calvin Clawson, *Mathematical Mysteries*, 1996).

- - - - - -

Mills's constant. The following formula yields only primes for integers n:

$$F(n) = \left\lceil \alpha^{\left(3^n\right)} \right\rceil \text{ where } \alpha =$$

1.30637788386308069046861449260260571
2916784585156713644368053759966434 . . .

The constant α is sometimes called the Mills's constant. The brackets indicate truncation to an integer—sometimes called the "floor" operation. For example, for $n = 3$, we get the prime number 1,361. Alas, because α is only an approximation and the precise value has not been calculated, at some point this formula breaks down (Calvin Clawson, *Mathematical Mysteries*, 1996).

On prime numbers: "There are tetradic, pandigit, and prime-factorial plus one primes. And there are Cullen, multifactorial, beastly palindrome, and antipalindrome primes. Add to these the strobogrammatic, subscript, internal repdigit, and the elliptic primes. In fact, a whole new branch of mathematics seems to be evolving that deals specifically with the attributes of the various kinds of prime numbers. Yet understanding primes is only part of our quest to fully understand the number sequence and all of its delightful peculiarities" (Calvin Clawson, *Mathematical Mysteries*, 1996).

Definition of a Sophie Germain prime. A prime number, p, is a Sophie Germain prime if you can double it, add 1, and produce another prime. One of the largest such primes is $2{,}540{,}041{,}185 \times 2^{114729} - 1$ with 34,547 digits, discovered in 2003 by David Underbakke.

Euler's prime formula. In 1772, Leonhard Euler discovered that $f(x) = x^2 + x + 41$ yields prime numbers for $x = 0$ through 39. Perhaps Euler was very excited until he tried $x = 40$, which, sadly, yields a composite number. Still, it appears to produce primes at a high rate. For

---

example, values of x from 40 through 79 yield 33 primes.

It turns out that $2x^2 - 199$ is the second-degree polynomial that yields the greatest number of primes for the first 1,000 values of x, which produces 598 primes.

Definition of a Fibonacci prime. A Fibonacci prime is a Fibonacci number that is prime. The 81,839th Fibonacci number is prime. It has 17,103 digits and is one of the largest known prime Fibonacci numbers.

On crazy numerical coincidences. "The ratio of the height of the Sears Building in Chicago to the height of the Woolworth Building in New York is the same to four significant digits (1.816 vs. 1816) as the ratio of the mass of a proton to the mass of an electron" (John Paulos, *Innumeracy*, 1988).

Four squares. In 1770, Joseph-Louis Lagrange showed that every positive integer can be written as the sum of four squares. For example, we can write the number 31 as $2^2 + 3^2 + 3^2 + 3^2$.

The smallest star-congruent prime containing all four prime digits. The number 7,777,277,227,777,772,327,777,772,222, 332,222,772,333,533,327,723,555,532,772,352,532,772,355, 553,277,233,353,332,772,222,332,222,777,777,232,777,777, 227,727,777 is the smallest star-congruent prime containing all four prime digits—and no other digits. This means that the 121-digit number can be arranged in the form of a six-pointed star:

```
                      7
                     7 7
                    7 2 7
                   7 2 2 7
      7 7 7 7 7 2 3 2 7 7 7 7 7
        7 2 2 2 2 3 3 2 2 2 2 7
          7 2 3 3 3 5 3 3 3 2 7
            7 2 3 5 5 5 5 3 2 7
              7 2 3 5 2 5 3 2 7
            7 2 3 5 5 5 5 3 2 7
          7 2 3 3 3 5 3 3 3 2 7
        7 2 2 2 2 3 3 2 2 2 2 7
      7 7 7 7 7 2 3 2 7 7 7 7 7
                   7 2 2 7
                    7 2 7
                     7 7
                      7
```

This curious number was discovered by the mathematician Dr. Michael Hartley in 2003 and was featured at "Prime Pages' Prime Curios!" managed by G. L. Honaker Jr. and Chris Caldwell (primes.utm.edu/curios/page.php?curio_id=5284). Hartley is an associate professor at the Malaysia campus of the University of Nottingham. A "congruent prime" contains "shapes" of identical digits nested about the center when drawn in the form of squares, stars, hexagons, and so forth. For example,

```
7 7 7
7 6 7   is a square-congruent prime of order 3.
7 7 7
```

Definition of round numbers. Round numbers are numbers that, when factored, contain a large number of primes. The greater the number of prime factors, the rounder the number. If you were to select a number at random, you might be surprised at just how rare "very round numbers" are.

Although this definition is rather vague, it is easy to rank numbers in order of roundness; for example, from least round to most round we have

$$1,957 = 19 \times 103$$
$$\text{(2 factors)}$$

$$1,958 = 2 \times 11 \times 89$$
$$\text{(3 factors)}$$

$$1,960 = 2^3 \times 5 \times 7^2$$
$$\text{(6 factors)}$$

$$362,880 = 2^7 \times 3^4 \times 5 \times 7$$
$$\text{(13 factors)}$$

- - - - - -

Ramanujan's goddess. Ramanujan credits his mathematical discoveries to Namagiri of Namakkal, a Hindu goddess who appeared in his dreams. (Narasimha was the male consort of the goddess Namagiri and also helped Ramanujan with his visions.)

- - - - - -

Five factors for big numbers. The number of different primes occurring in a number may be denoted ω(n), which is approximately ln(ln(n)). Here, "ln" is the natural logarithm. Calvin Clawson comments in *Mathematical Mysteries*, "If we consider the numbers in the range of 10^{80}, which is the approximate number of protons in the universe, we can determine that most such numbers have around five factors since $\ln(\ln 10^{80}) = 5.22$."

- - - - - -

Ramanujan's problem. Find integer solutions to

$$2^N - 7 = X^2$$

This equation was searched to about $N = 10^{40}$, and humanity has found only Ramanujan's original solutions ($N = 3, 4, 5, 7, 15$). It has recently been proved that these are the only ones.

- - - - - -

Ramanujan's beauty. "When we comprehend some of Ramanujan's equations, we realize that he was a true artist, expressing deep and beautiful mathematical truth in familiar symbols" (Calvin Clawson, *Mathematical Mysteries*, 1996).

The smallest titanic hex-congruent prime. Dr. Michael Hartley discovered this magnificent 1,027-digit prime number that can be arranged in the form of nested hexagrams (see the "smallest star-congruent prime containing all four prime digits" on page 99 for a definition of a square-congruent prime). To construct the original prime from this figure, simply concatenate each row of the hexagonal figure (primes.utm.edu/curios/page.php?number_id=2657).

```
        1 1 1 1 1 1 1 1 1 1 1 1 1 1 1 1 1 1 1 1 1
        1 0 0 0 0 0 0 0 0 0 0 0 0 0 0 0 0 0 0 0 1
        1 0 0 0 0 0 0 0 0 0 0 0 0 0 0 0 0 0 0 0 0 1
        1 0 0 0 0 0 0 0 0 0 0 0 0 0 0 0 0 0 0 0 0 0 1
       1 0 0 0 0 0 0 0 0 0 0 0 0 0 0 0 0 0 0 0 0 0 0 1
      1 0 0 0 0 0 0 0 0 0 0 0 0 0 0 0 0 0 0 0 0 0 0 0 1
     1 0 0 0 0 0 0 0 0 0 0 0 0 0 0 0 0 0 0 0 0 0 0 0 0 1
    1 0 0 0 0 0 0 0 0 0 0 0 0 0 0 0 0 0 0 0 0 0 0 0 0 0 1
   1 0 0 0 0 0 0 0 0 0 0 0 0 0 0 0 0 0 0 0 0 0 0 0 0 0 0 1
  1 0 0 0 0 0 0 0 0 0 0 0 0 0 0 0 0 0 0 0 0 0 0 0 0 0 0 0 1
 1 0 0 0 0 0 0 0 0 0 0 0 0 0 0 0 0 0 0 0 0 0 0 0 0 0 0 0 0 1
1 0 0 0 0 0 0 0 0 0 0 0 0 0 0 0 0 0 0 0 0 0 0 0 0 0 0 0 0 0 1
1 0 0 0 0 0 0 0 0 0 0 0 0 0 0 0 0 0 0 0 0 0 0 0 0 0 0 0 0 1
1 0 0 0 0 0 0 0 0 0 0 0 0 0 0 0 0 0 0 0 0 0 0 0 0 0 0 0 1
1 0 0 0 0 0 0 0 0 0 0 0 0 0 0 0 0 0 0 0 0 0 0 0 0 0 0 1
1 0 0 0 0 0 0 0 0 0 0 0 0 0 0 0 0 0 0 0 0 0 0 0 0 0 1
1 0 0 0 0 0 0 0 0 0 0 0 0 0 0 0 0 0 0 0 0 0 0 0 0 0 1
1 0 0 0 0 0 0 0 0 0 0 0 0 0 0 8 8 0 0 0 0 0 0 0 0 0 0 1
1 0 0 0 0 0 0 0 0 0 0 0 0 0 8 7 8 0 0 0 0 0 0 0 0 0 0 0 1
1 0 0 0 0 0 0 0 0 0 0 0 0 0 0 8 8 0 0 0 0 0 0 0 0 0 0 0 1
1 0 0 0 0 0 0 0 0 0 0 0 0 0 0 0 0 0 0 0 0 0 0 0 0 0 0 1
1 0 0 0 0 0 0 0 0 0 0 0 0 0 0 0 0 0 0 0 0 0 0 0 0 0 1
1 0 0 0 0 0 0 0 0 0 0 0 0 0 0 0 0 0 0 0 0 0 0 0 0 1
 1 0 0 0 0 0 0 0 0 0 0 0 0 0 0 0 0 0 0 0 0 0 0 0 0 1
  1 0 0 0 0 0 0 0 0 0 0 0 0 0 0 0 0 0 0 0 0 0 0 0 1
   1 0 0 0 0 0 0 0 0 0 0 0 0 0 0 0 0 0 0 0 0 0 0 1
    1 0 0 0 0 0 0 0 0 0 0 0 0 0 0 0 0 0 0 0 0 0 1
     1 0 0 0 0 0 0 0 0 0 0 0 0 0 0 0 0 0 0 0 0 1
      1 0 0 0 0 0 0 0 0 0 0 0 0 0 0 0 0 0 0 0 1
       1 0 0 0 0 0 0 0 0 0 0 0 0 0 0 0 0 0 0 1
        1 0 0 0 0 0 0 0 0 0 0 0 0 0 0 0 0 0 1
         1 0 0 0 0 0 0 0 0 0 0 0 0 0 0 0 0 1
          1 0 0 0 0 0 0 0 0 0 0 0 0 0 0 0 1
           1 1 1 1 1 1 1 1 1 1 1 1 1 1 1 1 1
```

In the mid-1980s, Samuel Yates started to compile a list known as the "Largest Known Primes" and coined the name "titanic prime" for any prime with 1,000 or more decimal digits.

- - - - - - - - - - - - - - -

Ramanujan synchronicity. "It is not guaranteed that a great mind will automatically find the nurturing support required for success. If Ramanujan had not been a Brahmin, or his mother had not been as patient, or if Hardy had ignored his letter, then Ramanujan might have slipped into obscurity, and his wonderful notebooks and equations would have been lost forever" (Calvin Clawson, *Mathematical Mysteries*, 1996).

Math in bed. "It is amazing to see the quantity of mathematics that Ramanujan communicated to Hardy from hospital beds in England!" (Krishnaswami Alladi, "Review of Ramanujan: Letters and Commentary," *The American Mathematical Monthly* 103, no. 8 [1996]: 708–11).

Ramanujan's fractions. Ramanujan excelled at producing infinite fractions. Here is an example.

$$\frac{1}{e-1} = \cfrac{1}{1+\cfrac{2}{2+\cfrac{3}{3+\cfrac{4}{4+\dots}}}}$$

Definition of the Landau-Ramanujan constant. The Landau-Ramanujan constant is 0.764223653 . . . Let $N(x)$ denote the number of positive integers not exceeding x that can be expressed as a sum of two squares. Edmund Landau and Ramanujan independently proved that

$$\lim_{x\to\infty} \frac{\sqrt{\ln(x)}}{x} \cdot N(x) = K$$

For example, $M(8) = 5$ because $1 = 0^2 + 1^2$, $2 = 1^2 + 1^2$, $4 = 0^2 + 2^2$, $5 = 1^2 + 2^2$, and $8 = 2^2 + 2^2$. Here, K is given by

$$K = \sqrt{\frac{1}{2} \prod_{p=4k+3} \frac{1}{1-p^{-2}}} = 0.764223653 \dots$$

The product is taken over all primes p congruent to 3 modulo 4. The convergence to the constant K, known as the Landau-Ramanujan constant, is very slow. Here it is to a few more digits, to impress your friends:

0.7642236535892206629906987312500923281167905413934095147216866737496146416587328588384015050131312337219372691207925926341874206467808432306331543462938 0531

Many more digits of the Landau-Ramanujan constant have been calculated by Philippe Flajolet at INRIA Paris (Institut National de Recherche en Informatique et Automatique) and Paul Zimmermann.

- -

Ramanujan and π. Ramanujan enjoyed finding approximations to π. Here's a nice one:

$$\pi \approx \frac{99^2}{2206\sqrt{2}} \text{ (correct to 8 places)}$$

- -

Hilbert sleeping. "If I were to awaken after having slept for a thousand years, my first question would be: Has the Riemann hypothesis been proven?" (David Hilbert, 1862–1943).

Zebra numbers. In my book *The Mathematics of Oz* (2002), I defined a class of irrational numbers that have remarkable patterns in their strings of digits. Most of the usual irrational numbers we think of, like $\sqrt{2} = 1.4142135623\ldots$, seem to be patternless. Zebra numbers are a class of irrational numbers with patterns, and are so named because the zebra's skin displays obvious patterns. Given the Zebra formula:

$$f(n) = \sqrt{9/121 \times 100^n + (112 - 44n)/121} =$$

we find that the 50th Zebra is

$$f(50) = \sqrt{9/121 \times 100^{50} + (112 - 44 \times 50)/121} =$$

2 72727 27272 72727 27272 72727 27272 72727 27272 72727 2727.272727 27272 72727 27272 72727 27272 72727 27272 72727 26956 36363 63636 36363 63636 36363 63636 36363 63636 36363 63636 36363 63636 36363 63636 36363 63636 36363 63636 36363 45287 27272 72727 27272 72727 27272 72727 27272 72727 27272 72727 27272 72727 27272 72727 27272 72727 27272 72727 27251 44232 72727 27272 72727 27272 72727 27272 72727 27272 72727 27272 72727 27272 72727 27272 72727 27272 72727 27272 69640 9556363636 36363 63636 36363 63636 36363 63636 36363 63636 3636363636 36363 63636 36363 63636 36363 63636 36358 62418 4680727272 72727 27272 72727 27272 72727 27272 72727 27272 7272727272 72727 27272 72727 27272 72727 27272 71855 15358 8991999999 99999 99999 99999 99999 99999 99999 99999 99999 9999999999 99999 99999 99999 99999 99999 99998 41025 13993 62559 99999 99999 99999 99999 99999 99999 99999 99999 99999 99999 99999 99999 99999 99999 99999 99999 99700 33238 87798 42559 99999 99999 99999 99999 99999 99999 99999 99999 99999 99999 99999 99999 99999 99999 99999 42064 26183 07695 61599 99999 99999 99999 99999 99999 99999 99999 99999 99999 99999 99999 99999 99999 99999 99885 75072 43302 7758 . . .

My book explains these numbers in greater detail.

- -

Ethereal plane. "Because it lies on a cool, ethereal plane beyond the everyday passions of human life, and because it can be fully grasped only through a language in which most people are unschooled, Ramanujan's work grants direct pleasure to only a few hundred pure mathematicians around the world, perhaps a few thousand . . . " (Robert Kanigel, *The Man Who Knew Infinity*, 1991).

- - - - - -

From Ramanujan.

$$\frac{e^{\pi} - 1}{e^{\pi} + 1} = \cfrac{\pi}{2 + \cfrac{\pi^2}{6 + \cfrac{\pi^2}{10 + \cfrac{\pi^2}{14 + \ldots}}}}$$

and, more generally,

$$\frac{e^{x} - 1}{e^{x} + 1} = \cfrac{x}{2 + \cfrac{x^2}{6 + \cfrac{x^2}{10 + \cfrac{x^2}{14 + \ldots}}}}$$

- - - - - - -

The foundation of math. "If you disregard the very simplest cases, there is in all of mathematics not a single infinite series whose sum has been rigorously determined. In other words, the most important parts of mathematics stand without a foundation" (Niels H. Abel, in G. F. Simmons, *Calculus Gems*, 1992).

Alien ships. Alien ships descend on America. The president shivers in fear. On the first day, 3 ships come. On the second day, 6 ships come. Each day, the number of ships increases as

3; 6; 18; 72; 144; 432; 1,728; ?

Find the eighth number in this sequence to determine how many ships come on the last day. (See Answer 2.84.)

Trains with crystal balls. My friend is superstitious and is riding in a train with seven cars. Below are the numbers of crystal balls he places in each car for good luck.

42, 63, 94, 135, 96, 157, ?

These numbers form a sequence. How many crystal balls are in the seventh car? (See Answer 2.85.)

An infinite product. From Ramanujan:

$$\prod_{p}^{\infty}\left(\frac{p^2+1}{p^2-1}\right)=\frac{5}{2}$$

Here, the infinite product symbolized by $\prod$, steps through every prime number; for example,

$$\left(\frac{2^2+1}{2^2-1}\right)\cdot\left(\frac{3^2+1}{3^2-1}\right)\cdot\left(\frac{5^2+1}{5^2-1}\right)\cdot\left(\frac{7^2+1}{7^2-1}\right)\ldots=\frac{5}{2}$$

Prime glue. In 1737, Euler discovered the following relationship, which appears to glue together the prime numbers and the sequence of natural numbers in a fascinating way:

$$\xi(s)=\sum_{n=1}^{\infty}\frac{1}{n^s}=\prod_{p=primes}\frac{1}{1-\frac{1}{p^s}}$$

For Euler, the variable s could be any real number! We can rewrite this formula as

$$\xi(s)=\frac{1}{1^s}+\frac{1}{2^s}+\frac{1}{3^s}+\frac{1}{4^s}+\ldots=$$

$$\left(\frac{1}{1-\frac{1}{2^s}}\right)\left(\frac{1}{1-\frac{1}{3^s}}\right)\left(\frac{1}{1-\frac{1}{5^s}}\right)\left(\frac{1}{1-\frac{1}{7^s}}\right)\ldots$$

$\xi(s)$ is called the zeta function. For values of $s>1$, the zeta function converges to a limit. For $s=1$, the zeta function is just the harmonic series, which diverges.

From Ramanujan.

$$\frac{1}{2}=\sqrt{1-\sqrt{1-\frac{1}{2}\sqrt{1-\frac{1}{4}\sqrt{1-\frac{1}{8}\sqrt{1-\ldots}}}}}$$

Mr. Zanti's ants. My friend, Mr. Zanti, collects ants, which he places in 7 ant farms. In the first ant farm are only 6 ants. The next ant farm contains 26 ants. What number of ants comes last in the following sequence?

6; 26; 106; 426; 1,706; 6,826; ?

(See Answer 2.86.)

- - - - - -

Integer representation as sums. An integer can be represented as the sum of the squares of two rational numbers if, and only if, it can be represented as the sum of the squares of two integers. This means, for example, that there are no rational numbers, r, s, such that $r^2 + s^2 = 3$.

- - - - - -

Factorial. Stirling's formula can be used to approximate n factorial ($n!$):

$$n! \approx \sqrt{2\pi n}\,(n/e)^n$$

It's interesting that factorial should be related so intimately with π and e.

- - - - - -

Prime king. The kings in a deck of cards abscond with all the cards that show prime numbers on their faces. How many cards does your deck now contain? (See Answer 2.87.)

A delight from Ramanujan.

$$1 - 5\left(\tfrac{1}{2}\right)^3 + 9\left(\tfrac{1\cdot3}{2\cdot4}\right)^3 - 13\left(\tfrac{1\cdot3\cdot5}{2\cdot4\cdot6}\right)^3 + \ldots = \tfrac{2}{\pi}$$

(Ramanujan; James R. Newman, *The World of Mathematics*, Volume 1 [New York: Simon and Schuster, 1956], pp. 371–72).

- - - - - - - - - - -

Another delight from Ramanujan.

$$\int_0^a e^{-x^2}\,dx = \tfrac{1}{2}\pi^{1/2} - \cfrac{e^{-a^2}}{2a + \cfrac{1}{a + \cfrac{2}{2a + \cfrac{3}{a + \cfrac{4}{2a + \ldots}}}}}$$

- - - - - - - - - - -

From Ramanujan.

$$\cfrac{1}{1 + \cfrac{e^{-2\pi}}{1 + \cfrac{e^{-4\pi}}{1 + \ldots}}} = \left(\sqrt{\frac{5+\sqrt{5}}{2}} - \frac{\sqrt{5}+1}{2}\right)\cdot e^{\frac{2}{5}\pi}$$

Note that the term in parentheses can be reduced to $\sqrt{2 + \phi} - \phi$, where ϕ is the golden ratio.

- - - - - - - - - - -

From Ramanujan.

$$\ln x + \gamma = \sum_{k=1}^{\infty} \frac{(-1)^{k-1} x^k}{k!\,k} = x - \frac{x^2}{4} + \frac{x^3}{18} - \frac{x^4}{96} + \ldots$$

where $\gamma = 0.5772157\ldots$ is the Euler-Mascheroni constant.

- - - - - - - - - - -

From Ramanujan.

$$-2 = \sqrt[3]{-6 + \sqrt[3]{-6 + \sqrt[3]{-6 + \sqrt[3]{-6 + \ldots}}}}$$

From Ramanujan.

$$10 - \pi^2 = \sum_{k=1}^{\infty} \frac{1}{k^3 (k+1)^3} = \frac{1}{1^3 \cdot 2^3} + \frac{1}{2^3 \cdot 3^3} +$$

$$\frac{1}{3^3 \cdot 4^3} + \frac{1}{4^3 \cdot 5^3} + \dots$$

From Ramanujan.

$$\ln 2 = \frac{1}{2} + \sum_{k=1}^{\infty} \frac{1}{(2k)^3 - 2k} = \frac{1}{2} + \frac{1}{2^3 - 2} + \frac{1}{4^3 - 4} +$$

$$\cdot \frac{1}{6^3 - 6} + \frac{1}{8^3 - 8}$$

A formula that unites. Here's a gem from Ramanujan relating π, a natural logarithm, and the golden ratio. (Can you find the golden ratio?)

$$\frac{\pi^2}{6} - 3 \ln \left(\frac{\sqrt{5} + 1}{2} \right)^2 = \sum_{k=0}^{\infty} \frac{(-1)^k (k!)^2}{(2k)! (2k+1)^2} =$$

$$1 - \frac{1}{2! \cdot 3^2} + \frac{(2!)^2}{4! \cdot 5^2} - \frac{(3!)^2}{6! \cdot 7^2} + \frac{(4!)^2}{8! \cdot 9^2} \dots$$

From Ramanujan.

$$\frac{x}{2x-1} = 1 - \frac{x-1}{x+1} + \frac{(x-1)(x-2)}{(x+1)(x+2)} -$$

$$\frac{(x-1)(x-2)(x-3)}{(x+1)(x+2)(x+3)} + \dots$$

Euler's conjecture on powers. Euler once conjectured that an nth power cannot be expressed as the sum of fewer than n smaller nth powers. Today we know that the simplest known counter example to Euler's conjecture is 31,858,749,840,007,945,920, $321 = 422,481^4 = 95,800^4 + 217,519^4 + 414,560^4$.

Pi deck. How many consecutive digits of pi (3.1415 . . .) can you display with a deck of cards, using the numbers on the cards and starting at any point you like in the digit string of pi? You can omit the cards with no numbers, like the jacks and the aces. (See Answer 2.88.)

Mathematical profundity. "In the company of friends, writers can discuss their books, economists the state of the economy, lawyers their latest cases, and businessmen their latest acquisitions, but mathematicians cannot discuss their mathematics at all. And the more profound their work, the less understandable it is" (Alfred Adler, "Reflections: Mathematics and Creativity," *New Yorker*, 1972).

A prime connection. Here is an approximation from Ramanujan that connects the constant e, logarithms, and primes. As x approaches zero, we have

$$\sum_{k=1}^{\infty} \frac{P_k}{e^{kx}} = \frac{2}{e^x} + \frac{3}{e^{2x}} + \frac{5}{e^{3x}} + \frac{7}{e^{4x}} + \ldots \approx -\frac{\ln x}{x^2}$$

- - - - - - - - - - - - - - - -

From Ramanujan.

$$\pi = 4\left(\sum_{n=0}^{\infty} \frac{(6n+1)\left(\frac{1}{2}\right)^3_{Poch(n)}}{4^n (n!)^3}\right)^{-1}$$

where the subscript n refers to the Pochhammer notation that is explained in the definition on page 107.

- - - - - - - - - - - - - - - -

From Ramanujan.

$$\pi =$$

$$\frac{27}{4}\left(\sum_{n=0}^{\infty} \frac{(15n+2)\left(\frac{1}{2}\right)_{Poch(n)}\left(\frac{1}{3}\right)_{Poch(n)}\left(\frac{2}{3}\right)_{Poch(n)}}{(n!)^3}\left(\frac{2}{27}\right)^n\right)^{-1}$$

where the subscript $Poch(n)$ refers to the Pochhammer notation.

- - - - - - - - - - - - - - - -

Nested radicals. Ramanujan presented this nested radical in *The Indian Journal of Mathematics*:

$$3 = \sqrt{1 + 2\sqrt{1 + 3\sqrt{1 + 4\sqrt{1 + \ldots}}}}$$

On the disorder of π. In 1997, the mathematicians Steve Pincus and Rudolf Kalman applied a statistical method called "approximate entropy" to assess the disorder of the digits in transcendental numbers like π and algebraic numbers like the square root of 2. Pincus initially had a gut feel that the digits in transcendental numbers were more disorderly than in algebraic numbers, but he was wrong. Although π had the most irregular sequence of digits, the square root of 2 was the second-most disorderly digit string tested. In order of randomness, Pincus and Kalman found

$$\pi = 3.141592\ldots$$
$$\sqrt{2} = 1.414235\ldots$$
$$e = 2.718281\ldots$$
$$\sqrt{3} = 1.732050\ldots$$

I do not know if more comprehensive entropy lists of transcendental and algebraic numbers continue to show this mixing. For further reading, see Charles Seife, "New Test Sizes Up Randomness," *Science* 276, no. 5312 (April 25, 1997): 532.

- - - - - -

The wonderful number 37. The numbers 111, 222, 333, 444, 555, 666, 777, 888, and 999 are all evenly divisible by 37, leaving no remainder.

Vibonacci numbers. We've already discussed Fibonacci numbers, in which each term (except for the first two) is the sum of the two preceding terms. We can produce the sequence by $F_n = F_{n-2} + F_{n-1}$ to yield 1, 1, 2, 3, 5, 8, 13, ... The mathematician Divakar Viswanath has studied the *Vibonacci sequence*—a new variation on the Fibonacci sequence, in which, instead of always adding two terms to produce the next term, we either add or subtract, depending on the flip of a coin at each stage in the calculation. This can be expressed as $V_n = V_{n-2} \pm V_{n-1}$, where we choose to add or subtract with equal probability.

An amazing degree of order can be found in the "randomized" sequences. In particular, the absolute value of V_n grows exponentially as n increases, with a growth rate controlled by a newly discovered constant called Viswanath's constant, $C = 1.13198824 \ldots$ (Divakar Viswanath, "Random Fibonacci Sequences and the Number 1.13198824 ... ," *Mathematics of Computation* 69 [2000]: 1131–55).

Large palindions and hyper-palindions. As already discussed, a palindrome is a natural number that does not change when its digits are reversed. Examples of palindromes include 101; 33; 1,234,321; and 2. *Palindions* are natural numbers that have more palindromic divisors than any smaller number. In other words, if a natural number p is a palindion, then no natural number n has as many palindromic divisors as p where $0 < n < p$. In 2004, Jason Earls discovered the palindion 2,666,664, which has 50 palindromic divisors. It satisfies the definition of a palindion because no smaller numbers exist with so many palindromic divisors. Its palindromic divisors include 80,808; 333,333; and 888,888. Later in 2004, Mark

Pochhammer notation defined. The Pochhammer symbol provides a way to simplify mathematical formulas and is used in some of the notations for the Ramanujan formulas in this book. In particular,

$$(x)_n \text{ or } (x)_{Poch(n)} = x \times (x + 1) \times (x + 2) \times \ldots \times (x + n - 1)$$

Sometimes, the subscript n is written "*Poch(n)*" to visually differentiate it from an ordinary subscript. Let's try it with an example that uses "5" as the Pochhammer subscript:

$$\left(\frac{1}{2}\right)_5^3 = \left[\left(\frac{1}{2}\right) \times \left(\frac{3}{2}\right) \times \left(\frac{5}{2}\right) \times \left(\frac{7}{2}\right) \times \left(\frac{9}{2}\right)\right]^3$$

The subscript value effectively tells you how many terms to include in the product. The Pochhammer symbol should have higher priority than the exponent, so that the Pochhammer symbol is applied first, and then the result is cubed.

Ganson discovered a few larger palindions, including 134,666,532, which has 80 palindromic divisors. All multidigit palindions that have been discovered so far are multiples of 6. (The reason for this is currently a mystery.)

Hyperpalindions are palindions that themselves are palindromic. The largest known hyperpalindion is 2,772, with 16 palindromic divisors: 1, 2, 3, 4, 6, 7, 9, 11, 22, 33, 44, 66, 77, 99, 252, and 2,772.

– – – – – –

🌐 **11,410,337,850,553.** Arithmetic progressions are number sequences in which each term differs from the preceding one by a constant. For example, the sequence 5, 11, 17, 23, and 29 happens to contain prime numbers in which successive primes differ by 6. The longest known arithmetic progression of prime numbers contains just 22 terms. One 22-term progression starts at 11,410,337,850,553, and the difference between successive terms is 4,609,098,694,200. In 2004,

🌐 **Palindromic primes.** This list of palindromic primes was created by G. L. Honaker Jr. in 1999:

2

30203

133020331

1713302033171

12171330203317121

151217133020331712151

1815121713302033171215181

16181512171330203317121518161

3316181512171330203317121518161 33

9333161815121713302033171215181613339

11933316181512171330203317121518161333911

The pyramid is unique, in that each palindromic prime is related to the previous by appending two digits to the start and the end of the previous palindromic prime. For further reading, see "On-Line Encyclopedia of Integer Sequences," primes.utm.edu/curios/page.php?number_id=2393.

Ben Green, of the University of British Columbia, and Terence Tao, of the University of California, offered a proof that demonstrates that there are infinitely many prime progressions of every finite length (Erica Klarreich, "Primal Progress: Pattern Hunters Spy Order among Prime Numbers," *Science*

News 165, no. 17 [April 24, 2004]: 260–61).

– – – – – –

🌀 **Math scribblings.** "After Ramanujan's death, Hardy received the loose sheets of paper on which Ramanujan had scribbled mathematical formulas in his dying moments. Contained in these sheets were several deep formulas for the mock-theta functions" (Krishnaswami Alladi, "Review of Ramanujan: Letters and Commentary," *The American Mathematical Monthly* 103, no. 8 [1996]: 708–11).

Legion's number. In my book *The Mathematics of Oz*, I wondered whether humanity could ever compute the first ten digits of the number $666!^{666!}$, which I called "Legion's number of the second kind." (Legion's number of the first kind was a mere 666^{666}.) The number $666!^{666!}$ is so large that it has roughly 10^{1596} digits. The answer in my previous book turned out to be incorrect for various technical reasons, and I would like to update readers with the correct number. After reading about Legion's number in *The Mathematics of Oz*, B. Ravikumar, of California State University, Sonoma, used various mathematical tricks to determine the first ten digits, 2765838869, using a personal computer that ran for just two minutes. In particular, he used computational tricks known as repeated squaring, addition chains, and Karatsuba's algorithm (a fast algorithm for multiplying integers). As part of this project, Ravikumar also determined the exact number of digits in $666!^{666!}$. Readers may write to me for details of his approach.

π-palindromic prime. The number 31,415,926,535,897, 932,384,626,433,833,462,648,323,979,853,562,951,413 is a palindromic prime formed from the reflected decimal expansion of π. (G. L. Honaker Jr., primes.utm.edu/curios/page.php?number_id=130). Carlos B. Rivera reports that the next two larger π-palindromic primes have 301 and 921 digits. Similarly, Jason Earls has discovered the following φ-palindromic prime, based on the first few digits of the golden ratio, φ:

161803398874989484820458683436563811772030
302771183656343868540284849894788933308161

Baxter-Hickerson prime numbers. Mathematicians of all levels of sophistication continue to search for elusive Baxter-Hickerson prime numbers. As background, in 1999, Lew Baxter discovered that $(2 \times 10^{5n} - 10^{4n} + 2 \times 10^{3n} + 10^{2n} + 10^n + 1)/3$ produces numbers whose cubes lack zeros. The first few terms for $n = 0, 1, 2, \ldots$ are 2, 64037, 6634003367, 666334000333667, \ldots *Baxter-Hickerson prime numbers* occur for $n = 0, 1, 7, 133, \ldots$ yielding $f(0) = 2, f(1) = 64037, f(7) = 666666633333340000000333333366666667, \ldots$ The largest known Baxter-Hickerson prime number is 66666 . . . 66667 (665-digits) and was discovered by G. L. Honaker Jr.

For your interest, cubing $f(7)$ produces a number that is zero free and equal to 29629625185186296296248148159259 25981481487777779111111177777788148148925925981 481486296296518518529629963.

On prime numbers. If an integer p is a prime number, then for all integers j, dividing both j^p and j by p gives a result with the same remainder. For example, if $p = 7$ (a prime) and $j = 9$, dividing 9^7 by 7 gives a remainder of 2, as does dividing 9 by 7. Any integer p that fails this test is not prime. For further reading, see Ivars Peterson, "Prime Pursuit," *Science News* 162 (October 26, 2002): 266–67.

Rare sums. Only 31 numbers exist that cannot be expressed as the sum of distinct squares. Only one prime number exists that is equal to the sum of the decimal digits of its 7th power.

Algebra, Percentages, Weird Puzzles, and Marvelous Mathematical Manipulations

I N WHICH WE ENCOUNTER TREASURE CHESTS OF ZANY AND EDUCATIONAL MATH problems that involve algebra, fractions, percentages, classic recreational puzzles, and various types of mathematical manipulation. Some are based on problems that are more than a thousand years old. Others are brand new. Get ready to sharpen your pencils and stretch your brains!

> Algebra is about constants and variables and
> relationships between them. If numbers are
> soldiers, algebra sets the rules of engagement.

⊛ Free association. "Rarely do I solve problems through a rationally deductive process. Instead I value a free association of ideas, a jumble of three or four ideas bouncing around in my mind. As the urge for resolution increases, the bouncing around stops, and I settle on just one idea or strategy" (Heinz Pagels, *Dreams of Reason*, 1988).

❓ ½ puzzle. What number gives the same result when it is added to ½ as when it is multiplied by ½? (See Answer 3.1.)

⊛ Math metaphysics. "Algebra is the metaphysics of arithmetic" (John Ray, 1627–1705).

❓ Shrunken heads puzzle. Gary and Joan collect lifelike shrunken heads made of leather. Gary said that if Joan gave him two shrunken heads, they would have an equal number, but if Gary gave Joan two of his, Joan would have twice as many as Gary. How many shrunken heads did they each have? (See Answer 3.2.)

❓ Loyd's Leaning Tower of Pisa. Sam Loyd, the famous nineteenth-century American puzzlemaster, proposed "The Leaning Tower of Pisa" problem, illustrated in figure 3.1. If an elastic ball is dropped from the Leaning Tower of Pisa at a height of 179 feet from the ground, and on each rebound the ball rises exactly one tenth of its previous height, what distance will it travel before it comes to rest? He gave a solution but, alas, never said how he solved this. How would you solve this? (See Answer 3.3.)

Figure 3.1   Sam Loyd's Leaning Tower of Pisa.

A real mathematician. "A person who can within a year solve $x^2 - 92y^2 = 1$ is a mathematician" (Brahmagupta [598–670], *Brahmasphutasiddhanta* [*The Opening of the Universe*], A.D. 628).

- - - - - -

Chimpanzees and gorillas. A chimpanzee, when asked by a gorilla how old it was, replied, "My age is now five times yours, but three years ago, it was seven times yours. If you can tell me my age, I will reward you by bringing you a banana every day." How old is the chimpanzee? (See Answer 3.4.)

- - - - - -

RED + BLUE. Mr. Antón Buol walks down the street and sees a red-and-blue-striped pole. "Great!" he shouts, when he realizes that

RED + BLUE = BUOL

forms an "alphametic," in which each letter is replaced by a digit. The same letter always stands for the same digit, and the same digit is always represented by the same letter. Can you solve the alphametic? (See Answer 3.5.)

Loyd's "teacher" puzzle. The teacher pictured in figure 3.2 is explaining to his class the remarkable fact that 2 times 2 gives the same answer as 2 plus 2. Although 2 is the only positive number with this property, there are many pairs of different numbers that can be substituted for a and b in the equations on the right of the blackboard, namely,

$$a \times b = y \qquad a + b = y$$

Can you find a value for a and b? For this puzzle, Sam Loyd asks us to give different values for a and b. They may be fractions, of course, but they must have a product that is exactly equal to their sum. (See Answer 3.6.)

Figure 3.2 Sam Loyd's classic $a \times b = y$, $a + b = y$.

New tools. "The great advances in science usually result from new tools rather than new doctrines. Each time we introduce a new tool, it always leads to new and unexpected discoveries, because Nature's imagination is richer than ours" (Freeman Dyson, in the introduction to *Nature's Imagination*, 1995).

- - - - - -

? Worms and water. You go to a bait store and buy 1,000 pounds of worms for your fishing business. This particular species of worm is 99 percent water.

The worms dry slightly in their hot, smelly enclosure, and an hour later they are 95 percent water. How much do the worms weigh now? (See Answer 3.7.)

? Loyd's mixed tea puzzle. According to Loyd, in Asia, the blending of teas is such an exact science that combining different kinds of teas is done with utmost care (figure 3.3). In order to illustrate the complications that arise in the science of blending teas, he calls attention to a simple puzzle that is based upon two tea blends only.

Mr. Han, the human mixer, has received two cases, each cubical but of a different size. The larger cube is completely full of black tea. The smaller cube is completely full of green tea. Mr. Han has mixed together the contents and found that the mixture exactly fills 22 cubical chests of equal size. Assuming that the interior dimensions of all the boxes and the chests can be expressed as exact decimals (i.e., numbers with decimal parts that don't repeat forever), can you determine the proportion of green tea to black? (See Answer 3.8.)

Figure 3.3 Sam Loyd's classic mixed teas puzzle.

? Fragile fractions. Is it possible to construct the fraction ½ by summing other fractions of the form $1/x^2$? For example, you can choose various denominators, as in: $\frac{1}{3^2} + \frac{1}{5^2} + \frac{1}{10^2} + \ldots$ (but this is not an answer!). The solution must have a finite number of terms, and no value of x can exceed 100 or be repeated. (See Answer 3.9.)

? Square pizzas. Abraham has a stack of square pizzas, each with the numbers 1, 2, 3, and 10 at the corners, as shown in figure 3.4.

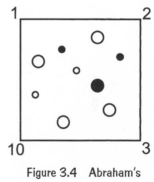

Figure 3.4 Abraham's square pizza.

Abraham tosses the pizzas into a pile at random so that they are rotated haphazardly and create quite a mess. An angel of the Lord comes to Abraham and asks him if there is a way to stack the pizzas so that the numbers in a column through each corner in the stack equal 67. In other words, Abraham can use as many pizzas as he likes, and he is free to rotate them however he likes, if he can ensure that each corner stack sums to 67. Is this possible? (See Answer 3.10.)

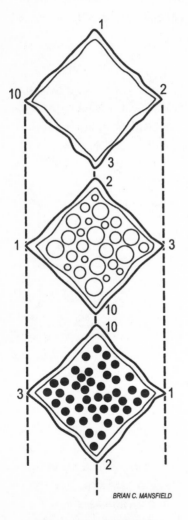

BRIAN C. MANSFIELD

Figure 3.5   A stack of square pizzas.

The problem of the rich jeweler. Here's a difficult problem I invented to test my brainy colleagues. Your friend, a jeweler in New York City, walks with you to a large room filled with three kinds of valuable objects: gems, cubical chunks of special alloys, and bottles of rare spices. There are four kinds of gems: one pink, another beige, a third yellow, and the last green. Among the gems, the number of pink gems is equal to one-fourth plus one-third the number of beige gems plus the number of green gems. The number of beige gems is equal to one-seventh plus one-third the green gems plus half the number of yellow gems. The number of green gems is equal to one-fifth plus one-eleventh the pink plus the beige gems.

We are also gazing at four colors of precious alloys.

Among the alloys, the number of pink alloys is equal to one-half plus one-fifth of the pink objects in the room plus twice the number of yellow alloys. The number of beige alloys is equal to one-third plus one-half the total of the green alloys. The number of green alloys is equal to one-fifth the number of pink alloys plus one-sixth the total number of the yellow alloys, and the number of yellow alloys is equal to one-eighth plus one-third the total of the green objects in the room.

There are four colors of spice bottles. The number of pink spice bottles is equal to one-half the number of beige objects in the room. The total number of green and yellow spice bottles is equal to the total number of pink and beige spice bottles.

What is the least number of precious objects that are in the room? (See Answer 3.11.)

A huge number? You have 99 constants labeled a_1 to a_{99}. Let $a_1 = 1$. The value of each successive constant a_n is equal to n raised to the value of the previous constant a_{n-1}. For example, $a_2 = 2^{a_1} = 2$ and $a_3 = 3^{a_2} = 9$, and so forth. What is the exact value of

$$(a_{97} - a_1) \times (a_{97} - a_2) \times (a_{97} - a_3) \times \ldots \times \ldots (a_{97} - a_{99})?$$

(See Answer 3.12.)

🌀 **Phoenix and e^x.** "Who has not been amazed to learn that the function $y = e^x$, like a phoenix rising again from its own ashes, is its own derivative?" (Francois le Lionnais, *Great Currents of Mathematical Thought*, 1962).

- - - - - -

❓ **x^0?** Can you show why $x^0 = 1$? In other words, your task is to informally demonstrate why you think that any number raised to the power of zero is 1. (See Answer 3.13.)

- - - - - -

🌀 **Mathematics and romance.** "Mathematicians are like lovers. Grant a mathematician the least principle, and he will draw from it a consequence which you must also grant him, and from this consequence another" (Bernard Le Bovier Fontenelle, quoted in V. H. Larney, *Abstract Algebra: A First Course*, 1975).

- - - - - -

🌀 **The fraction of life.** "A man is like a fraction whose numerator is what he is and whose denominator is what he thinks of himself. The larger the denominator the smaller the fraction" (Count

❓ **Winged robot.** Danielle is eating ice cream with her father and her friend Kate. Danielle tells Kate, "My grandfather is exactly the same age as my father."

"No way!" says Kate.

"It's true!" Danielle says, bringing out a photo of her grandfather and showing it to Kate.

"You're a liar," Kate says.

Suddenly, a winged robot descends and perches next to them. "I assure you that Danielle is telling the truth."

Danielle nods. "Kate, you can have my ice cream if I'm lying."

Could the robot be telling the truth? (See Answer 3.14.)

- - - - - - - - - - - - - - - - - -

❓ **Fraction family.** Which fraction is the odd one out?

$$\frac{17}{74}, \frac{29}{98}, \frac{35}{152}, \frac{42}{162}, \frac{87}{372}, \frac{74}{372}$$

(See Answer 3.15.)

Lev Nikolgevich Tolstoy [1828–1920], in Howard Eves, *Return to Mathematical Circles*, 1987).

- - - - - -

❓ **Tarantulas in bottles.** Tiffany places three opaque bottles on the table before you. One bottle contains a dead tarantula. The other two bottles contain live tarantulas. Tiffany knows what is in each bottle, but you do not.

You can ask Tiffany one yes-or-no question, but when you do, you have to point to one of the bottles. If you point to a live tarantula, she will tell the truth. If you point to the dead tarantula, she will randomly say "yes" or "no." Your mission is to find one of the live tarantulas by asking a single question. What question should you ask? (See Answer 3.16.)

- - - - - -

❓ **Large ape.** A large ape enters your kitchen, along with a zookeeper. The zookeeper says to you, "I will remove this ape from your kitchen four days after two days before the day before tomorrow." What day will you be free of the ape? (See Answer 3.17.)

- - - - - -

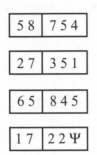

Ψ substitution. What number should replace the Ψ?

58	754

27	351

65	845

17	22 Ψ

(See Answer 3.18.)

- - - - - -

Angel number-guessing game. An angel materializes on your doorstep and speaks. "I am thinking of two whole numbers that represent the number of apples and grapes in the Garden of Eden. The product of these numbers is 1,000 times larger than their sum. What are the two numbers?"

You frown. "Wait, that is not enough information."

The angel nods. "You are right. There is more than one solution, and you have 24 hours to give me *any* solution. If you are correct, you

will live the rest of your life in bliss." (See Answer 3.19.)

- - - - - -

Starship journey. Captain Kirk's starship leaves Earth for Mars. Captain Eck's starship leaves from Mars for Earth. Their ships start at the same time and travel at uniform speeds, but one is faster than the other. After meeting and passing, Kirk requires 17 hours and Eck requires 10 hours to complete the journey. Approximately what total time did each starship require for its interplanetary journey? Assume stationary planets. (See Answer 3.20.)

Journey. An old car starts at a city in New Jersey and travels at a speed of 21 miles per hour toward New York. After reaching New York, the car returns, traveling over exactly the same distance, at only 3 miles per hour. What is the car's average speed over the entire journey? (Assume that the car turned around instantly once it reached New York.) (See Answer 3.22.)

- - - - - -

Four intelligent gorillas. Four intelligent gorillas in the forests of Africa are holding an election to determine who should be king of

Missing numbers. Insert the consecutive numbers 1 through 11 in the 11 empty cells. Each number in the gray cells is the sum of the adjacent empty cells (right, left, up and down) that touch the gray cell.

13			
		33	
	19		19
		10	

(See Answer 3.21.)

the gorillas. At the recent election, a total of 8,888 votes were received for the four candidate gorillas, the winner beating its opponents by 888, 88, and 8 votes, respectively. How many votes did the successful gorilla receive? (See Answer 3.23.)

- - - - - -

? George Nobl: A true story. In 2003, the math teacher George Nobl often posed math problems to passersby in New York's Time Square. He laid out different math problems on a table and offered Snickers bars to anyone who got them right. Here is one of my favorite Nobl problems.

Noah is a candy store owner. He has 20 pounds of cashews, costing $3.55 a pound. And he has peanuts that cost $2.50 a pound. How many pounds of peanuts would Noah have to mix with all the cashews to get a mixture that costs $3.20 per pound? (See Answer 3.24.)

- - - - - -

? Human hands. Mike collects lifelike models of human hands. He wants to put them on shelves in his den. His first thought is to put each one on a shelf, but when he tries this, two hands have to share one shelf. Next, Mike tries to place the hands on the shelves so that every shelf contains two hands, but when he tries this, one shelf is left empty. How many hands does Mike own? How many shelves is he using? (See Answer 3.25.)

? Ages. I have five wonderful friends, one of whom is quite young. The sum of all their ages is 109. If I add pairs of ages, I get the following:

Name	Name	Age Sum
Teja	Danielle	16
Danielle	Nick	32
Nick	Pete	52
Pete	Mark	72

How old are my friends? (See Answer 3.26.)

? Three arrays. I have filled the following three arrays with consecutive numbers from 1 to 12. The sum of the numbers in the first array is 24; in the second array, 34; and the third array, 20. Fill in the missing numbers. (See Answer 3.27.)

1	?
?	11

8	?
?	5

6	?
?	3

- - - - - -

? Perfect cubes? Let n be a positive integer. Can both $n + 3$ and $n^2 + 3$ be perfect cubes? (A perfect cube is a number that is the cube of an integer, like $2^3 = 8$ or $3^3 = 27$.)

It would seem as if I gave you too little information to solve this, and that is why the problem is so fascinating. How can you answer this problem with such meager information? (See Answer 3.28.)

- - - - - -

? Harmonic series. The harmonic series

$$1 + 1/2 + 1/3 + 1/4 + \ldots$$

grows without bound, but so slowly that it requires 12,367 terms to make the sum greater than 10.

A super slow-growing series. The series

$$1/(3\log3\log\log3) + 1/(4\log4\log\log4) + \ldots$$

grows so slowly that it requires $10^{10^{100}}$ terms to make the sum greater than 10.

- - - - - - - - - - - - - - - - - - -

A very slow-growing series. The series

$$1/(2\log2) + 1/(3\log3) + 1/(4\log4) + \ldots$$

grows so slowly that it requires $10^{4,300}$ terms to make the sum greater than 10.

Four pets. I have four pets: an iguana, a cat, a bird, and a large ape. Today, I loaned all the pets to friends who enjoy the company of pets. The iguana is returned every 6 days. The cat comes home every 4 days. The bird comes home every 3 days and the ape every 7 days. The pets are always returned at noon and stay with me for an hour during my lunch break so that I can enjoy their company before loaning them out again. In how many days will all my pets be together again in my cozy home? (See Answer 3.29.)

- - - - - -

Aesthetic math. "Blindness to the aesthetic element in mathematics is widespread and can account for a feeling that mathematics is dry as dust, as exciting as a telephone book. . . . Contrariwise, appreciation of this element makes the subject live in a wonderful manner and burn as no other creation of the human mind seems to do" (Philip J. Davis and Reuben Hersh, *The Mathematical Experience*, 1981).

- - - - - -

Fractional swap. Swap two numbers in the numerator with two numbers in the denominator to form a fraction equaling ⅓.

$$\frac{1630}{4542}$$

(See Answer 3.30.)

Rational roots. Can you find a number, n, such that n, $n - 7$, and $n + 7$ all have rational square roots? In other words, the square root of each of these three numbers can be expressed as an integer or a fraction. (See Answer 3.31.)

- - - - - -

Consecutive integers. In 1769, Leonhard Euler conjectured that the following equation has no solution for positive integers:

$$A^4 + B^4 + C^4 = D^4$$

However, 218 years later, Noam Elkies and Roger Frye found solutions like $(A, B, C, D) = (95,800; 217,519; 414,560; 422,481)$. Note also that $27^5 + 84^5 + 110^5 + 133^5 = 144^5$. Recall that in chapter 1, we discussed the equation $a^n + b^n = c^n$, $n > 2$ as having no positive integer solutions. This statement is called Fermat's Last Theorem.

- - - - - -

Donald Trumpet. Donald Trumpet died, leaving a peculiar will. His will states that he will leave one million dollars to be split between his son William and his daughter Hillary. Hillary, his favorite child, gets four times the amount of William. If Hillary takes less than 30 seconds to determine how much William will get, the money is distributed immediately; otherwise, Hillary gets nothing. Can you help her? What did William get? (See Answer 3.32.)

- - - - - -

🐾 **Rubies and emeralds.** Three huge rubies and two emeralds weigh 32 pounds. Four rubies and three emeralds together weigh 44 pounds. Assume that the three rubies have an identical weight, as do the emeralds. What is the weight of two rubies and one emerald? (See Answer 3.33.)

- - - - - -

🐾 **Crow and eagle.** A crow and an eagle are gazing at 100 worms. The crow says, "Because I am smaller than you, you will get six times the number of worms I will get. If you tell me how many worms you will get in this fine deal, I will build a beautiful nest for you. If you are not that smart, I will get all the worms."

The eagle said, "It's a deal!" How many worms did the eagle get? (Hint: You may have to cut some of the worms to be fair and accurate.) (See Answer 3.34.)

- - - - - -

🌀 **Puzzle solving.** "Mathematics began to seem too much like puzzle solving. Physics is puzzle solving, too, but of puzzles created by nature, not by the mind of man" (Maria Goeppert Mayer, in Joan Dash's *Maria Goeppert-Mayer: A Life of One's Own*, 1973).

🐾 **Four digits.** Substitute four different digits for A, B, C, and D to make the following mathematical expression correct. (AB is a two-digit number, and DAC is a three-digit number. A, B, C, and D are single-digit numbers.)

$$(AB + A) \times C = DAC$$

(See Answer 3.36.)

- - - - - -

🐾 **Wizard's card.** A wizard in a long white robe approaches you with a single card that

🐾 **Number grid.** The grid specifies 6 mathematical formulas, 3 horizontally and 3 vertically. (Recall that multiplication and division are performed before addition and subtraction.) Fill in the missing integer numbers. (See Answer 3.35.)

	÷		+		=	6
×		+		×		
	×		+		=	34
+		+		−		
	×		+		=	24
=		=		=		
59		11		3		

has a 9 on one side and a 4 on the other. When the wizard tosses the card, it can land on either side.

The wizard can toss the card as often as you like, and each time he tosses the card, you add the number to the running sum. For example, he might toss a 9 and 4 and 4, which gives us a score of 17. What is the highest whole number score that is impossible to obtain while playing this strange game? (See Answer 3.37.)

- - - - - -

Missing numbers. The missing numbers in the following grid are one-digit integers. The sums for each row and each column, and one diagonal, are listed outside the 4-by-4 array. How quickly can you find the missing numbers?

?	3	?	6	16
4	?	4	?	17
?	0	?	8	15
2	?	5	?	17
10	6	19	30	18

(See Answer 3.39.)

Insert symbols. Insert + or – between the numbers to find the total.

$$8\ 7\ 6\ 5\ 4\ 3\ 2\ 1 = 88$$

(If you don't place a symbol between the numbers, they merge to form a multidigit number. For example, here is one possibility that, alas, turns out to be incorrect: $8,765 - 4,321 = 88$.) (See Answer 3.38.)

- - - - - -

New Guinea sea turtle. My New Guinea sea turtle is four times older than I am. As it ages, its shell will turn a bright green. As I ponder its beautiful color, I suddenly realize that in 20 years, the turtle will be only twice as old as I will be then. How old is my turtle, and how old am I? (See Answer 3.40.)

- - - - - -

Find A and B. What are A and B in this mathematical expression? (BA is a two-digit number, and $176B$ is a four-digit number.)

$$BA^A = 176B$$

(See Answer 3.41.)

- - - - - -

Prosthetic ulnas. Suzy Samson is a world-champion weight lifter with arms strengthened by prosthetic ulnas shaped like helices. Today, she is on a TV show, demonstrating her mental and physical strength. She gazes at some barbells, and a question forms in her mind. If she weighs 120 pounds plus a fourth of her own weight, how much does she weigh? (See Answer 3.42.)

- - - - - -

Ladybugs. Teja has a number of ladybugs in a jar. The number of ladybugs plus 10 grasshoppers is 2 less than 5 times the number of ladybugs.

In addition, Teja has ten times the number of butterflies as she has grasshoppers. If you wish, denote the number of ladybugs by L and the number of grasshoppers by G. How many ladybugs does Teja have? (See Answer 3.43.)

- - - - - -

Nebula aliens. Aliens from the Trifid Nebula descend to New Jersey. The surface area and the volume of their spherical spaceship are both four-digit integers times π, expressed in units of feet. Assume that the strange ship contains only air, and its volume is $(4/3)\pi r^3$. What is the radius r of the alien sphere? If you solve this puzzle, the aliens will give you their ship so that you can explore the universe. (See Answer 3.44.)

- - - - - -

Martian females. In a Martian crater, three-sevenths of the females are married to one-half of the males. What fraction of the crater's Martians are married? (Assume that no Martian is married to more than one Martian.) What is the least number of Martians who could live in the crater? (See Answer 3.45.)

- - - - - -

Dimension X. You are transported to a nearby dimension where half of 6 is 4, not 3, as you expected. What would a third of 12 be? (See Answer 3.46.)

- - - - - -

Wine and vinegar. Monica walks along the New Jersey Turnpike, carrying ten 3-cup bottles of vinegar that are one-quarter full. Her friend William is carrying five 4-cup bottles of red wine that are one-quarter empty. How much more liquid does William carry than Monica? (See Answer 3.47.)

- - - - - -

Unholy experiment. There are several humans and rabbits in a dirty cage (with no other types of animal). Perhaps they are trapped there for some kind of unholy experimentation.

All that we know is this: there are 70 heads and 200 feet inside the cage. Do you have the gut feeling that there are more rabbits than humans? Exactly how many humans are there, and how many rabbits? (See Answer 3.48.)

- - - - - -

Space race. America and Russia are in an important race from Earth to Saturn. Both spaceships start at Moscow and end at Saturn. They start the race traveling at the same speed and neither of them speeds up or slows down. The result is not a tie. How is this possible? (See Answer 3.49.)

- - - - - -

Number journey. "Mathematics is not a careful march down a well-cleared highway,

Hyperpower towers. Hyperpowers of the form

$$x = g(z) = z^{z^z}$$

for real positive z comprise a fascinating reservoir for computer study. You can define the sequence $\{z_n\}$ by $z_1 = z$, $z_{n+1} = z^{z_n}$, $n = 1, 2, 3, \ldots$ You can repeat this mathematical feedback loop over and over again. Whenever the sequence $\{z(n)\}$ converges, we write

$$g(z) = \lim_{n \to \infty} z(n)$$

You will find that the hyperpower tower diverges (gets larger and larger) for starting values greater than $e^{1/e}$. You'll also find some interesting surprises when the starting value is less than e^{-e}. Try it.

but a journey into a strange wilderness, where the explorers often get lost. Rigor should be a signal to the historian that the maps have been made, and the real explorers have gone elsewhere" (W. S. Anglin, "Mathematics and History," *The Mathematical Intelligencer*, 1992).

❓ Odin sequence. The Norse god Odin tells you to pick any two-digit number. Multiply by 3, and use Odin's mighty sword to sever the number so that you retain the last two digits of this result, and multiply by 3 again. Repeat the process. For example, 13 becomes 39, then 117, which we cleave to 17. Thus, starting with 13, we produce $13 \rightarrow 39 \rightarrow 17 \rightarrow 51 \rightarrow 53 \rightarrow 59, \ldots$ How many steps does the starting number 13 take to return back to 13? Do such sequences always return to their starting numbers? If so, how many steps are usually needed? (See Answer 3.50.)

❓ Tanzanian zoo. Monica is visiting her zookeeper friend Bill in the rain forests of Tanzania. Bill loves long-necked animals, and his zoo is a strange one, for it consists of just two types of animal: giraffes and ostriches. Monica gazes across the wooded area. "How many animals do you have?" Bill replies, "Among my animals, I have 22 heads and 80 legs in all. The number of ostriches is less than half the number of giraffe eyes minus 1. From this little information, can you tell me how many giraffes and ostriches I have?"

For the second problem, consider Bill's other zoo in Kenya. This zoo is filled only with long-necked birds. Monica is strolling with Bill through the Kenya zoo. "How many birds do you have altogether?" she asks.

"In my vast collection of birds, all but two of them are geese, all but two of them are swans, and all but two of them are ostriches. From my meager information, you should be able to find the answer."

"Are you some kind of nut?"

"Not at all. Tell me the answer, and we'll have a fine goose for dinner." (See Answer 3.51.)

❓ Fluid pool. Creatures from a nearby dimension penetrate our reality and want to fill a large swimming pool with their lime-scented nutritional fluid. From Hose A pours a green slime that would, by itself, take 30 minutes to fill the pool. From Hose B surges a crimson slime that, by itself, would take 20 minutes to fill the pool. How long would it take to fill the pool if both hoses poured at the same time?

The police will arrive at the scene in 15 minutes. If the creatures can fill the pool in under 15 minutes, they will deposit their spores in the liquid, multiply at fantastic rates, and take over Earth. Will the creatures succeed in their plan for world conquest? (See Answer 3.52.)

❓ Madagascar death snail. Bill is racing his poisonous Madagascar snail on a circular track that is 1 foot in circumference. The first time around, the snail travels at 30 feet per hour. How fast must the snail go the second time around to average 60 feet per hour for the two laps together? (See Answer 3.53.)

Pulsating brain. Dr. Matrix is a godlike being with a brain so active that its glowing pulsations can be seen through his glassine skull. He is also able to create miniature universes.

Dr. Matrix points to a shiny glass jar filled with black holes and glowing stars, and nothing else. In other words, there are only two kinds of astronomical objects to consider.

Dr. Matrix tells you that the percentage of black holes in his jar is more than 70 percent but less than 75 percent. Can there be as few as seven astronomical objects in the jar? (See Answer 3.54.)

- - - - - -

Grotesque vessel for total human harvest. My colleague Ignis Fatuus Jaymz once posed the following question to my discussion group. The year is 2030. Several alien poachers, while flying through space, discover Earth on their way home from hunting. They have unused space in their cargo hold. Starting in New York City, the aliens begin to examine all life on Earth. After a cursory taste test, they decide that they can easily market humans as exotic cuisine, so the aliens set out to harvest all living humanity from Earth.

Their standard process involves dissolving the harvest through a process of enzymatic deliquescence and pumping the noxious exudate into a single titanium spherical vessel. Their standard spherical vessels come in diameter sizes of 1, 2, 4, 8, 16, 32, and 64 kilometers.

Queequeg, their most senior hunter, instantly estimates the correct size of vessel to use, as the others begin setting up the rendering pumps and vats. He assumes that an average human weighs about 75 kilograms (165 pounds). What size vessel did Queequeg specify? (See Answer 3.55.)

Kama Sutra puzzle. In the *Kama Sutra*, an ancient Indian sex guide, we find a man who is tired of having sex, pausing and asking his lover:

Oh beautiful maiden with beaming eyes, tell me, since you understand the method of inversion, what number multiplied by 3, then increased by three-quarters of the product, then divided by 7, then diminished by one-third of the result, then multiplied by itself, then diminished by 52, whose square root is then extracted before 8 is added and then divided by 10, gives the final result of 2?

This is apparently a kind of mathematical foreplay. Can you solve the puzzle? (See Answer 3.56.)

Alien robot insect. A tiny disabled alien robot insect is attempting to climb over the edge of its spaceship, which is 40 feet tall. The creature starts at the base of the ship wall and takes a day to crawl 8 feet upward. The insect needs to recharge its fuel cells and so rests.

A month later, the insect awakens and realizes that it has slipped down 4 feet while sleeping. It begins its upward journey, and 8 feet later it sleeps and falls down by 4 feet. If this happens every month, in about how many months will the insect reach the top of a 40-foot wall of its craft? (See Answer 3.57.)

- - - - - -

Goa party. You are partying in Goa, India—listening to Goa trance music—and suddenly see a group of Sikhs riding a total of thirteen tricycles and bicycles. You also see thirty-five wheels. How many tricycles do you see? (See Answer 3.58.)

- - - - - -

Gelatinous octopoid. Dr. Eck is a gelatinous octopoid living within a gas pocket ten miles beneath the surface of Planet Uranus. He spent one-third of his life as a baby, one-fifth as a youth, and one-seventh as an active octopoid. If Dr. Eck finally spent 10 centuries as an old octopoid, then how many centuries did he spend as an active octopoid? (See Answer 3.59.)

- - - - - -

Cloned Jefferson in Rome. Thomas Jefferson has been "resurrected" from the dead by a cloning technology and sent back in time to visit ancient Rome. He comes across a 0.3-by-0.3-mile-square marble floor with columns equally spaced along the periphery of the floor. He counts 15 columns along each side of the wondrous floor. Each column is the same distance from neighboring columns. The columns at each corner of the square floor contain beautiful Doric caps. How many columns did the ancient Romans use? How did you solve this? (See Answer 3.60.)

- - - - - -

Hamburgers in space. Britney Spiracle is a blonde pop singer and a rocket hobbyist. She wants to make the *Guinness Book of World Records* by being the first person to launch a hamburger into the air. A small, specially designed rocket holds the hamburger. Britney's crazy rocket has a complicated set of rocket boosters. She believes that the distance of the hamburger in miles from Earth can be represented as a function of time in seconds: $f(t) = 3t^2 + 4t^3$.

How fast is the hamburger moving after it has traveled 100 seconds? Do you think Britney can actually get her hamburger to travel this long and this fast? (See Answer 3.61.)

- - - - - -

Spores of death and madness. From my pocket, I withdraw a small amber vial. Inside this vial are ten spores of an alien bacterium. I open the vial and place the ten spores in the center of a glass of milk. The bacteria multiply quickly. The bacteria emit a foul stench, but we need not worry about that for now. This is, however, a dangerous experiment, because the spores are said to cause wild hallucinations if inhaled.

By tomorrow at this time, there will be 2 million bacteria in the milk. The milk can contain a total of 6 billion bacteria, and our goal is to determine when the milk will contain this number of alien bacteria.

To solve this problem, we need to know that exponential growth makes use of the formula $N(t) = N_0 e^{kt}$. If this formula means nothing to you, skip to the answer. Exponential growth is growth that increases at a rate proportional to the current population. Or, to put it crudely: the more bacteria, the faster the population grows. Or, more specifically: given twice the number of bacteria, the population grows twice as fast. (See Answer 3.62.)

❓ An ancient problem of Mahavira. Mahavira (A.D. 800–A.D. 870) (or Mahaviracharya, meaning Mahavira the Teacher) lived in southern India. The only known book by Mahavira is *Ganita Sara Samgraha*, dated A.D. 850. Here is one of his problems.

A young lady has a quarrel with her husband and damages her necklace. One-third of the necklace's pearls scatter toward the lady. One-sixth fell on the bed. One-half of what remained (and one-half of what remained thereafter and again one-half of what remained thereafter) and so on, counting six times in all, fell everywhere else. [Thus] 1,161 pearls were found to remain unscattered. How many pearls did the girl originally have in total? (See Answer 3.63.)

🌏 Passionate math. "Contrary to popular belief, mathematics is a passionate subject. Mathematicians are driven by creative passions that are difficult to describe, but are no less forceful than those that compel a musician to compose or an artist to paint. The mathematician, the composer, the artist succumb to the same foibles as any human—love, hate, addictions, revenge, jealousies, desires for fame, and money" (Theoni Pappas, *Mathematical Scandals*, 1997).

❓ Alien slug. Dr. Oz is an alien slug. "Isn't it interesting," he said to his jumentous wife, "that 6 years ago, I would have been 10 years older than you were 3 years before I was half the age I am now."

"Impressive that you should say such a thing," his wife replied, with a slurping sound. "As for me, 12 years ago, I would have been 3 years older than you were 6 years before I was a third of my present age."

How old are Dr. Oz and his lovely wife? (See Answer 3.64.)

❓ Lucite pyramids. Dr. Eck sets before you two Lucite pyramids, one labeled alpha, the other labeled omega. Both pyramids contain a number of beetles from the Peruvian rain forests.

From alpha, Dr. Eck steals a number of beetles equal to one-third of the number of beetles already in omega and puts these beetles into omega. Next, you retrieve the stolen beetles from omega, a number equal to one-third of the number remaining in alpha, and return them to alpha. Together, alpha and omega now have 70 beetles. How many beetles does each pyramid contain? (See Answer 3.65.)

- - - - - -

❗ Doomsday occurs in November 2026. One day, while combing through a trash pile of old *Science* magazines, I found a gem from the year 1960. The title is "Doomsday: Friday, 13 November, A.D. 2026." That's quite a title for a serious magazine! The authors claimed that on this date, human population will approach infinity—if it grows as it has grown in the last two millennia. The researchers' work contains the following formula for world population N as a function of time t:

$$N = \frac{1.79 \times 10^{11}}{(2026.87 - t)^{0.99}}$$

where time is measured in years A.D. Just plug in a year, t, and you can calculate the population for that year. The researchers derive their model using a combination of empirical and theoretical reasoning that deals with fertility and mortality rates. The surprise is that the formula was so accurate. The formula gave remarkably close figures for human population between the years 1750 and 1960. It was even in agreement with world population estimates when Christ was born.

There is just one problem. In the year 2026 the U.S. population is infinite. N goes to infinity. Some religious extremists have taken this to mean that in 2026 Armageddon comes—Doomsday. We all die.

Let me quote from their paper where they describe a parameter in their model called t_0: "For obvious reasons, t_0 [A.D. 2026] shall be called 'doomsday,' since it is on that date, $t = t_0$, that N goes to infinity and that the clever population annihilates itself."

The authors' methods were so good that if Charlemagne had their initial equation, $N = K/\tau^k$, and also several estimates of the world population available to him when he lived, he could have predicted Doomsday accurately within 300 years. Elizabeth I of England could have predicted the critical date within 110 years, and Napoleon within 30 years. Various technological revolutions in human history show that food hasn't been a limiting factor to human growth. The Doomsday authors suggest our "great-great-grandchildren will not starve to death, they will be squeezed to death."

If this little description has attracted your interest, you can consult the original paper for all the details: H. von Foerster, P. Mora, and L. Amiot, "Doomsday: Friday, 13 November, A.D. 2026," *Science* 132 (November 1960): 1291–95.

The following are controversial responses to the von Foerster paper:

- Robertson, J., V. Bond, and E. Cronkite. "Doomsday (Letter to the Editor)," *Science* 133 (March 1961): 936–37.

- Hutton, W. "Doomsday (Letter to the Editor)," *Science* 133 (March 1961): 937–39.

- Howland, W. "Doomsday (Letter to the Editor)," *Science* 133 (March 1961): 939–40.

- Shinbrot, M. "Doomsday (Letter to the Editor)," *Science* 133 (March 1961): 940–41.

- von Foerster, H., P. Mora, and L. Amiot. "Doomsday (Letter to the Editor)," *Science* 133 (March 1961): 941–52.

- Serrin, J. "Is 'Doomsday' on Target?" *Science* 189 (1975): 86–88.

- - - - - -

Strange paper title. In 1988, Steven Strogatz published "Love Affairs and Differential Equations" in the prestigious *Mathematics Magazine* (61, no. 1: 35). The paper analyzes the time-evolution of the love affair between Romeo and Juliet.

- - - - - -

Bakhshali **manuscript: A true story.** The famous *Bakhshali* manuscript was found in 1881, in a stone enclosure in northwest India, and it may date as far back as the third century. When it was discovered, a large part of the manuscript had been destroyed and only about 70 leaves of birch bark, of which a few were only scraps, survived to the time of its discovery. Here is one problem from the manuscript:

> Before you are a group of twenty people comprising men, women, and children. They earn 20 coins between them. Each man earns 3 coins, each woman 1.5 coins, and each child 0.5 coin. How many men, women, and children are there?

(See Answer 3.66.)

- - - - - -

⊘ **Mathematics and truth.** "The higher arithmetic presents us with an inexhaustible storehouse of interesting truths—of truths, too, which are not isolated, but stand in the closest relation to one another, and between which, with each successive advance of the science, we continually discover new and wholly unexpected points of contact. A great part of the theories of arithmetic derive an additional charm from the peculiarity that we easily arrive by induction at important propositions, which have the stamp of simplicity upon them, but

the demonstration of which lies so deep as not to be discovered until after many fruitless efforts; and even then it is obtained by some tedious and artificial process, while the simpler methods of proof long remain hidden from us" (Carl Friedrich Gauss, 1849, quoted in Ivars Peterson, "Waring Experiments," 2004).

- - - - - -

❶ **Calculus of love.** In 2004, the psychologist John Gottman and the mathematician James Murray, both of the University of Washington, created a set of equations that accurately predicts whether a marriage will end in divorce. They have tested their equations on videotaped interviews made years earlier. In

particular, using their formulas, the researchers can examine just one interview and almost always correctly predict whether couples will divorce. One key input to the formulas is the degree to which one subject displays a contemptuous facial expression as the partner speaks. Happily, the researchers can improve relationships by studying their love formulas to guide several days of therapy. For further reading, see Erica Klarreich, "The Calculus of Love," *Science News* 165, no. 9 (February 28, 2004): 142.

- - - - - -

❶ **Strange paper title.** In 1983, Bruce Reznik published "Continued Fractions and an Annelidic PDE" in the

❶ **Defying imagination with a small formula.** If you were asked to find the smallest rational number x (smallest in the sense of smallest numerator and denominator) such that there exist rational numbers y and z and

$$x^2 - 157 = y^2, x^2 + 157 = z^2$$

the numerator and the denominator would be so large as to defy imagination. We know that a solution does exist, and you may contact me for the details. However, no one has ever found a rational number solution for this:

$$x^2 - x - 193 = y^2, x^2 + 193 = z^2$$

prestigious *Mathematical Intelligencer* (5, no. 4: 61–63). The term *annelidic* means "earthwormlike." The paper starts, "If you cut an earthworm (annelid) in two, each half will regenerate its missing part and become a new earthworm."

- - - - - -

🌀 **Riemann hypothesis.** Over 5,000 volunteers around the world are working on the Riemann hypothesis, using a distributed computer software package at Zetagrid.Net to search for the zeros of the Riemann zeta function.

The Riemann zeta function is a very wiggly curve defined for Re(s) > 1 by

$$\zeta(s) := \sum_{k=1}^{\infty} \frac{1}{k^s}$$

According to the Riemann hypothesis, all nontrivial zeros of the zeta function are on the critical line $1/2 + it$, where t is a real number. More than 300 billion zeros have so far been studied using the software at Zetagrid.Net. The verification of Riemann's hypothesis (formulated in 1859) is considered to be one of mathematics' most challenging problems.

🌀 **Abu'l Wafa.** The Muslim astronomer and mathematician Abu'l Wafa (940–998) established this relation: sin $(a + b)$ = sin a cos b + cos a sin b. Abu'l-Wafa was a distinguished researcher at the caliph's court in Baghdad.

🌀 **Harmonic series and Euler's γ.** We've previously mentioned that the harmonic series grows very slowly. In fact, the following "harmonic series" is divergent; that is, it approaches infinity as n does.

$$H = \sum_{n=1}^{\infty} \frac{1}{n} = 1 + \frac{1}{2} + \frac{1}{3} + \frac{1}{4} \ldots$$

For centuries, mathematicians mistakenly believed that the harmonic series converges because each new term gets continually smaller. Nicole Orseme (1323–1382) finally proved that it diverges. Interestingly, Leonhard Euler (1707–1783) discovered a formula to approximate the value of the sum of a finite number of its terms:

$$\sum_{n=1}^{m} \frac{1}{n} \approx \ln m + \gamma$$

where ln m is the natural logarithm of the number of terms in the series, and γ is Euler's constant. The formula becomes more accurate as m increases. Thus γ =

0.57721566 4901532860 6065120900 8240243104
2159335939 9235988057 6723488486 7726777664
6709369470 6329174674 9514631447 2498070824
8096050401 4486542836 2241739976 4492353625
3500333742 9373377376 7394279259 5258247094
9160087352 0394816567 . . . .

In chapter 2, I mentioned how this constant is also known as the *Euler-Mascheroni* constant or *Euler's gamma*. Humanity knows more than a million digits of γ, but we don't know whether γ is a rational number (the ratio of two integers a/b). Yet if it is rational, the denominator (b) must have more than 244,663 digits!

Al-Battani. The Muslim mathematician Al-Battani (850–929) discovered and promoted various trigonometric relationships and, as far as we know, was the first person on Earth to produce a table of cotangents that corresponded to every degree. His family had been members of the Sabians, a religious group of star worshippers from Harran.

Luminescent being. A luminescent being from Alpha Centauri comes to your home to test humanity's intelligence. If you answer correctly, humanity may become part of a large intergalactic federation. If you answer incorrectly, the alien will shroud Earth in perpetual darkness. Here's the problem that the alien writes on your front door, so that your neighbors may help.

1. Consider an integer N that is greater than 1.

2. Consider the integer M, which is the square root of N.

3. Given that M is the sum of the digits in N, what is N?

(Hint: Only one value of N can possibly satisfy these properties.) (See Answer 3.67.)

- - - - - -

Al-Khwarizmi and algebra. The word *algebra* comes from the title of Al-Khwarizmi's book *Kitab Al-Mukhasar fi Hisab Al-jabr Wa'l Muqabala* (*The Book of Summary Concerning Calculating by Transposition and Reduction*). Al-Khwarizmi (c. 790–847) was an Arab mathematician, and his book used no symbols but expressed mathematics as words. In fact, he explained how to find solutions to quadratic equations of the form $ax^2 + bx + c = 0$, using only words, instead of writing out the solutions by using symbolic notations as we would today: $x = [-b \pm \sqrt{b^2 - 4ac}] / 2a$.

- -

Indefinitely divergent. The following series is "indefinitely divergent" because it does not diverge, in the sense of being unbounded, or converge to a limit.

$$S = 1 - 1 + 1 - 1 + 1 - 1 + \ldots$$

We can group terms one way:

$$S = (1 - 1) + (1 - 1) + (1 - 1) + \ldots$$

or another:

$$S = 1 + (-1 + 1) + (-1 + 1) \ldots$$

to get seemingly different results, that is, 0 or 1, respectively.

Mathematical romance. One romantic evening in front of a warm fireplace, your sweetheart turns to you and says, "I will marry you if you can solve my mathematical challenge. Consider a multidigit integer N that is a power of 2. Moreover, each digit of N is also a power of 2. What is N?"

You think about it, but no solutions pop into your head.

Your lover sighs, massages the back of your neck, and finally whispers, "If you do not answer correctly, I will seek a more intelligent mate."

Can you satisfy your potential spouse? (See Answer 3.68.)

- - - - - -

A single solution? According to *The Inquisitive Problem Solver*, the only positive integer solution to

$$A \times B \times C = C \times D \times E = E \times F \times G$$

is $8 \times 1 \times 9 = 9 \times 2 \times 4 = 4 \times 6 \times 3$, if we assume that each variable must be a single digit.

Harmonic series. Can you show that the harmonic series H diverges, that is, it sums to infinity?

$$H = \sum_{n=1}^{\infty} \frac{1}{n} = 1 + \frac{1}{2} + \frac{1}{3} + \frac{1}{4} + \ldots$$

(See Answer 3.69.)

Harmonic series on a diet. Show that the harmonic series H, which normally diverges, does not diverge if we remove all the terms with at least one 9 digit in the denominator of the terms.

$$H = \sum_{n=1}^{\infty} \frac{1}{n} = 1 + \frac{1}{2} + \frac{1}{3} + \frac{1}{4} + \ldots$$

(See Answer 3.70.)

Variable notation. The philosopher René Descartes (1596–1650) established a convention of using letters at the end of the alphabet for variables (like x, y, and z) and letters at the beginning for constants (like a, b, and c). Example $f(x) = ax + by + c$. We still use this convention today.

Shopping mall puzzle. You are in a large New Jersey shopping mall. Over the mall's public address speaker comes a message:

"Consider an integer N consisting of different digits. Next consider M_i—all the two-digit numbers that can be made by selecting two digits from N. If I told you that the sum of all M_i equals N, what is the smallest possible value for N?"

All the shoppers stop dead in their tracks. You look up and shout, "We don't understand!"

The announcement continues, "Okay, here's a hint. As an example that does not work, consider 215. If we take every two-digit subset and add, we get $21 + 25 + 12 + 15 + 52 + 51 = 176$, which does not equal 215. If you can solve my problem, you can have all of the items in the mall that you can bring to your car on a one-hour shopping spree."

The people in the mall roar with delight, yet no one can solve the problem. Can you? (See Answer 3.71.)

Castles and strings. You are in a damp castle with two lengths of string and a book of matches. If you touch the flame to the end of one string, the string will burn for exactly 10 minutes. Similarly, the shorter string will burn for exactly 1 minute. How do you measure 5 minutes and 30 seconds using the matches and the strings? Assume that the strings burn at the same rate all along their lengths. (See Answer 3.72.)

- - - - - -

❓ Target practice. A tiny green elf hands you a bow and an arrow and points to a square target hanging on an ancient oak tree. "With four shots, hit four different numbers on the target that total 100." Can you do it?

22	15	62	61
87	23	13	63
17	9	24	19
55	20	7	51

(See Answer 3.73.)

- - - - - -

⚠ A nonlinear recurrence yielding binary digits. Consider the enigmatic properties of the sequence 1, 2, 3, 4, 6, 9, 13, 19, 27, 38, 54, 77, . . . defined by the recurrence $u_1 = 1$, $u_{n+1} = [\sqrt{2}\, u_n + \frac{1}{2}]$, $n \geq 1$, where $[x]$ denotes the floor of x, the largest integer not larger than x. This formula was discussed in the June 1991 issue of *Mathematics Magazine* (Stanley Rabinowitz and Peter Gilbert,

"A Nonlinear Recurrence Yielding Binary Digits," 64, no. 3: 168–71). They note the unusual property that $u_{2n+1} - 2u_{2n-1}$ is just the nth digit in the binary expansion of $\sqrt{2}$.

- - - - - -

❓ Hobson gambit. My colleague Nick Hobson blithely tossed out this problem to my discussion group: "If the sum of three numbers is 1, the sum of their squares is 15, and the sum of their cubes is 3, what is the sum of their fourth powers?"

What is the answer, and, more interestingly, how much time do you think it would take someone to provide an answer? (See Answer 3.74.)

- - - - - -

❓ Magic light board. You are in an incense-filled psychedelic shop in San Francisco. Black light posters adorn the walls. Music from Jefferson Airplane pours from an old stereo.

Before you is a magic light board. Each cell in this board of eight cells is one of eight colors: red, yellow, orange, green, blue, indigo, violet, or tan. No color is repeated and all colors are used. Using the clues given, can you determine the color of each cell?

1. One column has a red cell over a blue cell. Another column has an orange cell over a violet cell.

2. Number 8 is tan. Number 1 is not yellow.

3. Green is to the right of orange, and violet is to the right of indigo.

4. Blue is on an odd number, and indigo is on an even number.

5. Blue is to the left of indigo, and both orange and green are in sections with higher numbers than both red and yellow.

1	2	3	4
5	6	7	8

(See Answer 3.75.)

Aqueduct. You are transported back in time and to another dimension and find yourself on a thin stone path on top of a Roman aqueduct between Italy and Greece. The aqueduct is guarded by a Roman soldier whose orders are to kill anyone trying to leave Italy.

Anyone trying to come into Italy from Greece will be turned back. The Roman soldier is on the Italian side of the aqueduct inside a small hut. Every 10 minutes he comes out of the hut to check the aqueduct path. Your mission is to escape from the barbaric Romans and flee to Greece. It takes 20 minutes to cross the aqueduct. There is no place to hide, and you cannot go under the aqueduct. How do you escape? (See Answer 3.76.)

Students and math. "I advise my students to listen carefully the moment they decide to take no more mathematics courses. They might be able to hear the sound of closing doors" (James Caballero, "Everybody a Mathematician?" *CAIP Quarterly* 2 [Fall 1989]).

Geometry,
Games, and
Beyond

I N WHICH WE EXPLORE TILES, PATTERNS, POSITION PROBLEMS, ARRAYS, Venn diagrams, tic-tac-toe, other games played on boards, Königsberg bridges, catenaries, Loyd's and Dudeney's puzzles, Sherck's surface, magic squares, lituuses, inside-out Mandelbrot sets, the Quadratrix of Hippias, hyperspheres, fractal geese, Pappus's Arbelos, Escher patterns, Schmidhuber circles, and chess knights.

> Geometry is about spatial relationships and
> glistening shapes that span dimensions.
> It's the Silly Putty of mathematics.

Königsberg bridges. One of my favorite branches of mathematics deals with graph theory, or the mathematics of how objects are connected, which might be diagrammed as linkages of dots connected by lines. One of the oldest problems in graph theory involves the seven famous Königsberg bridges of Germany (now part of Russia), schematically illustrated in figure 4.1.

People in old Königsberg loved to take walks along the river and the island, and it had become a Sunday tradition to take the walk of the seven bridges. Over a few centuries ago, people debated whether it was possible to take a journey across all seven bridges without having to cross any bridge more than once.

In other words, could you take a complete tour of the town and return to the starting point by crossing all of the bridges just once? This problem had plagued them for years, because no one had ever been able to devise such a tour. No one knew for sure until the Swiss mathematician Leonhard Euler in 1736 was able to prove absolutely that this was impossible.

Euler represented the bridges by a graph in which land areas are represented by dots and bridges by lines. Figure 4.2 shows a simplified diagram of the Königsberg bridges.

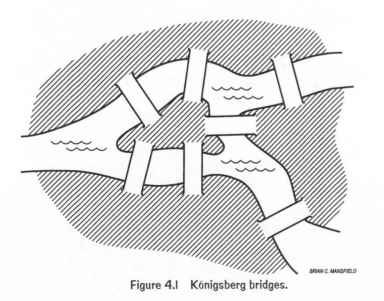

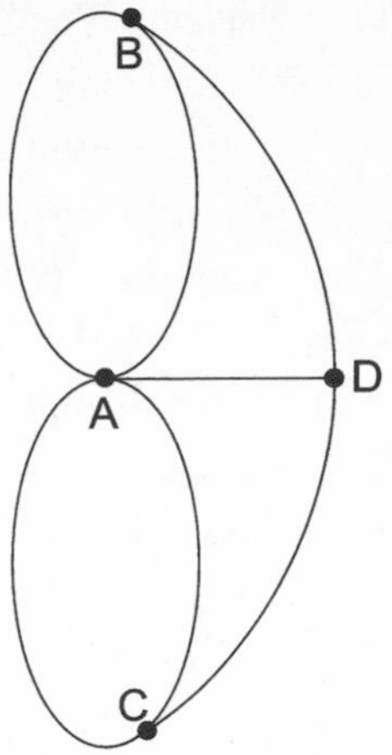

Figure 4.2 Simplified graph of the Königsberg bridges.

Euler showed that one could traverse the graph by going through every segment just once only if the graph had fewer than three vertices of odd "valence." The valence of a vertex is the number of lines that start or stop at the vertex. For example, Point A in the Königsberg graph has a valence of 5, and Point B has a valence of 3. Because all of the vertices of the Königsberg graph have odd valences, it is not possible to traverse the graph without going through a line more than once.

You can verify this yourself by trying to draw figure 4.2 with a pencil and not lifting the pencil from the paper. You cannot draw the Königsberg graph without repeating a line.

Figure 4.1 Königsberg bridges.

BRIAN C. MANSFIELD

Venn diagrams. The "Martian bodies" puzzle that follows shows the usefulness of simple *Venn diagrams* for practical problem solving. John Venn (1834–1923), a cleric in the Anglican Church, devised a scheme for visualizing elements, sets, and logical relationships. A Venn diagram usually contains circular areas representing groups of items that share common properties. For instance, within the universe of all real and legendary creatures (the bounding rectangle in figure 4.3), Region *H* represents the humans, Region *W* the winged creatures, and Region *A* the angels. A glance at the diagram reveals that

- All angels are winged creatures (Region A lies entirely within Region *W*).

- No humans are winged creatures (Regions *H* and *W* are nonintersecting).

- No humans are angels (Regions *H* and A are nonintersecting).

This is a depiction of a basic rule of logic—namely, that from the statements "all *A* is *W*" and "no *H* is *W*," it follows that "no *H* is *A*." The conclusion is evident when we look at the diagram's circles.

Venn struggled with generalizing his diagrams for visualizing many sets with intersecting areas. For example, he got as far as four sets

Figure 4.4   John Venn's diagram for four sets, using ellipses.

by using ellipses (figure 4.4): Venn tried to ensure that his diagrams would always be "symmetrical figures . . . elegant in themselves."

A century passed before various means that satisfied Venn's "elegance" statement were devised for larger numbers of sets. For example, Branko Grünbaum, a mathematician at the University of Washington, was the first to show that there are rotationally symmetric Venn diagrams made from five congruent ellipses.

Mathematicians gradually realized that rotationally symmetric diagrams can be drawn only with prime numbers of petals. Many different symmetrical Venn diagrams exist for five sets, including the one shown in figure 4.5

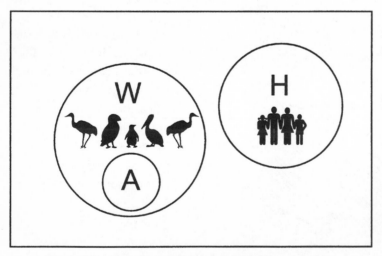

Figure 4.3   Venn diagram for real and legendary creatures, which include humans, winged creatures, and angels.

Figure 4.5 Branko Grünbaum's rotationally symmetric Venn diagram made from five congruent ellipses.

that was made by rotating an ellipse. However, symmetrical diagrams with seven petals were so hard to find that mathematicians initially doubted their existence.

In 2001, Dr. Peter Hamburger, of Indiana-Purdue University in Fort Wayne, constructed an example for eleven petals. The diagram is so complicated that it is difficult to appreciate without using color. Figure 4.6 shows Hamburger's diagram after removing the exterior of one of the curves. The original large and beautiful color image was created by Edit Hepp, following the methods of Dr. Hamburger. You can read more about these complicated objects and artistic renditions at Dr. Hamburger's

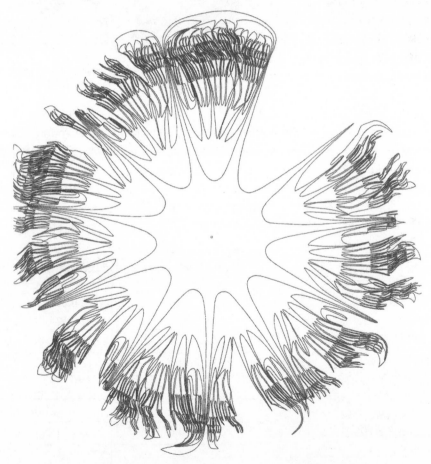

Figure 4.6 Symmetric 11-Venn diagram, after removal of one of the curves, courtesy of Dr. Peter Hamburger of Indiana-Purdue University and Edit Hepp.

Web site, www.ipfw.edu/math/Hamburger/, and in Barry Cipra, "Diagram Masters Cry 'Venn-i, Vidi, Vici'," *Science* 299, no. 5607 (January 31, 2003): 651.

The mathematician Anthony Edwards, of Cambridge University, came up with "cogwheel representations" for Venn diagrams. For example, figure 4.7 is Edwards's version of a Venn diagram that shows intersections of five sets.

Edwards has extended his representations to as many intersecting sets as we may wish to show.

Edwards has written extensively on the subject, including such books and papers as *Cogwheels of the Mind* (Baltimore, Md.: Johns Hopkins University Press, 2004); "Venn Diagrams for Many Sets," *Bulletin of the International Statistical Institute*, 47th Session, Paris, 1989 (contributed papers, Book 1, 311–12); "Venn Diagrams for Many Sets," *New Scientist* (7 January 1989): 51–56; "Rotatable Venn Diagrams," *Mathematics Review* 2 (February 1992): 19–21; and "Seven-Set Venn Diagrams with Rotational and Polar Symmetry,"

Combinatorics, Probability, and Computing 7 (1998): 149–52. You can read more about the Branko Grünbaum work here: Branko Grün-baum, "Venn Diagrams and Independent Families of Sets," *Mathematics Magazine* 48 (January–February 1975): 12–23.

Figure 4.7   Anthony Edwards's "cogwheel representations" for displaying the intersections of five sets.

Martian bodies and Venn diagrams. A group of theologians and scientists has discovered a humanoid race of creatures living in a crevice on Mars. Nine hundred Martians were examined for pointed ears (a_1), fangs (a_2), and forehead horns (a_3). The number of Martians with the various characteristics can be summarized as

Body Characterics	Number of Martians
a_1	600
a_2	390
a_3	400
a_1 and a_2	250
a_2 and a_3	150
a_1 and a_3	200
a_1, a_2, and a_3	20

How many Martians have none of these body characteristics? (See Answer 4.1.)

Geometry and beyond. "It was formerly supposed that Geometry was the study of the nature of the space in which we live, and accordingly it was urged, by those who held that what exists can only be known empirically, that Geometry should really be regarded as belonging to applied mathematics. But it has gradually appeared, by the increase of non-Euclidean systems, that Geometry throws no more light upon the nature of space than arithmetic throws upon the population of the United States" (Bertrand Russell, "Mathematics and Metaphysicians," *Mysticism and Logic and Other Essays*, 1918).

— — — — — —

Tic-tac-toe patents. I collect U.S. patents dealing with tic-tac-toe and Rubik's Cube–like games. For example, in 1995, U.S. Patent 5,433,448 was issued for a three-dimensional tic-tac-toe game (pictured in figure 4.8) that uses Velcro patches on the movable pieces.

Star of David. Is it possible to draw figure 4.9 without lifting your pencil and without repeating lines? (Recall what we learned about graphs and vertices in the section on the Königsberg bridges.) (See Answer 4.2.)

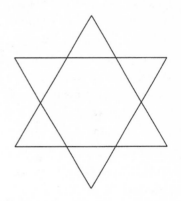

Figure **4.9** Draw the star without lifting your pencil and without repeating lines.

— — — — — —

Mathematics as the universe's language. "The universe cannot be read until we have learnt the language and become familiar with the characters in which it is written. It is written in mathematical language, and the letters are triangles, circles and other geometrical figures, without which means it is humanly impossible to comprehend a single word" (Galileo Galilei, *Opere Il Saggiatore*, 1623).

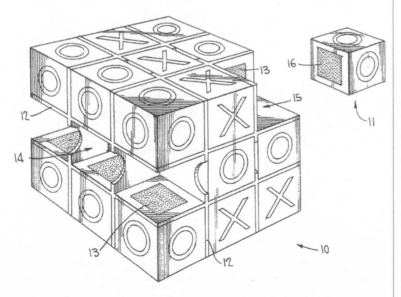

Figure **4.8** U.S. Patent **5,433,448**: a three-dimensional tic-tac-toe game.

Tic-tac-toe and *The Simpsons*. In 1996, tic-tac-toe was prominently featured on the hit TV show *The Simpsons*. In an episode titled "Much Apu about Nothing," Apu Nahasapeemapetilon, the convenience store clerk, describes how he wrote the first computerized version of tic-tac-toe for his doctoral thesis, using several thousand punch cards. It took Apu nine years to complete his thesis. Supposedly, only the top players in the world could beat his computer program, which is ridiculous because when played correctly, tic-tac-toe is always a tie!

Mathematician as poet. "A mathematician, like a painter or poet, is a maker of patterns. If his patterns are more permanent than theirs, it is because they are made with ideas" (G. H. Hardy, *A Mathematician's Apology*, 1941).

Catenary. Many hanging shapes in nature (such as a rope suspended at two points and sagging in the middle) follow a catenary curve defined by $(a/2)(e^{x/a} + e^{-x/a})$.

Mondrian puzzle. In this hypothetical meeting, the Dutch painter Piet Mondrian is showing Albert Einstein the painter's latest work (figure 4.10). The painting is made up of 18 rectangles, which are drawn identically in the schematic figure for simplicity of presentation. All intersecting lines form right angles with one another. The areas of several of the rectangles are shown inside the rectangles.

Mondrian wants to know the area of one rectangle in particular, denoted by the question mark. Einstein looks at the painting, wondering whether it is possible to calculate the particular area, given the meager amount of information we have. Can you help Einstein? (See Answer 4.3.)

Figure 4.10   Mondrian puzzle.

Galileo once erroneously believed that the curve of a chain hanging under gravity would be a parabola. The curve is also called the alysoid or the chainette, and its formula was discovered in 1691.

❓ Compact formula. Is there a compact formula relating e, π, i, and ϕ, the golden ratio? (See Answer 4.4.)

- - - - - -

⚡ Tic-tac-toe robot. In 1998, researchers and students at the University of Toronto created a robot to play three-dimensional ($4 \times 4 \times 4$) tic-tac-toe with a human. The robot used an arm powered by three motors to achieve three degrees of freedom (x direction, y direction, z direction) in its movement. An electromagnet was used to pick up the playing pieces, which were steel ball bearings. According to students Mark Ebden, Wilfred Lam, and Ryan Lausman, the final version of the tic-tac-toe robot won approximately 80 percent of the time when it went first, playing against a second copy of itself.

- - - - - -

⊛ Mathematicians and immortality. "Archimedes will be remembered when Aeschylus is forgotten, because languages die and mathematical ideas do not. 'Immortality' may be a silly word, but probably a mathematician has the best chance of whatever it

⚡ Tic-Tac-Chec. Tic-Tac-Chec is played with four chess pieces on a 4×4 board and is sold by Dream Green (Weirton, West Virginia). Your goal is to get all four of your pieces in a row (like tic-tac-toe). The board starts with no pieces, and players alternate taking turns. During each turn, you place one of your unused pieces on any empty square or move your piece that's already on the board. Pieces move as normal chess pieces do, except that pawns reverse direction once they reach the edge of the board. Pieces attack as normal chess pieces do. Once a piece is taken, it is returned to its owner for later placement.

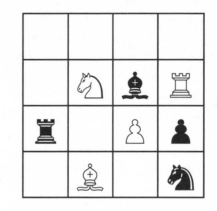

may mean" (G. H. Hardy, *A Mathematician's Apology*, 1941).

- - - - - -

⊛ Geometry and God. "Geometry is unique and eternal, and it shines in the mind of God. The share of it which has been granted to man is one of the reasons why he is the image of God" (Johannes Kepler, "Conversation with the Sidereal Messenger," 1610).

❓ Remarkable formulas involving 1. What is special about these two remarkable formulas that involve a single digit? What numbers do you think they represent? (See Answer 4.5.)

$$\sqrt{1+\sqrt{1+\sqrt{1+\sqrt{1+\ldots}}}}$$

$$1+\cfrac{1}{1+\cfrac{1}{1+\cfrac{1}{1+\cfrac{1}{\ldots}}}}$$

Goly. What is a golygon? (See Answer 4.6.)

- - - - - -

The false theorems of Archimedes. Archimedes (287 B.C.–212 B.C.), the ancient Greek geometer, is often regarded as the greatest mathematician and scientist of antiquity and one of the three greatest mathematicians to have walked on Earth—together with Isaac Newton and Carl Friedrich Gauss. But did you know that he sometimes sent his colleagues false theorems in order to trap them when they frequently stole his ideas?

In his book *The Sand Reckoner*, Archimedes estimates that 8×10^{63} grains of sand would fill the universe. We will discuss Archimedes further in chapter 5—specifically, his famous "Cattle Problem," which involves tremendously large numbers.

- - - - - -

Special number. What is special about the number $\phi = (1 + \sqrt{5})/2$? (See Answer 4.7.)

- - - - - -

Mathematics of toilet paper. In 1990, Don Thatcher of Leicester Polytechnic published a paper titled "The Length of a Roll of Toilet Paper" in a scientific book on mathematical modeling (*Mathematical Modeling*, Oxford University Press). He asks, "Given a roll of paper, find, without unwrapping it, the total length of paper on the roll."

He discusses the formulas $(n = r_2 - r_1)/t$ and $l = \pi(r_2^2 - r_1^2)/t$, where n is the number of turns, l is the total length of paper, t is the thickness of a sheet of paper, r_1 is the distance from the center of the roll to the cardboard tube within the roll, and r_2 is the distance from the center of the roll to the outer edge of the toilet paper.

- - - - - -

Mystery pattern. This mathematical object in figure 4.11 is my favorite of all shapes. What is it? Why is it special? (See Answer 4.8.)

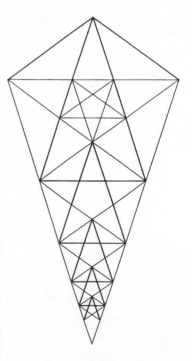

Figure 4.11   Mystery pattern.

Science and chess. "Even if the rules of nature are finite, like those of chess, might not science still prove to be an infinitely rich, rewarding game?" (John Horgan, *Scientific American* 6 [1992]).

- -

Math application. "There is no branch of mathematics, however abstract, which may not someday be applied to the phenomena of the real world" (Nikolai Lobachevsky, quoted in N. Rose, *Maxims and Minims*, 1988).

❓ Cut the crescent to make a cross. Here's another favorite from the nineteenth-century puzzlemaster Sam Loyd. Astonishing as it may seem, it is possible to cut the crescent moon shown here into as few as six pieces that can be fitted together to make a perfect Greek cross (figure 4.12). The shape of the symmetrical cross is shown in miniature on the head of the goddess. In forming the cross, it is necessary that one piece be turned over. (Notice that Loyd has put a straight edge on the crescent at the top and the bottom of the figure and that the two arcs of the crescent are arcs of a circle with the same-sized circumference.) Can you create the cross? (See Answer 4.9.)

Figure 4.12   Sam Loyd's "Cross and Crescent" puzzle.

⚠ Knight-tac-toe. David Howe invented a clever tic-tac-toe-like game in the late 1990s that is played on a 5 × 5 board. The goal is to either checkmate the opponent's king or arrange one's pieces to form three in a row in the center 3 × 3 area, as in the game of tic-tac-toe. The normal rules of chess apply. For example, the king may still not move into check. We believe that black has the advantage, although by modifying the starting positions, the game may be made more interesting.

Initial Board Setup for
Knight-Tac-Toe

- - - - - -

⚠ Three kinds of tori. A three-dimensional torus (ordinarily thought of as a doughnut shape) actually comes in three forms. If we let the radius from the center of the hole to the center of the torus "tube" be c and the radius of the tube be a, we can enumerate the three shapes. First, we have the common *ring torus* (shaped like a doughnut, in which $c > a$), a *horn torus* (in which the hole in the middle has zero diameter, i.e., $c = a$), and a *spindle torus* (where the doughnut walls intersect, and $c < a$). I would enjoy hearing from those of you who have eaten doughnut tori in all three forms.

This introduction to the torus leads to the next question.

– – – – – –

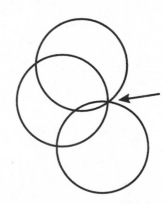 **Torus versus marmoset.** A rich person offers to give you a free, spacious house in the shape of either a large torus or a large marmoset. Which would you choose? Which do you think most people would choose? (Assume that both homes have the same ample living space and that the land on which they stand is similar. Also assume that the surface area of the torus, an object defined in the previous 〈!〉, is that for the usual ring torus: $S = 4\pi^2 ac$.) (See Answer 4.10.)

– – – – – –

〈?〉 **Green cheese moon puzzle.** In the 1800s, Sam Loyd asked, "If the moon were made of green cheese, into how many pieces could you divide it with five straight cuts of a knife?" (figure 4.13). In other words, what is the maximum number of pieces that you can cut a two-dimensional crescent, using straight lines? (See Answer 4.11.)

Figure 4.13   Sam Loyd's green cheese moon puzzle.

〈🎵〉 **Music and math.** "Music is the pleasure the human mind experiences from counting without being aware that it is counting" (Gottfried Leibniz, quoted in Marcus du Sautoy, *The Music of the Primes*, 2003).

– – – – – –

〈?〉 **Circle crossing.** If three circles with the same radii pass through a common point (see arrow), what interesting observation can we make

Figure 4.14   Circle crossing.

about the other three intersection points? (See Answer 4.12.)

Sam Loyd's fifteen puzzle. Sam Loyd's "fifteen puzzle" is the equivalent of today's sliding square puzzles that you may have seen in novelty stores. These plastic puzzles have 15 squares (tiles) and one vacant spot in a 4 × 4 frame or box. At startup, the squares sequentially contain the numbers 1 through 13, followed by 15, then 14.

1	2	3	4
5	6	7	8
9	10	11	12
13	15	14	

Loyd's Fifteen Puzzle
(Starting Position)

The idea is to "slide" the squares up, down, right, and left to arrive at the sequence 1 through 15. In other words, the goal is to rearrange the squares from a given starting arrangement by sliding squares one at a time into the configuration shown as follows. Loyd offered $1,000 for the correct solution. Can you see a way to solve this?

1	2	3	4
5	6	7	8
9	10	11	12
13	14	15	

Loyd's Fifteen Puzzle
(Ending Position)

(See Answer 4.13.)

Cow pi. I believe that π^3 appears very rarely in real physics or geometry problems. (One notable exception was discussed in chapter 2, in which π^3 appears in antenna physics.) However, here's a nice example of π^3 in a practical geometry problem. A circular column is 20 feet in diameter. A cow is tied to a point on the column wall. The rope is 10π feet long. Assuming that the ground is flat, the cow can cover an area $(250/3)\, \pi^3$. Could you determine how this solution was derived? John Derbyshire discusses these kinds of problems at his "Dog on a Leash" Web page, olimu.com/Notes/DogOnLeash.htm.

- - - - - -

Find the bugs! A devious spy has planted bugging devices in a secret government facility composed of a 4-by-4 array of rooms. Here's a top view of the facility showing the 16 little square rooms. Your mission is to determine the number of bugs in each room, given the following information. Two of the rooms contain three bugs each. Two of the rooms contain two bugs each. Four of the rooms contain one bug each. The remaining rooms are bugless. Can you find the number of bugs in each room? The numbers at the ends of each row and each column and one diagonal indicate the total number of bugs in that row, that column, and that diagonal.

				2
				5
				1
				6
5	2	3	4	4

(See Answer 4.14.)

— — — — — —

Strange dimension. You are in a strange dimension in another galaxy, overseeing the construction of a new multidenominational township with sectors for the three major local religions, whose houses of worship and religions are symbolized by 🏭, 🏛, and 🏠. Each of these religions is at war with the others. To make it more difficult for terrorists to bomb or shoot a laser through any one religious class, and to minimize religious conflicts, the

Tablet of Ezekiel. A tall, bearded man shows you three stone tablets divided into square tiles of beautiful colors. The first tablet is a 16×19 array:

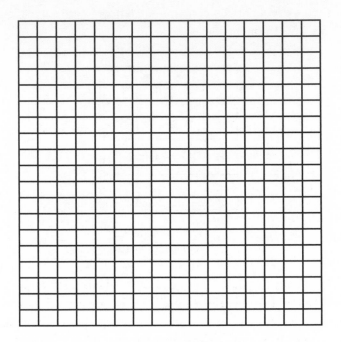

Tablet of Ezekiel

The other two are 16×24 and 16×18 arrays. The man draws a straight line on the rectangular tablets from one corner to another. On which of the three tablets does the diagonal line cross the most tiles? How could you solve this without drawing lines on each tablet and counting crossed tiles? (See Answer 4.15.)

architect is to design the township as a 3×3 matrix of sectors, so that (when viewed from above) each row and each column contain only one sector of a particular reli-

gious denomination. An aerial view of the religious center looks like a tic-tac-toe board, in which you are not permitted to have two of the same religion in any row or column. Is this arrangement possible? For a second problem, can you arrange the religions so that there are only two of the same religion in each row and each column?

The following is an arrangement prior to your attempt to minimize conflict:

(See Answer 4.16.)

A Dangerous Situation

Ant mathematics.
Aliens capture you and seat you in front of a large terrarium containing colonies of red ants (R), black ants (B), fire ants (F), and army ants (A), each species at one corner of a square (figure 4.15).

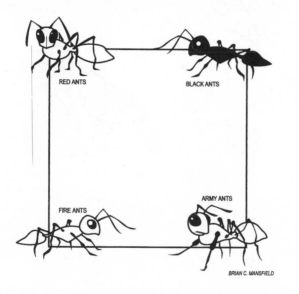

Figure 4.15   Ant puzzle.

Your captors ask you to create a tunnel out of plastic tubing that links all four species of ants together. You must use as little tubing as possible and still allow access from any colony to any other. The aliens provide you with a little glue to help stick pieces of tubes together, if necessary. What is the best solution? (See Answer 4.17.)

Nursery school geometry.
My nursery school has several unhappy children ☹. I want to position happy children ☺ so that every sad child has exactly one happy child next to him or her from a horizontally or vertically adjacent cell. To keep the room calm, two happy children can never be adjacent (not even diagonally). The numbers at the ends of each row and each column and one diagonal indicate the number of happy children in that row, that column, or that diagonal. Can you determine where the happy children are?

☹						1
☹	☹					0
	☹		☹			1
☹	☹	☹				1
☹		☹		☹		2
	☹			☹	☹	1
1	2	0	2	0	1	1

Nursery School Geometry

(See Answer 4.18.)

- - - - - -

Mathematical ballet.
"Math is a perfection in expression, like ballet or a shaolin class martial art" (V. Guruprasad, personal communication).

Frodo's magic squares.
Your small friend Frodo
loves magic squares. He asks
you to exchange four of the
numbers in one array with
four of the numbers in the
other to form two magic
squares in which each row,
each column, and each
diagonal have the same
sums.

16	3	2	13
5	10	11	8
9	6	20	3
4	10	14	22

15	18	1	14	1
11	24	7	7	12
17	5	13	21	9
23	6	19	2	15
4	12	25	8	16

How long does it take you to
fulfill Frodo's ambitious
request?

(See Answer 4.19.)

Lost in hyperspace. White cadaverous creatures from a parallel universe abduct you from your bedroom and place you in a long narrow tube. Assume that you are executing an infinite random walk within the tube; that is, you walk forever by moving randomly one step forward or one step backward in the tube (figure 4.16). Assume that the tube is infinitely long and that you don't get tired. What is the probability that the random walk will eventually take you back to your starting point? (See Answer 4.20.)

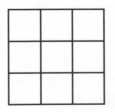

BRIAN C. MANSFIELD

Figure 4.16   Lost in hyperspace.

Heterosquares. Place the consecutive numbers 1 through 9, one number in each cell, so that the rows, the columns, and the main diagonals have *different* sums.

(See Answer 4.21.)

Charged array. The devious Doctor Brain has designed a memory array with 9 cells outlined in bold, as follows. A third of the cells have positive charges (+), a third have negative charges (−), and a third have no charge (0). Your mission is to decipher the secret arrangement of charges. The symbols at the ends of each row and each column and one diagonal indicate the sign of the charge for that row, that column, or that diagonal. Place a "+" symbol, a "−" symbol, or a "0" symbol in the appropriate cells. (See Answer 4.22.)

?	?	?	−
?	?	?	−
?	?	?	+
+	−	0	0

- - - - - -

Mathematical battle without conflict. "The tantalizing and compelling pursuit of mathematical problems offers mental absorption, peace of mind amid endless challenges, repose in activity, battle without conflict, 'refuge from the goading urgency of contingent happenings,' and the sort of beauty changeless mountains present to senses tried by the present-day kaleidoscope of events" (Morris Kline, *Mathematics in Western Culture*, 1953).

- - - - - -

Bouncing off the Continuum. An alien ship is located at position *S* in figure 4.17. The aliens wish to travel to a glimmering violet wall in space called the Continuum, to refuel their ship by using energy in the Continuum's plasmoid wall, and then travel back to a star called Aleph-Naught, denoted by *A*. They would like their trip to be as short a distance as possible. To what point on the edge of the Continuum should they travel? (See Answer 4.23.)

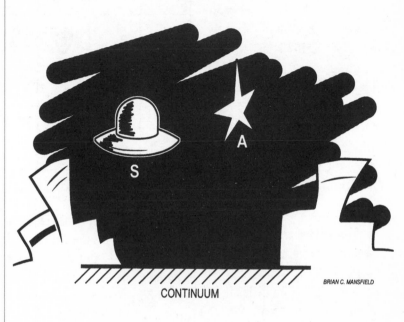

CONTINUUM

BRIAN C. MANSFIELD

Figure 4.17   Bouncing off the Continuum.

Alien heads. The following large square contains 49 little tiles and can be divided into a set of quadrilaterals (squares and rectangles) by drawing thick lines along the tile edges. I have already drawn one such quadrilateral in the upper left that "covers" 9 tiles. Can you draw exactly 10 more quadrilaterals, such that each quadrilateral contains exactly one alien head? Your 10 new quadrilaterals must "cover" the original bounding square completely, leaving no small tiles uncovered. (If useful to you, a single cell may be considered a quadrilateral.) (See Answer 4.24.)

Relationship. One of the gray cells should be white, and one of the white cells should be gray, in order to fit a simple relationship. Which cells should you change? (See Answer 4.25.)

1	2	28	15
4	3	50	70
5	6	18	30
8	7	27	48

Sticky faces. Cliff, Danielle, and Pete are dropping sticky smiley faces on the following chart. Each face they drop gets 10, 13, and 17 points. The chart shows how many of each face were dropped, but you don't know how many of each face each person dropped. You do know this: Danielle has 56 points. The number of Pete's points is a two-digit number with both digits the same. Pete has 3 more points than Cliff. What can we say about the number of each type of face each person dropped into the rows?

10 points	☺ ☺		☺ ☺
13 points	☺	☺☺	☺ ☺
17 points	☺☺☺	☺☺	☺

(See Answer 4.26.)

152 A Passion for Mathematics

Bioterrorist puzzle. The year is 2010, and a bioterrorist's bacterium endangers the planet. It is in this future world that we have some detective work to do. Romeo is located at one number in the following 5×5 square array, which represents an aerial view of plots of land. Juliet is located at another square. Romeo travels up, down, right, or left (not diagonally) to reach Juliet.

The numbers in the squares represent numbers of vaccine patches. Romeo starts with the number of vaccines that is indicated by the number in his starting cell and adds all the vaccines in a cell to his sack as he traverses the array until he lands on Juliet's cell and finally adds the number of vaccines in her cell. Once he adds the vaccines to his sack, the square no longer contains any vaccines.

1	2	3	4	5
6	7	8	9	10
11	12	13	14	15
16	17	18	19	20
21	22	23	24	25

If Romeo brings Juliet too many vaccines, her medical office will not be able to pay for them all. If he brings her too few, she will not be able to inoculate all of the people in her office. Romeo now has exactly 93 vaccines to give Juliet. Romeo starts on a square that is located immediately to the left of Juliet's square. On what cells do Romeo and Juliet initially sit? Note that Juliet is stationary as she waits for the 93 vaccines. (See Answer 4.27.)

- - - - - -

Rope capture. You are being held captive in Grand Central Station, New York, by five bearded drug dealers.

Next to you are two heroin-laced ropes suspended from the 100-foot ceiling. The two ropes are 2 feet apart. You have scissors in your pocket. If you can hand most of the rope to your captors, they will set you free. Otherwise, they will lock you in a coffin filled with monosodium glutamate and leave you to die.

You think you can climb to the top of the ceiling and cut one of the ropes, but you must climb down a rope until you are close to the ground before cutting it or you will die slamming into the hard floor. You think that there must be a better plan. What is your best strategy to get as

Magic square. To create this test, I started with a magic square that has the same sums for the rows, the columns, and two diagonals. In the following square, I've swapped a few tiles. Can you generate the original square, given the additional condition that tiles with ▶ or ◀ shapes must be adjacent horizontally and point to each other? For example, under this rule, 13 ▶ ◀ 7 might go side by side, but 13 ▶ 6 ▶ could not. (See Answer 4.28.)

17 ▶	6 ▶	◀ 7	23	19
13 ▶	◀ 10	15	11	◀ 12
16	2	8	14	20
21	◀ 24	9 ▶	0	◀ 3
22	18 ▶	◀ 4	5	1 ▶

much rope as possible? Assume that you have enough tools so that you can secure yourself to the rope while cutting pieces off. (See Answer 4.29.)

Talisman square. Fill in the squares with consecutive numbers 1 through 16 (one number to a cell) so that the difference between any one number and its neighbor is greater than some given constant. A *neighboring* number is defined as being in a cell that's horizontally, vertically, or diagonally adjacent to the current cell.

(See Answer 4.30.)

Platonic solids. A Platonic solid is a multifaceted 3-D object (or polyhedron) whose faces are all identical regular polygons. A "regular polygon" has sides of equal length and angles of equal degrees; two examples of regular polygons are the square and the equilateral triangle.

A Platonic solid also has the same number of faces meeting at every vertex. The best known example of a Platonic solid is the cube, whose faces are six identical squares. How many Platonic solids do you think we will ever discover? (See Answer 4.32.)

Circle madness. Draw Circle 1. Arrange six circles around it of any size, labeled 2 through 7 in the diagram. As you create your visual masterpiece, Circle 3 must touch 2, Circle 4 must touch 3, Circle 5 must touch 4, Circle 6 must touch 5, and Circle 7 must touch 6. Circles 2 through 7 must all touch the first circle. In this example, I happened to make Circle 7 enclose most of the others, but you can experiment with all kinds of arrangements. What can we say is interesting about the relationship between the circles' points of intersection with Circle 1? (See Answer 4.31.)

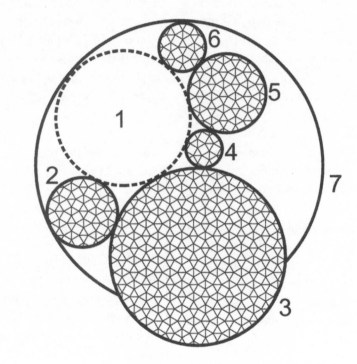

Figure 4.18 Circle madness.

Mystery pattern. What does figure 4.19 represent? (See Answer 4.33.)

Figure 4.19   Mystery pattern by Patrick Grim and Paul St. Denis.

Fractals and brains. "I wonder whether fractal images are not touching the very structure of our brains. Is there a clue in the infinitely regressing character of such images that illuminates our perception of art? Could it be that a fractal image is of such extraordinary richness, that it is bound to resonate with our neuronal circuits and stimulate the pleasure I infer we all feel" (Professor Peter W. Atkins, Lincoln College, Oxford University, "Art as Science," *The Daily Telegraph*, 1990).

Omega sphere. Dr. Brain places you in front of a rapidly spinning sphere called an omega sphere. The axis of rotation rapidly changes as the sphere rotates.

Dr. Brain approaches you with a dart. "Throw the dart at the sphere three times. If your three points of impact are all on the same half of the sphere (hemisphere), I will reward you with $25,000 and let you leave immediately. If your three points of impact are not all on the same hemisphere, I will remove a cubic inch of your brain. If you choose not to throw at the sphere, I will unleash a plague on the world that will kill 10,000 people." What is your best strategy for rapid escape from Dr. Brain's puzzle palace? (See Answer 4.34.)

- - - - - -

Twinkle, twinkle, little stars. You look up at a cluster of seven stars in the night sky. Your astronomer friend tells you that he believes that creatures from the seven stars are at war with one another. No two distances between any two pairs of stars are the same.

All at once, creatures from each star system launch a devastating nuclear arsenal at their nearest neighbor, and they wait to see what happens. None of the star systems has a defense against such awesome fury. Will all seven races be annihilated, or will at least one star system escape injury? (See Answer 4.36.)

- - - - - -

Mystery triangles. A giant places an array of ice crystals on the cold, snowy ground (figure 4.20). He then hands you a stick. "What is the maximum number of triangles without right angles that you can draw in the grid? None of the triangles can intersect or share the same vertices. Triangles may be nested inside one another. Each triangle must have its vertices on the ice crystals." (See Answer 4.37.)

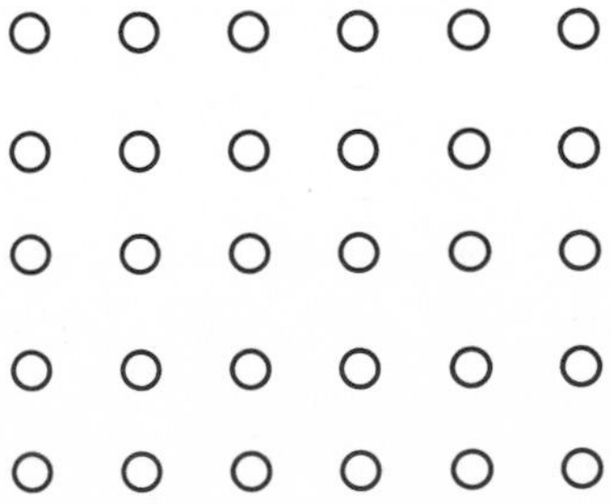

Figure 4.20 Mystery triangles.

Cutting the plane. Create two areas, both of exactly the same size and shape, so that both areas contain equal numbers of each symbol.

(See Answer 4.35.)

Space station jam. You are living in the new space station circling Earth when you find a painted metallic object floating through space (figure 4.21). Upon closer inspection, you see that the object is a collection of plates. Most of them are colored yellow (Y), green (G), red (R), or blue (B), but one of the plates is white (W). The collection was constructed according to a certain logic and will function as a teleportation device if you can solve a small mystery.

If you select the correct color that should replace the white plate, you and your crew will be transported to a wonderful universe where you will be rewarded with eternal bliss and will be able to solve humanity's most pressing problems. Can you work out which color should replace the missing (i.e., white) color? (See Answer 4.38.)

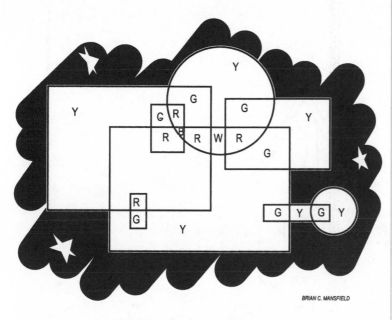

BRIAN C. MANSFIELD

Figure 4.21   Space station jam.

Grid of Gebeleizis. Gebeleizis, the ancient Dacian god of lightning, comes to you with a particu-larly vexing problem. He asks you to consider a grid of infinitesimal dots spaced one millimeter apart in a gargantuan cube having an edge that is equal to the north-south length of Greece. That's a lot of dots!

For conceptual purposes, you can think of the dots as having unit spacing, being precisely placed at 1.00000 . . . , 2.00000 . . . , 3.00000 . . . , and so on. When Gebeleizis uses the term *infinitesimal*, he means to imply that the dots are not bulbous objects that cover a range of locations in space. They really are located exactly at 1.00000 . . . , 2.00000 . . . , and so forth, and do not have a thickness that would make them extend to, for example, 1.010000 . . . , 2.010000 . . . , and so on.

Now, pick a dot, any dot, in this densely packed cube of dots. After you have made your selection, draw a straight line through the dot and extend it from that dot to the edge of the cube in both directions.

Gebeleizis asks, "What is the probability that your line will intersect another dot in the fine grid of dots within the cube the size of Greece?" (See Answer 4.39.)

- - - - - -

In an ancient tomb. You are exploring an ancient Egyptian tomb in search of a group of six mummies that supposedly have rested there for several thousand years. After a few minutes, you come upon a large triangular room that contains the six mummies lying on the cold, dusty floor. The room is covered with hieroglyphic symbols (figure 4.22). Because you are a world-renowned expert in hieroglyphics, you quickly decipher their meaning:

Three mummies are marked with three different symbols, and each of these three mummies is touching the south wall of the tomb. Three other mummies, marked with the same three symbols, are positioned so that one is touching the tomb walls that meet at the north corner. Is it possible to connect mummies bearing the same symbols using lines that you draw on the floor? The lines may not cross or touch the tomb walls. (For example, you must try to draw a line from the northmost mummy with the hooklike symbol to the other mummy with the hooklike symbol. You must also draw a line from the mummy with the eye symbol to the other mummy with the eye symbol, etc.) Your lines may be curvy, but they cannot touch or cross each other, nor can they go "through" the mummies. If you can solve this problem within five minutes, you will be granted great powers, wealth, and a long life.

(See Answer 4.40.)

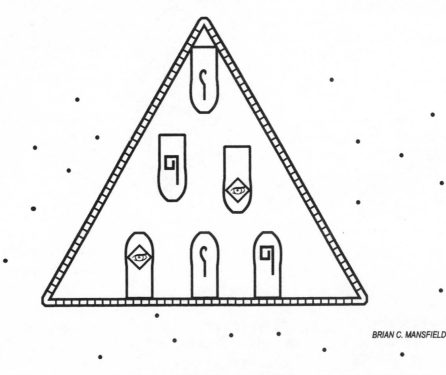

BRIAN C. MANSFIELD

Figure 4.22   In an ancient tomb.

Poor Pythagoras. Zeus steps out of the chariot and walks toward Pythagoras. Spirals of scarlet, thin as spider webs, float from Zeus's hypnotic eyes.

Pythagoras shakes his sword. "Don't come any closer. Last time you visited me, you turned my wife into stone."

"Listen, Pythagoras. I have a test for you. If you supply the proper answer, I will return your wife to normal."

Zeus places a marble table before Pythagoras. On the table are two circular disks that resemble Frisbees (Figure 4.23).

Zeus motions to the table. "We gods call them Omega disks, and we use them to test all intelligent mortals with whom we come in contact. As you see, one disk is red, the other gold. Both disks are the same size. The red disk is stuck to the table. The gold disk rotates around the red disk, touching it without

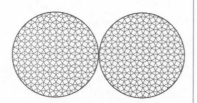

Figure 4.23   Poor Pythagoras.

Alien colonies. On planet Zarf in the Zeta Reticuli star system live alien bacterial colonies that come in many colors. Each bacterium is circular and touches at least one other bacterium to exchange nutrients. Figure 4.24 shows one example, in which different colors are represented by the numbers –3, –2, –1, 0, 1, and 2.

Scientists on Earth have obtained a few specimens and are trying to determine how each bacterium gets its color. Can you determine the rules by which the bacterial colonies get their colors? (See Answer 4.42.)

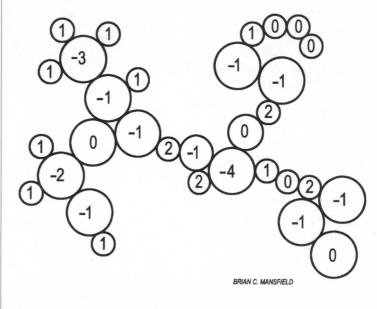

BRIAN C. MANSFIELD

Figure 4.24   Alien colonies.

slipping. When the gold disk has completed a turn around the red one, how many turns has it made around its own axis?"

Pythagoras does not answer, but he begins to charge Zeus with his sword. Can you help Pythagoras?

What is your answer? (a) 1 revolution, (b) 1.5 revolutions, (c) 2 revolutions, (d) 2.5 revolutions, (e) 3 revolutions, or (f) not on the list.

What do you think is the most common answer given to this problem? (See Answer 4.41.)

Ant planet. Lisa is all alone in her bedroom, playing with bugs. Her long blond hair trails to her knees. She laughs.

Lisa enjoys making maze-like structures with leftover wires from her electrical experiment. If an object touches the wire, it rings a buzzer. Today, Lisa is experimenting with ants. She places ants in these structures, and, depending upon where the insect is located, it is possible or impossible to escape without ringing the buzzer.

The ant prison mazes are of a peculiar type. Topologically speaking, they are *Jordan curves*, such as the ones shown here, which are merely circles that have been twisted out of shape (figure 4.25). Recall that a circle divides any flat surface into two areas—inside and outside. Like a circle, Jordan curves have an inside and an outside, and to get from one to the other, at least one line (wire) must be crossed.

Let's return to the ant story, in which Lisa is fantasizing about intelligent ants. One day, a "prisoner" ant named Mr. Nadroj is able to accurately determine, simply by poking its head over the wires and looking in one direction, whether or not it is on the inside or the outside of the maze. What's the quickest way that a creature can determine whether it is inside or outside the Jordan prison? How can you easily tell if the ant in the drawing can escape without actually trying to trace a path to the outside? (See Answer 4.43.)

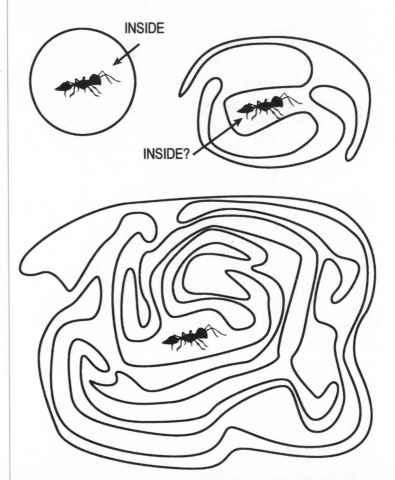

IS THE CREATURE INSIDE OR OUTSIDE ITS CURVY PRISON?

BRIAN C. MANSFIELD

Figure 4.25   Ant planet.

Aesculapian mazes. Aesculapius, the Roman god of medicine and health, devised the following problem for Romans so that they could improve both their minds and their bodies. The object is to travel from start to finish by taking three steps at a time and then turning right or left. Only those locations containing a vitamin pill (denoted by a dot) are valid points to land on and make turns (figure 4.26). If you desire, you may eat the pills along the way.

Start

2 1   3

End

Figure 4.26   Aesculapian mazes.

For example, when starting the maze, you land on the vitamin marked "1," and then must turn to the vitamin marked either "3" or "2." You are not allowed to retrace or go back along your path during any part of your journey. How many vitamins can you collect along your way? (See Answer 4.44.)

Contact from Aldebaran. At 3:00 A.M. on August 15, the Norwegian fashion model Britney Bjørlykke was walking near her home in Oslo. The air blew the spiral curls of Britney's blond hair out of place just as she saw a silvery, disk-shaped craft in the sky. The UFO swooped down and a figure emerged.

"We are from Aldebaran, a star 60 light-years from Earth," the alien said. "We wish to assess your intelligence."

The alien gave Britney instructions and then placed her in a transparent spiral tube one mile in length. The diameter was so small that she had to crawl through the tube. She started at the center of the spiral at 5 A.M. and crawled until she reached the outlet of the spiral at 5 P.M. She traveled at varying speeds, and every now and then paused, rested, and ate from meager food rations strapped to her belt.

When she arrived at the outlet of the spiral, she rested and then began her journey back into the spiral the next day at 5 A.M., reaching the center of the spiral at 5 P.M.

"You have completed your mission," the alien said.

"Now answer a question. What are the chances that there is a location along the spiral that you passed at exactly the same time both days?" (See Answer 4.45.)

- - - - - -

🌀 **Equation swirl.** "Perhaps an angel of the Lord surveyed an endless sea of chaos, then troubled it gently with his finger. In this tiny and temporary swirl of equations, our cosmos took shape" (Martin Gardner, "Order and Surprise," 1950).

- - - - - -

🌀 **Toilet paper and the infinite.** My favorite toilet paper geometries are not the simple, realistic Archimedean spiral kinds, but rather the squashed Archimedean variety. These exotic nonlinear forms of toilet paper rolls are pleasing to look at and can hold an amazing length of paper.

In particular, imagine a roll of toilet paper defined by the spiral $r = a\sqrt{\theta}$, where a is 1 inch. Since the compression is nonlinear, your theoretical toilet paper can grow wildly in length while being constrained to a small roll. For example, it's been estimated that a mile of "imaginary"

❓ **Detonation!** It is midnight, a cold Saturday night in New York City. A phone call comes to the precinct station informing the police that a mad bomber has planted an explosive device in the center of Grand Central Station.

The police rush to the scene and discover an intricate detonator that they must defuse. The detonator consists of a 6-by-6 wire mesh with white and black spheres containing two different explosive chemicals (figure 4.27). In order to defuse the bomb, the police must ever-so-carefully separate the white and the black balls by cutting through the wires and making two identically shaped cutouts—one containing the white spheres, the other containing the black spheres.

Your job is more difficult, because you know that one of the spheres has slipped and is in the wrong position. You must move it back to an adjacent grid position before cutting.

Tension mounts as the police bring out wire cutters. One false cut, and the bomb goes off! It seems impossible! Remember: the two cutouts must be identical in size and shape. Can you help? (See Answer 4.46.)

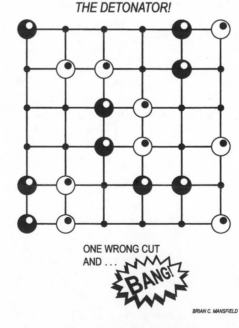

THE DETONATOR!

ONE WRONG CUT AND . . . **BANG!**

BRIAN C. MANSFIELD

Figure 4.27   Detonation!

toilet paper of this form could fit in a $a\sqrt{\theta}$ hypertoilet paper roll less than 50 inches in radius.

🛈 **Fractal Ford froth.** Imagine a frothy milkshake with an infinite number of bubbles, schematically illustrated in figure 4.28. As you stare into the cosmological milkshake, you notice bubbles of all sizes, touching one another but not interpenetrating. In this hypothetical foam, the bubbles become smaller and smaller, always filling in the cracks and the spaces between larger ones.

If you were to magnify the foam, tiny bubbles would always be interspersed with larger ones, but the overall structures would look the same at different magnifications. In other words, the froth would be called a fractal because it displays "self-similar" structures at different size scales. Do you have any guesses as to how this mathematical structure, first discussed in 1938, was created?

The figure shows the little-known Ford circles, named after L. R. Ford, who published on this topic in 1938. Ford circles provide an infinite treasure chest to explore, and the circles are among the most mind-numbing mathematical constructs to contemplate. In fact, it turns out that they describe the very fabric of our rational number system in an elegant way.

As a review, recall that rational numbers are numbers that can be expressed as fractions. For example, ½, ⅓,

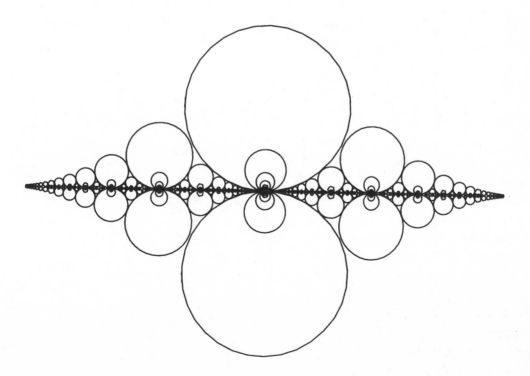

Figure 4.28   Fractal Ford froth.

and ⅔ are all rational numbers. As you might expect, there is an infinite number of such numbers.

What follows is a mathematical recipe for creating a Ford froth, which characterizes the location of rational numbers in our number system. You can use a compass and some graph paper to get started. No complicated mathematics is required for your journey.

Let us begin by choosing any two integers, h and k. Draw a circle with a radius of $1/(2k)^2$ and centered at $(h/k, 1/(2k)^2)$. For example, if you select $h = 1$ and $k = 2$, you draw a circle centered at $(0.5, 0.125)$ and with a radius of 0.125. Note that the larger the denominator of the fraction h/k, the smaller the radius of its Ford circle. Choose another two values for h and k, and draw another circle. Continue placing circles as many times as you like. As your picture becomes more dense, you'll notice something quite peculiar. None of your circles will intersect, although some will be tangent to one another (i.e., just kissing one another). Even if we place infinitely many Ford circles,

none will overlap, and each will be tangential to the x-axis. We can get a visual confirmation of this by magnifying the froth. For much more information on the Ford froth, see my book *Keys to Infinity*.

- - - - - -

Infinitely exploding circles. Draw a circle with a radius equal to 1 inch. Next, circumscribe (i.e., surround) the circle with an equilateral triangle. Next, circumscribe the triangle with another circle. Then circumscribe this second circle with a square. Continue with a third circle, circumscribing the square. Circumscribe the circle with a regular pentagon. Continue this procedure indefinitely,

Gear turns. Figure 4.29 shows a collection of intermeshed gears. The numbers of teeth on certain gears are indicated by the numbers within those circles. Does the gear marked "4" at the upper left spin faster, slower, or at the same speed as the gear marked "6" at the bottom right? (See Answer 4.47.)

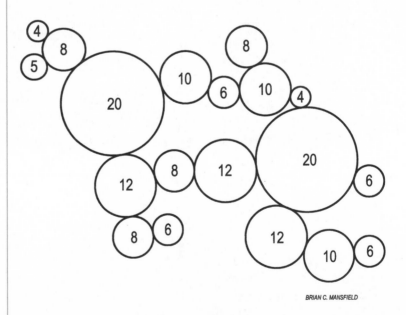

Figure 4.29   Gear turns.

each time increasing the number of sides of the regular polygon by one. Every other shape used is a circle that grows continually as it encloses the assembly of its predecessors (figure 4.30).

Figure 4.30   Infinitely exploding circles.

If you were to repeat this process, always adding larger circles at the rate of a circle a minute, how long would it take for the largest circle to have a radius equal to the radius of our universe, 10^{26} feet? (See Answer 4.48.)

– – – – – –

⊗ **Hexagonal challenge.** Dr. Brain walks with you in a forest and hands you a hexagonal piece of wood and a chisel. He wants you to cut a regular hexagon into eight congruent quadrilaterals (figure 4.31). In other words, each four-sided

piece must be identical to the others. Can this be done? (See Answer 4.49.)

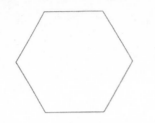

Figure 4.3I   Hexagonal challenge.

⊗ **Robotic worm.** A robotic worm named T'Pol creeps through cells (right, left, up, and down), starting with *A* and always repeating the pattern *ABCD, ABCD, ABCD, ABCD* as it crawls. The worm starts its journey with an *A* on a cell in the grid. Here are hints that show the worm's values in two cells. In the *B*

⊗ **Magic circles.** In figure 4.32, fill in the missing numbers so that numbers add up to 205 for each circle. The consecutive numbers 1 through 40 are used in the figure. (See Answer 4.50.)

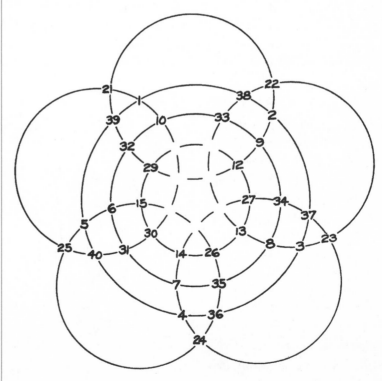

Figure 4.32   Magic circles.

and the *C* cells shown, T'Pol is moving upward. Can you fill in the rest of the letters? How did you go about solving this puzzle? (See Answer 4.51.)

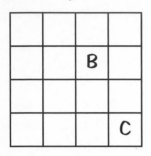

- - - - - -

❷ *y* = *x*sin(1/*x*). At night, I often dream of the function

$$y = x\sin\left(\frac{1}{x}\right)$$

The graph of this function for $|x| < 1$ shows smaller, yet more rapid, oscillations the nearer it approaches zero. The limit of this function as *x* approaches zero is zero (figure 4.33). We can define a very similar function $\psi(x)$ as follows:

$$\psi(x) = x\sin\left(\frac{1}{x}\right) \text{ when } x \neq 0$$

otherwise, $\psi(x) = 0$.

Like a Koch curve, this curve has infinitely many bumps, decreasing in size, in a finite

region of space. Is $\psi(x)$ continuous? (See Answer 4.52.)

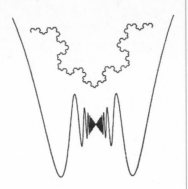

Figure 4.33 $y = x\sin(1/x)$

❷ Continuity. I ask you to draw a line on my hand, and you draw a flowing curve until I tell you to stop. Your curve is *continuous* because you never lifted the pen off my hand as you drew the curve. Now, I want you to imagine a crazier curve. Can a curve that has *infinitely* many turns or corners in a finite space still be continuous, or must it "break up" or become discontinuous due to

❷ Magic sphere. Figure 4.34 is a sphere containing the consecutive numbers 1 through 26, arranged in nine circles. Each circle has eight numbers, and their magic sum is 108. Can you fill in the missing numbers? (See Answer 4.53.)

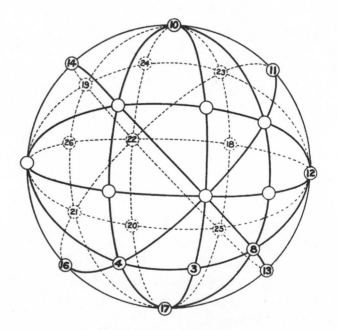

Figure 4.34 Magic sphere.

the infinity of turns? (See Answer 4.54.)

❓ Limits. When we want to take the "limit" of a function, this means we want to find out the function's behavior as we approach a particular value; for example, we might want to know the value of the function $f(x)$ as x approaches a number a.

Sometimes we can get a good idea about the limit of a function by examining its graph. We can also compute limits without graphing a function. Consider the simple parabola, $y = x^2$. As x approaches 0 from either the positive or the negative direction, the values of the function $f(x)$ approach the value of 0. For something simple like this, the limit of the function as x approaches some number a is just a^2. You just insert a into the function for x. We can write this as follows:

$$\lim_{x \to 0} x^2 = 0$$

For

$$\lim_{x \to 0} (x^2 - 3) = ?$$

we find a solution of -3. Simply insert the value for 0 for x, and you find that the limit is -3.

Simple insertion often works if we don't produce a number that has 0 in a denominator or a negative number inside a square root. But sometimes limits don't exist. For example, consider $y = 1/x^2$. As x approaches 0, y approaches infinity, and we usually say that "the limit does not exist."

Let us try a more complicated example. Can you find the limit of

$$\lim_{x \to 2} \frac{x - 2}{x^2 - 4}$$

(See Answer 4.55.)

❓ Space station scordatura. Four stars are located at the corners of a planar quadrilateral (figure 4.35). Musical creatures from each star want to build a space station where they can practice music. For convenience, the sum of the distances from the station to their stars must be as small as possible. Where should the creatures build the space station? (Assume that the stars are static in space.) (See Answer 4.56.)

BRIAN C. MANSFIELD

Figure 4.35   Space station scordatura.

The Procyon maneuver. A plantation owner is on a hunting trip south of Caracas, Venezuela, when at 2:30 A.M. he is paralyzed by a flash of light. He is lifted to a cylindrical craft by aliens in dusty silver suits and masks. In order to assess his intellectual prowess, they transport him to an Earthlike planet circling Procyon, a star that is 11 light-years from Earth.

In minutes, the plantation owner is facing a large in-ground funnel that has a circular opening 1,000 feet in diameter (figure 4.36). The walls of the funnel are quite slippery, and if the plantation owner attempts to enter the funnel, he will slip down into it. At the bottom of the funnel is a sleep-inducing liquid that will instantly put him to sleep for 8 hours if he touches it.

As shown in figure 4.36, there are two ankh-shaped towers. One stands upon a cylindrical platform in the center of the funnel. The platform's top surface is at ground level. The distance from the platform's top surface to the liquid is 500 feet. The other ankh tower is on land, at the edge of the funnel, as illustrated.

The aliens hand the plantation owner two objects: a rope 1,016.28 feet in length and the skull of a chicken.

The aliens turn to him and say, "If you are able to get to the central tower and touch it, we will give Earth the cure for cancer. We will also give Earthlings the ability to see in the ultraviolet range, thereby opening up a vast new arena of sensory experience. If you do not get to the tower, we will leave you on this planet after we have implanted a tracking device in your nose. Please note that with each passing hour, we will decrease the rope length by a foot."

How can the plantation owner reach the central ankh tower and touch it? (See Answer 4.57.)

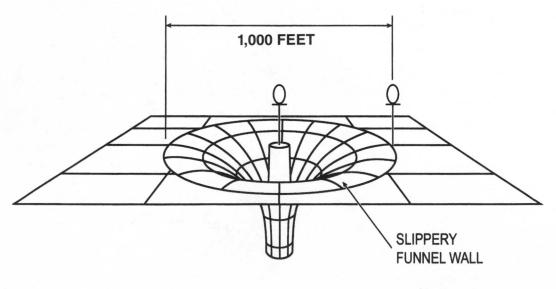

1,000 FEET

SLIPPERY
FUNNEL WALL

BRIAN C. MANSFIELD

Figure 4.36   The Procyon maneuver.

"I once asked Gregory Chudnovsky if a certain impression I had of mathematicians was true, that they spend immoderate amounts of time declaring each other's works trivial. 'It is true,' he admitted" (Richard Preston, *The New Yorker*, 1992).

❓ **Leaning tower of books.** One day while walking through the New York City Public Library, you notice a stack of dusty books leaning over the edge of a table (figure 4.37). It seems as if it is about to fall. A question comes immediately to mind: Would it be possible to stagger a stack of many books so that the top book would be far out into the room—say 10 or 20 feet? Or would such a stack fall under its own weight? You ask several friends, and each gives a different answer. What is your answer?

BRIAN C. MANSFIELD

Figure 4.37   Leaning tower of books.

(See Answer 4.58.)

❗ **Spirals.** *The Fermat* or *parabolic spiral* in figure 4.38 is pretty to ponder. Pierre de Fermat studied this form in 1636, and it can be created using the polar equation $r^2 = a^2\theta$. Fermat was only twenty-five when he studied this curve. Today, researchers sometimes use this form to model the arrangement of seedheads in flowers.

Figure 4.38   Fermat or parabolic spiral.

- - - - - -

🔄 **Good notation.** "A good notation has a subtlety and suggestiveness which, at times, make it almost seem like a live teacher" (Bertrand Russell [1872–1970], in J. R. Newman's *The World of Mathematics*, 1956).

- - - - - -

Pappus's Arbelos. I enjoy contemplating the mysteries of Pappus's Arbelos (figure 4.39). To create the figure, draw the two largest circles in figure 4.39a that have a horizontal line through their centers. These two circles, pointed to by arrows, create a crescent-shaped region between them. Next, place another circle with its center on the line and tangent to the two original circles. Continue to fill the crescent area with tangent circles, as shown. This chain of ever-diminishing circles is called a Pappus chain, and the arbelos is the unfilled region outside the circles.

The first small circle, which has its center on the line, can be denoted C_0, the next smaller circle is denoted C_1, and so on. Interestingly, the vertical distances from all the small circles' centers in the crescent to the line segment equal $2nr_n$, where r is the radius of each of the circles, C_n. Pappus's Arbelos seems to have been known to the early Greek mathematicians.

Figure 4.39b shows a Pappus chain with more circles filling the arbelos. Countless elegant and mind-boggling mathematical relations describe the relationships between all the circles in this infinite structure.

Loxodrome. A loxodromic sequence of tangent spheres in n-space is an infinite sequence of spheres having the property that every $n + 2$ consecutive members are mutually tangent (figure 4.40). H. S. M. Coxeter has shown that points of contact of consecutive pairs of spheres lie on a curve known as a *loxodrome*. If β is a constant angle, while θ and ϕ are the longitude and the latitude of a point on the loxodrome, the loxodrome's equation may be written $x = \sin \phi \cos \theta$, $y = \sin \phi \sin \theta$, $z = \cos \phi$, where $\theta = -\tan \beta \log \tan (\phi/2)$.

Doing a little historical

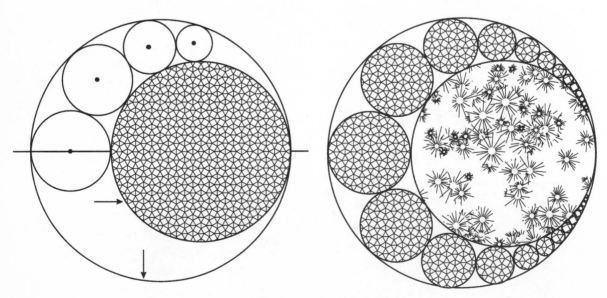

Figure 4.39a and b   Pappus's Arbelos.

research, you'll find out that the loxodrome curve was actually first conceived by Pedro Nunes around 1550. Note that a loxodrome is a curve on the surface of a sphere that makes a constant angle with the parallels of latitude—for example, a course with a constant compass bearing. It is the spherical analog of the logarithmic spiral in the plane, which makes a constant angle with concentric circles.

❓ Mystery swirly curve. Figure 4.41 is a beautiful and famous smoky-looking curve. Can you take a guess at its name or what class of shapes it represents? (See Answer 4.59.)

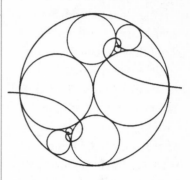

Figure 4.40   Loxodromic sequence of tangent spheres.

- - - - - -

❗ On the supernatural existence of Borromean rings. Borromean rings are three mutually interlocked rings named after the Italian Renaissance family whose members used these on their coat of arms. Ballantine Beer also uses this configuration in its logo (figure 4.42).

Notice that Borromean rings have no two rings that are linked, so if we cut any one of the rings, all three rings fall apart. Some historians speculate that the rings represent the three families of Visconti, Sforza, and Borromeo, which formed a

Figure 4.41   Mystery swirly curve.

tenuous union through inter-marriages.

Structures such as these fall within a mathematical discipline known as "knot theory," which is useful in polymer and theoretical physics. Mathematicians now know that we cannot actually construct a true set of Borromean rings with flat circles, and, in fact, you can see this for yourself if you try to create the interlocked rings out of wire, which requires some deformation or kinks in the wires. The theorem stating that Borromean rings are impossible to construct with flat circles is proved in Bernt Lindström and Hans-Olov Zetterström, "Borromean Circles Are Impossible," *The American Mathematical Monthly* 98, no. 4 (1991): 340–41.

In 2004, UCLA chemists created a breathtaking

beauty—a molecular counterpart of interlocked Borromean rings. Each molecule of the molecular Borromean ring compound was 2.5 nanometers across and contained an inner chamber that was a quarter of a cubic nanometer in volume and was lined by 12 oxygen atoms.

❓ **Kissing circles.** Figure 4.43 shows adjacent circles that are packed so that they *osculate* ("kiss" or just touch). I find that circle packing can provide a deep reservoir for striking images. How do you think I created this figure? Do you have even the slightest idea how to go about creating it yourself? The "bubbles" can be explored with a magnifying glass to yield details at increasingly tiny-sized scales. In fact, the arrow points to an inset figure that is a magnification of a tiny region of the froth in the larger figure. (See Answer 4.60.)

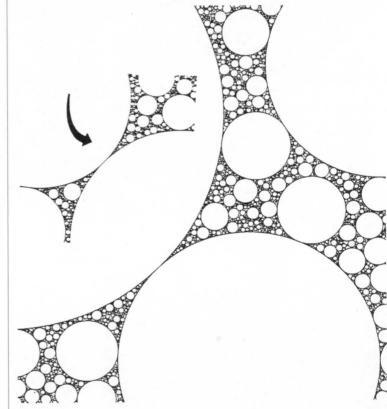

Figure 4.43 Osculatory circles.

Figure 4.42 Borromean rings.

Researchers speculate that molecular Borromean rings could be used as highly organized nanoclusters in a materials setting, such as spintronics, or in a biological context, such as medical imaging.

Thébault construction. In 1937, the French mathematician Victor Thébault (1882–1960) discovered that if you construct squares on the sides of any parallelogram, their centers form another square when connected (figure 4.44).

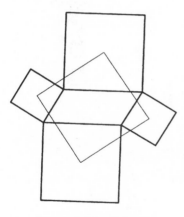

Figure 4.44   Thébault construction.

Schoenberg curve. "There is a discernible pattern [to the Schoenberg space-filling curve], albeit a very compli-

cated one. Attractive it is not. It would appear that what Arnold Schoenberg has done to music, I. J. Schoenberg has done to Peano-curves" (Hans Sagan, *The American Mathematical Monthly*, 1986).

Heron's problem. Heron (c. A.D. 500) asked whether we can find two rectangles, with integral sides, such that the area of the first is three times the area of the second, and the perimeter of the second is three times the

Withers's attractor. William Douglas Withers of the U.S. Naval Academy has described an interesting attractor for $A(z) = z^2 - 2z^*$, where z is a complex number and z^* is the complex conjugate. (See the answer to the earlier entry in this chapter titled "Mystery swirly curve" for more information on attractors.) The complex conjugate of a complex number is created by changing the sign of the imaginary part. Thus, the conjugate of the complex number $z = x + iy$ is $z^* = x - iy$. Starting with any initial value for z, you will finally be positioned somewhere on a triangular-shaped object with vertices at (3, $-3/2 \pm 3\sqrt{3}/2$) as you repeatedly apply Withers's formula in a mathematical feedback loop (figure 4.45). To implement this feedback loop or recursion, your new z value becomes input to the equation, and the mapping is repeated. For more information on this curve, see W. D. Withers, "Folding Polynomials and Their Dynamics," *The American Mathematical Monthly* 95, no. 5 (1987): 399–407.

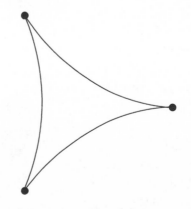

Figure 4.45   Withers's attractor.

perimeter of the first. Can you solve this? (See Answer 4.61.)

🌑 **Scherk's surface.** Scherk's surface, discovered by Heinrich Ferdinand Scherk in 1835, has the following form:

$$z = \ln\left(\frac{\cos y}{\cos x}\right)$$

where $-2\pi < x < 2\pi$ and $-2\pi < y < 2\pi$. Can you have your computer draw this strange minimal surface? Stewart Dickson has actually created a 3-D physical model of this saddlelike surface, using a process called stereolithography, which employs a laser-based tool and a photosensitive liquid resin that hardens as it forms the 3-D object.

Interestingly, Scherk's surface is a plausible model for the structure of interacting polymers that prefer to have as little contact as possible. Carlo H. Sequin has sculpted these surfaces out of wood, and you can use Google to search for various renditions. For further information, see Stewart Dickson, "Minimal Surfaces," *The Mathematica Journal* 1, no. 1 (1990): 38.

❓ **The game of elegant ellipses.** Many puzzles have been based on the problem of drawing straight lines in such a way that objects are segregated into separate regions on a plane. The game of elegant ellipses can be played with coins or ellipses that are thrown on a large piece of paper. In the example in figure 4.46, can you draw four straight lines that will divide the plane in such a way as to place each ellipse in a separate region? (See Answer 4.62.)

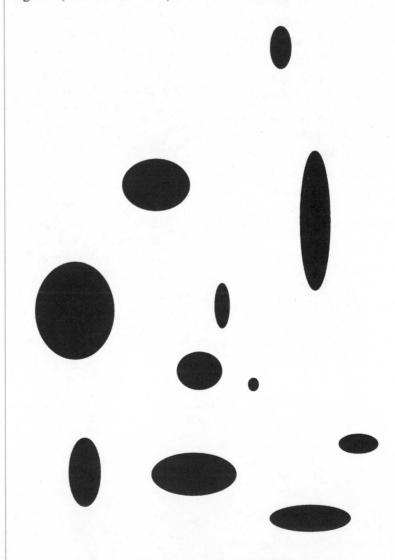

Figure 4.46   The game of elegant ellipses.

⟲ Fractal caves. "Out of the vast main aisle there opened here and there smaller caves, exactly, Sir Henry said, as chapels open out of great cathedrals. Some were large, but one or two—and this is a wonderful instance of how Nature carries out her handi-work by the same unvarying laws, utterly irrespective of size—were tiny. One little nook, for instance, was no larger than an unusually big doll's house, and yet it might

---

⚡ Schoenberg curves. The Schoenberg space-filling curve is one of the most intricate and exotic of all space-filling curves discussed in the world's scientific literature. *Space-filling curves* are interesting patterns that grow in length without limit while they fill the region in which they lie. The two most famous are the Hilbert and the Sierpinski curves. A more recent discovery is a self-similar curve, devel-oped by Mandelbrot, which fits exactly inside a Koch snowflake (see the April 1978 cover of *Scientific American*).

The Schoenberg curve was invented in 1938 by I. J. Schoenberg and further devel-oped in 1986 by Hans Sagan. It is certainly a challenge for mathematicians, programmers, and computer graphics specialists. Hans Sagan speaks of the curve's complexity: "To draw a 5th order Schoenberg curve, with no simple pattern to serve as guide, would tax the manipulative skills of a seventeenth-century mathematician, and the mere thought of going beyond that boggles the mind." Here I give the recipe for creating these curves and you can read more about the gory details in Sagan's paper. First, you must define a func-tion $p(t)$, which looks something like a chunky sine wave:

$p(t) = 0$ for $0 \leq t < 1/3$

$p(t) = 3t - 1$ for $1/3 \leq t < 2/3$

$p(t) = 1$ for $2/3 \leq t < 1$

This curve continues to infinity in both the $+t$ and $-t$ directions. Also, $p(-t) = p(t)$, $p(t + 2) = p(t)$. To create the Schoenberg monstrosity for different orders, connect each vertex to its predecessor by a straight line, using

$$x = f(t) = \sum_{k=1}^{\infty} p(3^{2k-2} t)/2^k$$

$$y = g(t) = \sum_{k=1}^{\infty} p(3^{2k-1} t)/2^k$$

The result is a complicated, chaotic assembly of zigzags. Schoenberg's curve has vertices for $t_{n,m} = n/3^m$, for $m = 1, 2, 3, \ldots, n = 0, 1, 2, \ldots$, and 3^m, where m is the "order" of the curve. My book *Computers and the Imagina-tion* shows a curve of order 4, and I believe that the highest-order Schoenberg curve ever to be plotted in a scientific journal is 7. You can read more about his curve in I. J. Schoenberg, "The Peano-Curve of Lebesgue," *Bulletin of the American Mathematics Society* 44 (1938): 519. Additional informa-tion can be found in H. Sagan, "Approximat-ing Polygons for Lebesgue's and Schoenberg's Space-Filling Curves," *The American Mathe-matical Monthly* 93, no. 5 (May 1986): 361. For a general description of space-filling curves, see F. Hill, *Computer Graphics* (New York: Macmillan, 1990).

have been the model of the whole place, for the water dropped, the tiny icicles hung, and the spar columns were forming in just the same way" (Sir Henry Rider Haggard, *King Solomon's Mines*, 1885).

- - - - - -

❓ Dudeney's circles. Here's a classic puzzle from British puzzlemaster Henry Ernest Dudeney (1857–1930).

Dudeney began his career at age nine, when he started composing puzzles that he published in a local paper. Later, he became angry when the American puzzle guru Sam Loyd published some of Dudeney's puzzles without giving sufficient credit. Some of Dudeney's puzzles continue to be of interest to modern mathematicians.

Here's one favorite Dudeney puzzle. Following

are 12 circles arranged to form 6 identical squares when perpendicular lines are drawn between the circles. Remove just 3 circles to leave just 3 identical squares. You might experiment with coins.

○　○　○　○

○　○　○　○

○　○　○　○

(See Answer 4.63.)

❓ Eschergrams.
I created figure 4.47 by randomly positioning a small square tile that contained 8 straight lines. In particular, the tile was placed in random orientations on a checkerboard to create this composite pattern. Can you draw the original tile? (See Answer 4.64.)

Figure 4.47   Eschergram.

Zenograms: Squashed worlds. You can compress all of mathematical space from $-\infty$ to $+\infty$ into a viewable cube that extends from -1 to 1. One way to do this makes use of the hyperbolic tangent function:

$$\tanh x = \frac{e^x - e^{-x}}{e^x + e^{-x}}$$

I call the resulting representation a Zenogram, after the ancient philosopher who studied various properties of infinity. My graphics program, called Zenospace, allows you to explore this strange squashed world using advanced computer graphics. No matter how large your numbers are, the tanh function can return only a maximum value of positive 1 or a minimum value of -1. Thus, any function, no matter how spatially extended, becomes viewable in the Zenogram.

Here are some observations on this weird space. In the Zenogram, diagonal parallel planes begin to curve, and they meet at infinity (the sides of the box). Paraboloids ($z = x^n + y^n$, $n = 2$) become squashed in interesting ways as they near the side of the box. (What happens as you increase n?) Spheres deform in interesting ways as they grow larger or are "pushed" toward the side of the box. My book *Computers and the Imagination* shows a Zenogram for a sphere as it is forced to a side wall of the Zenogram, at positive infinity. What do you think happens to the shape of a sphere centered at the origin as its radius grows to infinity? I discuss this novel method in my books *Computers and the Imagination* and *Mazes for the Mind*.

Dudeney's house. A charitable individual built a house in one corner of a square plot of ground and rented it to four persons. On the grounds were four cherry trees, and it was necessary to divide the grounds so that each person might have a tree and an equal portion of garden ground. Figure 4.48 is a sketch of the plot. How is it to be divided? (See Answer 4.65.)

Figure 4.48   Dudeney's house.

Dudeney's 12 counters.
Place 12 counters in 6 rows
(straight lines with any
orientation you like) so that
there are now 4 counters in
each row.

○ ○ ○ ○ ○ ○
○ ○ ○ ○ ○ ○

(See Answer 4.66.)

- - - - - -

Dudeney's cuts. How can
the board, marked as shown
in figure 4.49, be cut into four
identical pieces, so that each
piece contains three circles,
and no circle is cut. (See
Answer 4.67.)

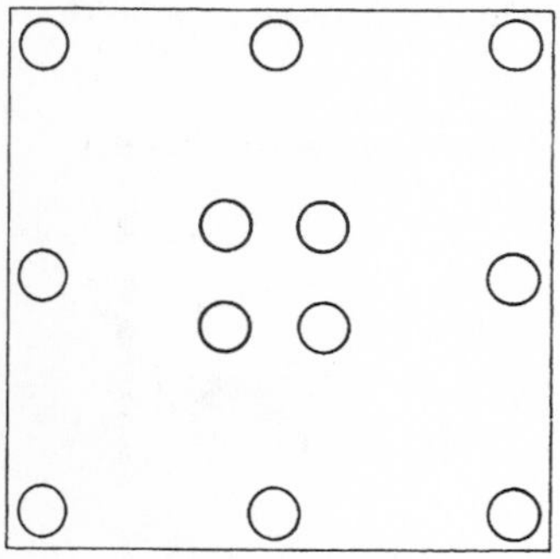

(Willi Hartner, in Annemarie
Schimmel's *The Mystery of
Numbers*, 1993).

- - - - - -

$e^{\frac{1}{e}}$. You go on a television
game show where the host
gives contestants the answers,
and the contestants have to
come up with the questions.
Today, the host gives this
answer:

$$e^{\frac{1}{e}}$$

What is the correct question?
(See Answer 4.68.)

**Tunnel through a cube:
A true story.** In the 1600s,
Prince Rupert of Bavaria
asked a fascinating and
famous geometrical question,
which incidentally won him
a sum of money. What is the
largest wooden cube that
can pass through a given
cube with a side length of 1
inch? More precisely, what is
the size of the edge of the
largest tunnel (with a square
cross section) that can be
made through a cube without
breaking the cube? (See
Answer 4.70.)

Dudeney's horseshoe. How can a horseshoe (figure 4.50) be
cut into six separate pieces with just two cuts? Your challenge
is to try to solve this in your head, without even using a pencil
and paper. (See Answer 4.69.)

Figure 4.49   Dudeney's cuts.

- - - - - -

Mathematical spirit. "The
mathematical spirit is a pri-
mordial human property that
reveals itself wherever human
beings live or material ves-
tiges of former life exist"

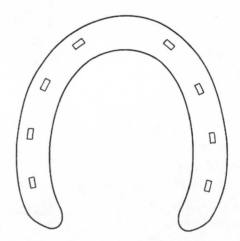

Figure 4.50   Dudeney's horseshoe.

❓ The mathematics of love. Mark and Bill are both in love with a TV actress named Shannon who is sitting in a park (figure 4.51). Alas, sometimes the men compete with each other to see who can meet her first, in order to engage in stimulating conversation and flirtation. Bill, the faster runner of the two, always runs at three times the speed of Mark.

Shannon now sits on a park bench exactly 200 feet directly to the east of Mark. Bill is some distance directly north of Mark's position.

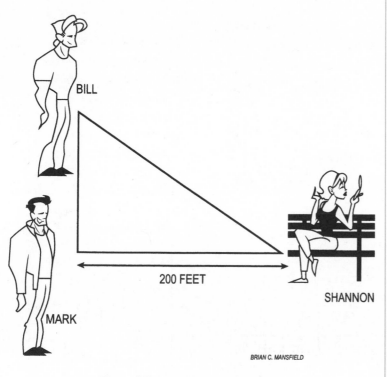

BRIAN C. MANSFIELD

Figure 4.51　The mathematics of love.

At the same time, Mark and Bill race to Shannon and arrive at the same moment. Mark and Bill are both fair, so they decide to share their conversations with Shannon. Just how far north of Mark was Bill at the start of their race to Shannon? (See Answer 4.71.)

- -

🛈 Albrecht Dürer. The Renaissance artist Albrecht Dürer created this wonderful 4 × 4 magic square in 1514:

16	3	2	13
5	10	11	8
9	6	7	12
4	15	14	1

Note that the two central numbers in the bottom row read "1514," the year of its construction. The rows, the columns, and the main diagonals sum to 34. In addition, 34 is the sum of the numbers of the corner squares (16 + 13 + 4 + 1) and of the central 2 × 2 square (10 + 11 + 6 + 7).

- - - - - -

🛈 Magic squares. There are 275,305,224 different 5 × 5 magic squares in existence. In 1973, Richard Schroeppel, a mathematician and a computer programer, used 100 hours of PDP-10 computer time to arrive at this large number.

For compactness, we can divide this large number by 4 and give the total as 68,826,306, because, in addition to the eight variants obtained by rotation and reflection, four other variants also preserve magical properties:

- Exchange the left and the right border columns, then exchange the top and the bottom border rows.

- Exchange Rows 1 and 2 and Rows 4 and 5. Then exchange Columns 1 and 2 and Columns 4 and 5.

As Martin Gardner has pointed out in *Time Travel and Other Mathematical Bewilderments*, when these two transformations are combined with the two reflections and the four rotations, the result is $2 \times 4 \times 2 \times 2 = 32$ forms that can be called isomorphic (having essentially the same structure). With this definition of isomorphic, the number of "unique" fifth-order magic squares drops to 68,826,306.

- - - - - -

Alphamagic square. According to Lee Sallows, an *alphamagic square* is a magic square for which the number of letters in the word for each number generates another magic square. For example,

5	22	18
28	15	2
12	8	25

Alphamagic Square

4	9	8
11	7	3
6	5	10

Resulting Square

where the magic square on the right corresponds to the number of letters in

five	twenty-two	eighteen
twenty-eight	fifteen	two
twelve	eight	twenty-five

Numbers Spelled Out

In other words, you can spell out the numbers in the first magic square. Then count the letters in the words. The integers make a second magic square. This second square contains the consecutive digits from 3 to 11.

Antimagic square. An antimagic square is an $N \times N$ array of numbers from 1 to N^2 such that each row, each column, and each main diagonal produces a different sum, and the *sums* form a consecutive series of integers. Fill in the missing numbers to complete this antimagic square. (See Answer 4.72.)

	8	9	
	12	15	
	1	14	
	15	4	

Antimagic Square

- - - - - -

Annihilation magic squares. According to Ivan Moscovich, the author of *Fiendishly Difficult Math*

Puzzles, "annihilation magic squares" are those in which the rows, the columns, and the two main diagonals sum to zero. For these squares, consecutive numbers from $-N/2$ to $N/2$ are required, and zero is excluded. Following is an example with several numbers removed. Can you fill in the missing numbers? (See Answer 4.73.)

-4	3	2	-1
1	-2	-3	4

Annihilation Magic Square

❓ Cramming humanity into small spots. Let us consider today's world population and give everyone a 3-by-3-foot piece of ground on which to stand. How large an area would humanity cover?

Put your answer in terms that are meant to delight us. For example, could they all squeeze onto an area the size of Manhattan Island? (See Answer 4.74.)

❓ Chocolate cake computation. You are walking down Fifth Avenue in New York City and enter a restaurant selling a delicious "Triple Chocolate Mousse Cake." The cake consists of chocolate chiffon layers, a rich chocolate mousse filling, and a glaze of bittersweet chocolate. Your mouth begins to water!

For $10, you can buy a circular cake of diameter 8.5 or a square cake of side length 7.5. Which is a better buy? In general, is a circular cake of diameter d always a better buy than a square cake with an edge of $d - 1$? (See Answer 4.75.)

⚠ Apocalyptic magic square. A rather beastly six-by-six magic square was invented by the mysterious A. W. Johnson. No one knows when this square was constructed, nor is there much information about Johnson, except that he has occasionally published in the *Journal of Recreational Mathematics*. All of its entries are prime numbers, and each row, each column, and each diagonal and broken diagonal sum to 666, the Number of the Beast. (A "broken diagonal" is the diagonal produced by wrapping from one side of the square to the other; for example, the outlined numbers 131, 83, 199, 113, 13, 127 form a broken diagonal.)

3	107	5	131	109	311
7	331	193	11	83	41
103	53	71	89	151	199
113	61	97	197	167	31
367	13	173	59	17	37
73	101	127	179	139	47

Apocalyptic Magic Square

Amateur findings. In the 1970s, Marjorie Rice, a San Diego housewife and a mother of five, was working at her kitchen table when she discovered numerous new geometrical patterns that professors had thought were impossible. Rice had no training beyond high school, but by 1976 she had discovered 58 special kinds of pentagonal tiles that would tile a plane, and most of the tiles had been previously unknown. Her most advanced diploma was a 1939 high school degree, for which she had taken only one general math course.

Mirror magic square. If you reverse each of the entries in this "mirror magic square," you obtain another magic square.

96	64	37	45
39	43	98	62
84	76	25	57
23	59	82	78

Mirror Magic Square

In both cases, the sums for the rows, the columns, and the diagonals are 242:

69	46	73	54
93	34	89	26
48	67	52	75
32	95	28	87

Rorrim Magic Square

π-square. To create T. E. Lobeck's magic square, which follows, start with a conventional 5-by-5 magic square and then substitute the nth digit of π (3.14159 . . .) for each number n in the square. This means that a 3 is substituted for a 1, a 1 is substituted for a 2, a 4 substituted for a 3, and so on. Amazingly, every column sum duplicates some row sum for the π-square. For example, the top row sums to 24, as does the 4th column.

17	24	1	8	15
23	5	7	14	16
4	6	13	20	22
10	12	19	21	3
11	18	25	2	9

5 × 5 Magic Square

2	4	3	6	9
6	5	2	7	3
1	9	9	4	2
3	8	8	6	4
5	3	3	1	5

π-Square

69-square. In the following magic square, notice how the digits 0, 1, 6, 8, 9, when rotated 180 degrees, become 6, 8, 9, 1, 0. This magic square is still magic when rotated 180 degrees. However, if these digits are simply turned upside down, the 6 becomes a backward 9 and the 9 a backward 6. If you turn the square upside down, then reverse the 6s and 9s so that they read correctly, you end up with different numbers, but the square is still magic! Notice that corners of any 2×2, 3×3, or 4×4 squares, as well as many other combinations, also sum to 264.

18	99	86	61
66	81	98	19
91	16	69	88
89	68	11	96

69-Square

- - - - - -

Existence. "The external world exists; the structure of the world is ordered; we know little about the nature of the order, nothing at all about why it should exist"

Upside-down magic squares. In the following pair of upside-down magic square pairs, the numbers in either square will form the other square by turning either square upside down. In either case, the sum of the numbers in the first-place digits (9 + 8 + 1 + 6) or the sum of the numbers in the second-place digits (9 + 1 + 6 + 8) in each row, each column, each two-by-two opposite short diagonal, or each main diagonal will be 24. The sum of the numbers in each row, each column, each two-by-two short diagonal, each main diagonal, the four corner squares, the four center squares, and any four adjacent corner squares equals 264.

99	81	16	68
18	66	91	89
61	19	88	96
86	98	69	11

Upside-Down of Right Square

89	68	96	11
91	16	88	69
18	99	61	86
66	81	19	98

Upside-Down of Left Square

(Martin Gardner, "Order and Surprise," 1950, paraphrasing Bertrand Russell).

- - - - - -

Strange paper title. In 1980, Dr. Forest W. Simmons published the unusual mathematical title "When Homogeneous Continua Are Hausdorff Circles (or Yes, We Hausdorff Bananas)" in the prestigious *Continua, Decompositions, Manifolds*

(Proceedings of Texas Topology Symposium), University of Texas Press. The figures are reminiscent of bananas!

- - - - - -

Prime number magic square. The following "small-constant" square at left has the smallest possible magic constant, 177, for an order-3 square composed only of prime numbers. The number 1 is usually not considered a prime number; however, if 1 is allowed in a cell as a prime, the only all-prime order-3 magic square has a constant of 111.

71	89	17
5	59	113
101	29	47

Prime Number Magic Square

- - - - - -

🔵 **8811 magic square.** The following magic square totals 19,998 in all directions in the square as is, upside down, or as reflected in a mirror. In every case, any 2 × 2 sub-square (e.g., one formed by 8,188; 1,111; 1,881; 8,818), as well as the four corner cells, totals 19,998.

8811	8188	1111	1888
1118	1881	8818	8181
8888	8111	1188	1811
1181	1818	8881	8118

8811 Magic Square

🔵 **Prime number magic square.** This square has the smallest possible magic constant, 3,117, for an order-3 square filled with primes in an arithmetic sequence. (In an arithmetic sequence, each term is equal to the sum of the preceding term and a constant.)

1669	199	1249
619	1039	1459
829	1879	409

Prime Number Magic Square

- - - - - - - - - - - - - - - - -

🔵 **Consecutive prime magic square.** Harry Nelson was the first person to produce the following 3 × 3 matrix containing only consecutive primes (for which he won a $100 prize offered by Martin Gardner).

1480028159	1480028153	1480028201
1480028213	1480028171	1480028129
1480028141	1480028189	1480028183

Consecutive Prime Magic Square

🔵 **Strange paper title.** In 1985, Tom Morley published "A Simple Proof That the World Is Three-Dimensional" in the prestigious *SIAM Review* 27: 69–71. The article starts, "The title is, of course, a fraud. We prove nothing of the sort. Instead we show that radially symmetric wave propagation is possible only in dimensions one and three."

Knight's tour. Figure 4.52 is a representation of a knight's tour, in which a chess knight jumps once to every square on the (8 × 8) chessboard in a complete tour. This is the earliest recorded solution and was found by Abraham De Moivre (1667–1754), the French mathematician who is better known for his theorems about complex numbers. Note that in De Moivre's solution, the knight ends his tour on a square that is not one move away from the starting square. The French mathematician Adrien-Marie Legendre (1752–1833) improved on this and found a solution in which the first and the last squares are a single move apart, so that the tour closes up on itself into a single loop of 64 knight's moves. Such a tour is said to be *reentrant*. Not to be outdone, the Swiss mathematician Leonhard Euler (1707–1783) found a reentrant tour that visits two halves of the board in turn. (The little squares show positions where the knight transits from one half to the other.)

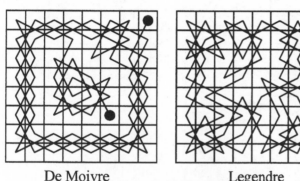

De Moivre   Legendre

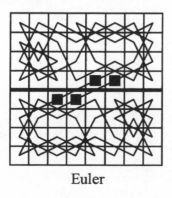

Euler

Figure 4.52   Knight's tours.

Chess knight exchange on a tiny board. Barry Cipra and John Conway have discovered the *fewest* possible moves it would take to exchange all the positions of the chess knights on this small chess board. The knights move as do traditional knights. The goal is to exchange the black knight at *A* with its white partner at *A*, *B* with *B*, *C* with *C*, and *D* with *D*. The fewest possible moves for such an exchange is 32.

Geometers of war. Mathematics, especially trigonometry, was vital in war and colonization during the sixteenth-century age of European expansion. Math was needed to navigate on the seas, to design fortifications, and for artillery tables of targeting cannons. The "Geometers of War" helped to devise instruments to measure the size of shot, determine the elevation of guns and mortars, and calculate the range of fire.

Knight's tour on a cube.
Here is a knight's tour over the six surfaces of a cube, each surface being a chessboard (figure 4.53). A spider moves as a chess knight and jumps once to every square on the (8 × 8) chessboards in a complete tour. H. E. Dudeney presented this in his book *Amusement in Mathematics*, and I believe that he based the solution (in which each face is toured in turn) on earlier work of the French mathematician Alexandre-Théophile Vandermonde (1735–1796).

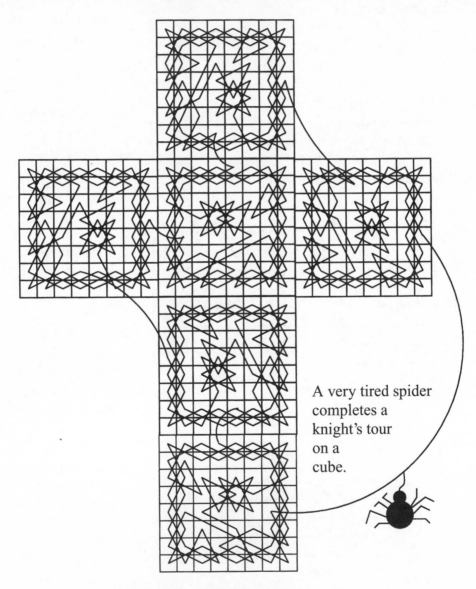

A very tired spider completes a knight's tour on a cube.

Figure 4.53   Knight's tour on a cube.

Pythagorean triangle with square sides. In 1643, the French mathematician Pierre de Fermat wrote a letter to his colleague Mersenne asking whether it were possible to find a Pythagorean triangle (a right triangle) whose hypotenuse and sums of its legs were squares. In other words, if the sides are labeled X, Y, and Z, this requires

$$X + Y = a^2; \; Z = b^2;$$
$$X^2 + Y^2 = Z^2 = b^4$$

The *smallest* three numbers satisfying these conditions are quite large: X = 4,565,486,027,761, Y = 1,061,652,293,520, and Z = 4,687,298,610,289.

- - - - - -

Euclid's postulates. Euclid (c. 330–275 B.C.) put forth five famous postulates in geometry:

1. Exactly one straight line can be drawn between two points.

2. Any straight line segment can be extended indefinitely in a straight line.

3. Given any straight line segment, a circle can be drawn having the segment as radius and one endpoint as center.

4. All right angles are congruent.

5. Through a given point outside a given straight line, there passes only one line parallel to the given line. Such a line does not intersect the given line.

In 1823, János Bolyai and Nikolai Lobachevsky realized that self-consistent "non-Euclidean geometries" could be created in which the "parallel lines" could intersect. More precisely, non-Euclidean geometries can be postulated using the first four Euclidean postulates, plus, instead of the fifth:

a. The postulate that more than one line can be drawn through a given point parallel to a given line (i.e., that every line has more than one parallel).

b. The postulate that no such lines can be drawn (i.e., there are no parallels).

Lituus and divination. The lituus is a cute-looking spiral that represents the equation $r^2 = a/\theta$ (figure 4.54), and it was first published by the English mathematician Rogert Cotes in 1722. Poor Cotes (1682–1716) died at age thirty-four, having published only two memoirs during his lifetime. He was appointed professor at Cambridge at age twenty-four, but his work was published only after his death. Note that the lituus gets increasingly closer to the origin but never reaches it.

The lituus got its name from an ancient Roman trumpet called the lituus. In art, the lituus spiral is a recurring shape called the volute. The lituus was also the curved wand used by the ancient Roman priesthood. Priests were called "augurs," who tried to predict the future, and the shape appears on Roman coins.

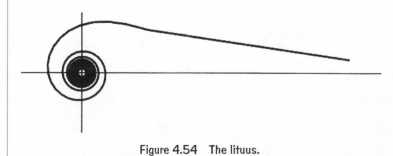

Figure 4.54   The lituus.

Butterfly curves. Incredible beauty can be found in the variety of forms that inhabit the algebraic and the transcendental curves on the plane. Many of these curves express beauty in their symmetry, leaves and lobes, and asymptotic behavior. Butterfly curves, developed by Temple Fay at the University of Southern Mississippi, are one such class of beautiful, intricate shapes (figure 4.55). These curves can be easily used for experimentation, even on personal computers. The equation for the butterfly curve can be expressed in polar coordinates by

$$\rho = e^{\cos\theta} - 2\cos 4\theta + \sin^5(\theta/12)$$

This formula describes the trajectory of a point as it traces out the butterfly's body. In the formula, ρ is the radial distance of the point to the origin.

Figure 4.55   Butterfly curves.

Helical curves. Consider $x = a\sin(t)$, $y = a\cos(t)$, and $z = at/(2\pi c)$, where a and c are constants. Try $a = 0.5$, $c = 5.0$, and $0 < t < 10\pi$. A plot of this *circular helix* curve resembles a wire spring. To draw a *conical helix*, try $x = az\sin(t)$, $y = az\cos(t)$, and $z = t/(2\pi c)$, where a and c are constants (figure 4.56). Conical helices are used today in certain kinds of antennas.

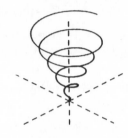

Figure 4.56   Conical helix.

- - - - - -

Mandelbrot magnification. "In principle . . . [the Mandelbrot set] could have been discovered as soon as men learned to count. But even if they never grew tired, and never made a mistake, all the human beings who have ever existed would not have sufficed to do the elementary arithmetic required to produce a Mandelbrot set of quite modest magnification" (Arthur C. Clarke, *The Ghost from the Grand Banks*, 1990).

Lorenz attractor. In 1962, the MIT meteorologist Edward Lorenz attempted to develop a model of the weather. Lorenz simplified a weather model until it consisted of only three differential equations.

$$dx/dt = 10(y - x)$$
$$dy/dt = -xz + 28x - y$$
$$dz/dt = xy - (8/3)z$$

Time is t, and d/dt is the rate of change with respect to time. If we plot the path that these equations describe using a computer, the trajectories seem to trace out a squashed pretzel (figure 4.57). The surprising thing is that if you start with two slightly different initial points, for example, (0.6, 0.6, 0.6) and (0.6, 0.6, 0.6001), the resulting curves first appear to coincide, but soon chaotic dynamics leads to independent, widely divergent trajectories. This is not to say that there is no pattern, although the trajectories do cycle, apparently at random, around the two lobes. In fact, the squashed pretzel shape always results, no matter what starting point is used. This is the behavior to which the system is attracted.

To create the Lorenz attractor, one needs to solve the system of differential equations, given previously. Several numerical techniques can be used that come up with an accurate value for x, y, and z as a function of time. The most straightforward approach, which I have used to get a rough idea about the Lorenz attractor, simply replaces dx with ($xnew - x$) and replaces dt by a time step, called h. Other higher-accuracy approaches, such as Runge Kutta methods, can be used but only with a consequent increase in computer time. To create a projection of this 3-D figure in the x-y plane, simply plot (x, y) pairs of points and omit the z value.

```
h = 0.01, npts = 4,000;
x, y, z = 0.6;
  frac = 8/3;
  do i = 1 to npts;
  xnew = x + h*10*(y - x);
  ynew = y + h*((-x*z) + 28*x - y);
  znew = z + h*(x*y - frac*z);
  x = xnew; y = ynew; z = znew;
  MovePenTo(x, y);
end;
```

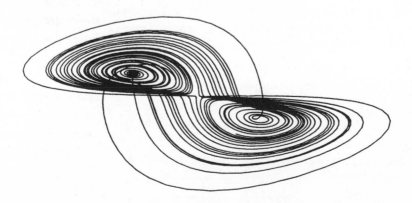

Figure 4.57   Lorenz attractor.

Build your own chaos machine. One of the best non-computer projects I know for observing chaos involves building a double pendulum—a pendulum suspended from another pendulum. The motion of the double pendulum is quite complicated. The second arm of the pendulum sometimes seems to dance about under its own will, occasionally executing graceful pirouettes, while at other times doing a wild tarantella. You can make the double pendulum from wood. At the pivot points, you might try to use ball bearings to ensure low friction. (Ball bearings can be obtained from hobby shops or from discarded motors and toys.)

Place a lead weight at the bottom of the first pendulum so that the pendulum will swing for a longer time. (The weight stores potential energy when the pendulum is lifted.) The second pendulum arm can be about half the length of the first. You can place a bright red dot, or even a light, on one end of the second pendulum so that your eye can better track its motion. Note that your pendulum will never trace the same path twice, because you can never precisely reposition it at the same starting location, due to slight inaccuracies in knowing where the starting point is. These small initial differences in position are magnified through time until the pendulum's motion and position become unpredictable. Can you predict where the lower pendulum will be after two or three swings? Could the most powerful supercomputer in the world predict the position of the pendulum after 30 seconds, even if the computer were given the pendulum's precise equations of motion? Unlike some of the strange attractor patterns in this book, your pendulum's pattern will eventually come to rest at a point due to friction.

— — — — — —

Fractal goose. In *The Story of an African Farm*, first published in 1883, Olive Schreiner gives an early, poignant description of the fractal geometry of nature: "A gander drowns itself in our dam. We take it out, and open it on the bank, and kneel looking at it. Above are the organs divided by delicate tissues; below are the intestines artistically curved in a spiral form, and each tier covered by a delicate network of blood vessels standing out red against the faint blue background. Each branch of the blood vessels is comprised of a trunk, bifurcating and rebifurcating into the most delicate, hair-like threads, symmetrically arranged. We are struck with its singular beauty. And, moreover—and here we drop from our kneeling into a sitting position—this also we remark: of that same exact shape and outline is our thorn-tree seen against the sky in midwinter; of that shape also is delicate metallic tracery between our rocks; in that exact path does our water flow when without a furrow we lead it from the dam; so shaped are the antlers of the horned beetle. How are these things related that such union should exist between them all? Is it chance? Or, are they not all the fine branches of one trunk, whose sap flows through us all? That would explain it. We nod over the gander's insides."

Turning a universe inside-out. Since its discovery around 1980, the Mandelbrot set has emerged as one of the most scintillating stars in the universe of popular mathematics and computer art. Figure 4.58 shows inside-out Mandelbrot sets produced by the iterative process

$$z \rightarrow z^p + (1/\mu)^p, \ p = 2.$$

For this figure, $|\text{Re}(\mu)| < 2$, $|\text{Im}(\mu)| < 2$. The convergence to infinity occurs at the center of the figure.

Bounded orbits, which do not explode, correspond to the black surrounding regions. Several contours (level sets) are plotted for the first few iterations. The contours indicate the rate at which the iteration explodes.

When most people talk about the Mandelbrot set, they often mean the area just outside the set near its infinitely complicated fractal boundary. But there is also action inside the set, as figure

Figure 4.58   Inside-out Mandelbrot set.

4.59 shows. I used a method called the "epsilon cross" technique for revealing this normally hidden internal structure—Mandelbrot stalks within the interior bounded region of the Mandelbrot set.

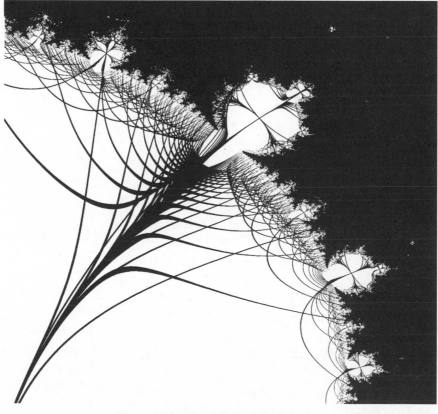

Figure 4.59   Mandelbrot stalks.

The Quadratrix of Hippias. Hippias of Elis, who came to Athens in the second half of the fifth century B.C., studied the behavior of the interesting formula $2r = \pi\rho(\sin\theta)/\theta$ in polar coordinates. Here, $\rho = \sqrt{x^2 + y^2}$ and $\theta = \arctan(y/x)$. After some manipulation, we have $y = x\cot(\pi x/2a)$.

Amazingly, Hippias and another Greek mathematician, Dinostratus (350 B.C.), saw only the tip of the iceberg for this curve—they knew only about the shape of the curve on the interval $-a < x < a$. (The curve looks like the top of a bald man's head on this interval.) Later, in the seventeenth century,

the full behavior of this curve became known. What amazing things happen to the curve at higher and lower values of x? Try plotting the curve yourself.

- - - - - -

Within a sybaritic sphere. "We sail within a vast sphere, ever drifting in uncertainty, driven from end to end" (Blaise Pascal, *Pensées*, 1660).

- - - - - -

Bears in hyperspace. The polar bear is the largest land carnivore alive in the world today. Adult males weigh from 400 to 600 kilograms (880 to 1,320 pounds). Would it be possible to cram a polar bear into an 11-dimensional sphere with a 6-inch radius? (See Answer 4.76.)

- - - - - -

Hypercube edges. The number of edges of a cube of dimension n is n times 2^{n-1}. For example, the number of corners of a seven-dimensional cube is $2^7 = 128$, and the number of edges is 7 times $2^6 = 7 \times 64 = 448$. Another interesting factoid: two perpendicular planes in 4-space can meet at a point.

Harborth configuration. How does one arrange identical sticks in a way such that four sticks meet end to end, without crossing each other, at every point in a geometrical figure on a flat surface? In the diagram in figure 4.60, four sticks meet at each vertex. This is the smallest arrangement known, but no one knows whether it's the smallest possible way to make a figure with four sticks meeting at each point!

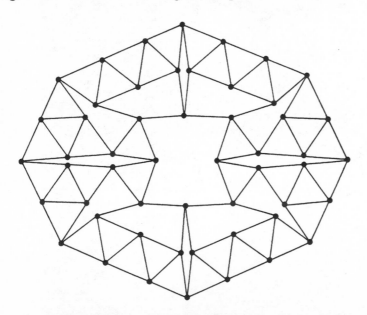

Figure 4.60   Harborth configuration. (Pattern discovered by Heiko Harborth; diagram after Ivars Peterson, *Islands of Truth.* [New York: Freeman, 1990.])

Hyperspheres and higher dimensions. The fourth dimension has fascinated me for a long time, and higher dimensional spheres, or hyperspheres, particularly delight me. Consider some exciting experiments you can conduct using a pencil and paper or a calculator. Just as a circle of radius r can be defined by the equation $x^2 + y^2 = r^2$, and a sphere can be defined by $x^2 + y^2 + z^2 = r^2$, a hypersphere in four dimensions can be defined simply by adding a 4th term: $x^2 + y^2 + z^2 + w^2 = r^2$, where w is the 4th dimension! Here are formulas that permit you to compute the volume of a sphere of any dimension, and you'll find that it's relatively easy to implement this formula using a computer or a hand calculator.

The volume of a k-dimensional sphere is

$$V = \frac{\pi^{k/2} r^k}{(k/2)!}$$

for *even* dimensions k. The exclamation point is the mathematical symbol for factorial. (Factorial is the product of all the positive integers from one to a given number. For example, $5! = 1 \times 2 \times 3 \times 4 \times 5 = 120$.) The volume of a 6-dimensional sphere of radius 1 is $\pi^3/3!$, which is roughly equal to 5.1. For odd dimensions, the formula is just a bit more intricate:

$$V = \frac{\pi^{(k-1)/2} m! 2^{k+1} r^k}{(k+1)!}$$

where $m = (k + 1)/2$. The formulas are really not too difficult to use. In fact, with these handy formulas, you can compute the volume for a 6-dimensional sphere just as easily as for a 4-dimensional one. For radius 2 and dimension 2, the previous equations yield the value 12.56, which is the area of a circle. A sphere of radius 2 has a volume of 33.51. A 4-dimensional

Hypersphere packing. I enjoy contemplating how hyperspheres might pack together—like pool balls in a rack or oranges in a box. On a plane, no more than four circles can be placed so that each circle touches all others, with every pair touching at a different point. Figure 4.61 shows two examples of four intersecting circles. In general, for n-space, the maximum number of mutually touching spheres is $n + 2$.

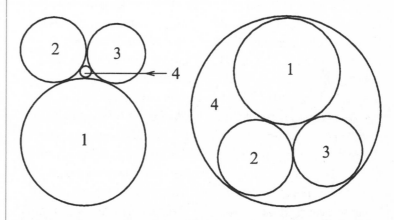

Figure 4.61   In two dimensions, no more than four circles can be placed so that each circle touches all the others, with every pair touching at a different point. What happens in higher dimensions?

What is the largest number of circles that can touch a *single* circle? (Assume that each circle has the same radius.) Also, answer the question for spheres. (See Answer 4.77.)

hypersphere of radius 2 has a hypervolume of 78.95. Intuitively, one might think that the volume should continue to rise as the number of dimensions increases. The volume, or perhaps we should use the term *hypervolume*, does grow larger and larger until it reaches a maximum—at which point the radius 2 sphere is in the 24th dimension. At dimensions higher than 24, the volume of this sphere begins to decrease gradually to 0, as the value for the dimension increases. An 80-dimensional sphere has a volume of only 0.0001. This apparent turnaround point occurs at different dimensions, depending on the sphere's radius, r. For all the sphere radii I tested, the sphere initially grows in volume and then begins to decline. (Is this true for all radii?) For example, for $r = 1$, the maximum hypervolume occurs in the 5th dimension. For $r = 1.1$, the peak hypervolume occurs in the 7th dimension. For $r = 1.2$, it occurs in the 8th dimension. (Incidentally, the hypersurface of a unit hypersphere reaches a maximum in the 7th dimension and then decreases toward zero as the

Hypercube diagonals. A cube has diagonals of two different lengths, the shorter one lying on the square faces and the longer one passing through the center of the cube. The length of the longest diagonal of an n-dimensional cube of side length m is $m\sqrt{n}$. This means that if I were to hand you a 3-foot-long thigh bone and ask you to stuff it in a 9-dimensional hypercube with edges 1 foot in length, the bone would just fit because $\sqrt{9} = 3$. A dinosaur bone 10 feet long could fit diagonally in a 100-dimensional cube with edges only 1 foot in length. A mile-long toothpick could fit inside an n-cube with edges the same length as those of an ordinary sugar cube, if n is large! On the other hand, a hypersphere behaves somewhat differently. An n-sphere can never contain a toothpick longer than twice its radius, no matter how large n becomes.

dimension increases.) If we examine the equations for volume more closely, we notice that this funny behavior shouldn't surprise us too much. The denominator contains a factorial term that grows much more quickly than any power does, so we get the curious result that an infinite dimensional sphere has no volume. Using the equations for volume given here, you'll find that an 11-dimensional sphere of radius 2 feet has a hypervolume of 333,763 feet[11]. Try plotting the *ratio* of a k-dimensional hypersphere's volume to the k-dimensional cube's volume that encloses the hypersphere. Plot this as a function of k.

Fractal swords and drums. In several of my books, such as *Sushi Never Sleeps*, I describe science-fiction warriors wielding swords with jagged Koch snowflake edges and tossing deadly Koch-curve throwing stars. These ideas are my own, but they are inspired by research in fractal drums. In 1991, Bernard Sapoval and his colleagues at the École Polytechnique in Paris found that fractally shaped drum heads are very quiet when struck. Instead of being round like an ordinary drum head, these heads resemble a jagged snowflake. Sapoval cut his fractal shape out of a piece of metal and stretched a thin membrane over it to make a drum. When a drummer bangs on an ordinary drum, the vibration spreads out to affect the entire drum head. With fractal drums, some vibrational modes are trapped within a branch of the fractal pattern. Faye Flam, in the December 13th, 1991, issue of *Science* (vol. 254, p. 1593), notes, "If fractals are better than other shapes at damping vibrations, as Sapoval's results suggest, they might also be more robust. And that special sturdiness could explain why in nature, the rule is survival of the fractal." Fractal shapes often occur in violent situations where powerful, turbulent forces need to be damped: the surf-pounded coastline, the blood vessels of the heart (a very violent pump), and the wind- and rain-buffeted mountain.

🔴 **Hyperpyramids.** A four-dimensional analog of a pyramid has a hypervolume one-fourth the volume of its three-dimensional base multiplied by its height in the fourth direction. An *n*-dimensional analog of a pyramid has a hypervolume

🔴 **Knight's move magic square.** To create a knight's move magic square, a chess knight has to be jumped once to every square on the (8×8) chessboard in a complete tour, while numbering the squares in the order they are visited, so that the magic sum is 260. The array of numbers must have the same sum in each row, each column, and the two main diagonals. One possible solution for the knight's movement is shown here. Start at the "1" at the bottom, jump to the "2," and so forth.

46	55	44	19	58	9	22	7
43	18	47	56	21	6	59	10
54	45	20	41	12	57	8	23
17	42	53	48	5	24	11	60
52	3	32	13	40	61	34	25
31	16	49	4	33	28	37	62
2	51	14	29	64	39	26	35
15	30	1	50	27	36	63	38

Jaenisch's Knight's Square

Notice that the knight can finally jump from 64 back to 1, a beautiful feature of this square. If the first and the last squares traversed are connected by a move, the tour is said to be *closed* (or *reentrant*); otherwise, it is *open*.

This magic square was created in 1862 by C. F. Jaenisch, in *Applications de l'Analyse Mathématique au Jeu des Echecs*. Sadly, it is not a perfect magic square because one diagonal sums to 264 and the other to 256. Therefore, the square is sometimes referred to as *semimagic*. I can imagine poor C. F. weeping. . . . So close, yet so far!

For centuries, the holy grail of magic squares was to find a perfect knight's move magic square, but, sadly, all have had minor flaws. (It is possible to produce a knight's move magic square for larger boards, such as for a 16×16 board.)

$1/n$ times the volume of its $(n-1)$-dimensional base multiplied by its height in the nth direction.

🛈 Seven-dimensional ice. Recently, scientists and mathematicians have researched the theoretical melting properties of ice in higher dimensions. In particular, the mathematicians Nassif Ghoussoub and Changfeng Gui, from the University of British Columbia, have developed mathematical models for how ice turns from solid into liquid in the seventh dimension and have proved that if such ice exists, it likely exhibits a different melting behavior than ice in lower dimensions. This dependence on dimension, although not very intuitive, often arises in the field of partial differential equations and minimal surfaces—recent results suggest that geometry depends on the underlying dimension in ways that were not suspected in the past. Other research suggests that there is something about eight-dimensional spaces that makes physical phase transitions inherently different from seven-dimensional spaces. If you want to read more about what happens when you lick a seven-dimensional popsicle, see I. Ekeland, "How to Melt If You Must," *Nature* 392, no. 6677 (April 16, 1998): 654–55.

🛈 On the existence of knight's tours. Does a knight's tour on an 8×8 magic square exist? On August 6, 2003, after 61 days of computation, this 150-year-old unsolved problem was finally answered. Günter Stertenbrink (Germany), Jean-Charles Meyrignac (France), and Hugues Mackay (a twenty-one-year-old military policeman in the Canadian Armed Forces) performed an exhaustive computer search of all the possibilities and finally demonstrated for the first time that no 8×8 magic knight's tour is possible. Meyrignac wrote the software for the computation, and, while doing the search, he found 140 distinct semimagic knight's tours. In other words, he found 140 "magic" knight's tours on the chessboard, but none of these is diagonally magic.

The computer checking program was executed as a

❓ Lattice growth. You and I are climbing Vinson Massif, the highest mountain of Antarctica, when I turn and show you the following diagram written in sepia ink on an ancient parchment. How many rectangles and squares are in this diagram?

Think about this for a minute, and then I'll give you an answer before you freeze. This is a simple lattice. There are the 4 small squares marked 1, 2, 3, and 4, plus 2 horizontal rectangles containing 1 and 2, and 3 and 4, plus 2 vertical rectangles, plus the 1 large surrounding border square. Altogether, therefore, there are nine 4-sided overlapping areas. The lattice number for a 2×2 lattice is therefore 9, or $L(2) = 9$.

Your challenge: What is $L(3)$, $L(4)$, $L(5)$, and $L(n)$? (See Answer 4.78.)

distributed project that operated for 61 days, 9 hours, and 36 minutes at an equivalent 2.25 Ghz (G. Stertenbrink, "Computing Magic Knight Tours," magictour.free.fr/, August 6, 2003).

In the garden of the knight. Jean-Charles Meyrignac, Mike Malak, and colleagues have wondered about the greatest number of knights that can be placed on a chess board so that each piece attacks exactly one other piece. Can you arrange 32 knights on a standard board so that each attacks only one other knight? (See Answer 4.79.)

In the garden of the knight. Place 80 knights on a 13×13 chessboard so that each attacks only four other knights. (See Answer 4.80.)

Enigma. "It may well be doubted whether human ingenuity can construct an enigma of the kind which human ingenuity may not, by proper application, resolve" (Edgar Allan Poe, "The Gold-Bug," 1843).

Knight's move magic square. The first person to attempt to draw a knight's move magic square was Leonhard Euler (1707–1783), a Swiss mathematician and the most prolific mathematician in history. Even when he was completely blind, Euler made great contributions to modern analytic geometry, trigonometry, calculus, and number theory.

1	48	31	50	33	16	63	18
30	51	46	3	62	19	14	35
47	2	49	32	15	34	17	64
52	29	4	45	20	61	36	13
5	44	25	56	9	40	21	60
28	53	8	41	24	57	12	37
43	6	55	26	39	10	59	22
54	27	42	7	58	23	38	11

Euler's Knight's Square

To traverse the square, start at the 1 in the upper left and leave the square at the 64 at the right. Figure 4.62 shows the actual path. Notice the nice symmetries in the path. The right side of the figure is a mirror image of the left.

The numbers in adjacent 2×2 subsquares add up to 130. Euler's knight's square adds up to 260 in the rows and the columns. The four 4×4 subsquares are also magic in their rows and columns, which sum to 130. (The Euler knight's square is actually four 4th-order magic squares put together.) Alas, like the Jaenisch square, the main diagonals do not add to 260.

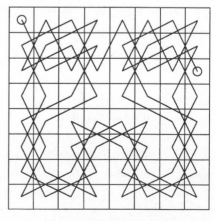

Figure 4.62   Knight's path.

In the garden of the knight. Here is the solution. What's the question? (See Answer 4.81.)

[A 10×10 grid containing knight chess pieces in various cells.]

at the start of a turn because all squares are occupied. For clarification, here is how each turn starts: (1) your new bishop is placed on the board at a position determined by your opponent; (2) next, you are free to move the new bishop or any other of your pieces as usual.

- - - - - -

Polyhedral universe. Is the universe more likely to be shaped like a cube, a dodeca-hedron, an icosahedron, an octahedron, or a tetrahedron? (See Answer 4.82.)

- - - - - -

Johnson functions. In his article "Approximating $\sqrt{n}$," which appeared in volume 19 of the journal *Mathematical Spectrum* (issue no. 2, page 40), Simon Johnson suggests that the iterative formula

$$z \to I(z) = \frac{z+n}{z^2 + 1}$$

applied to the rational num-ber z, converges to $n^{1/3}$. Later, Irving, Richards, and Sowley showed that this iteration converges to $n^{1/3} >$ if $0 < n \leq 2^{3/2}$, but that if $n > 2^{3/2}$, it converges only for certain

unusual initial values of z. Can you devise a graphics strategy to show these unusual values?

- - - - - -

Too many bishops. In this special version of chess that I invented, each player acquires an additional bishop before each turn. The bishop is placed on a position deter-mined by the opponent. The game ends when a king is captured, when a player can-not move due to the mob of bishops on the board, or when a player cannot add a bishop

Golden pentagram. The pentagram, or five-pointed star, contains many ratios that are the golden ratio. In figure 4.63, examples include $\phi = AB/BC = CH/BC = IC/HI = 2DE/EF = EG/2DE$.

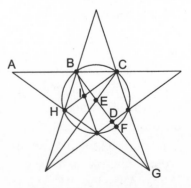

Figure 4.63   Golden pentagram.

Behold the hidden Stomachion. In 2003, math historians discovered long-lost information on the Stomachion of Archimedes. In particular, an ancient parchment, written over by monks nearly a thousand years ago, describes Archimedes' Stomachion, a puzzle involving combinatorics. Combinatorics is a field of math that deals with the number of ways a given problem can be solved. The goal of the Stomachion (pronounced sto-MOCK-yon) is to determine in how many ways the 14 pieces shown in figure 4.64 can be put together to make a square. In 2003, four mathematicians required six weeks to determine that the number is 17,152.

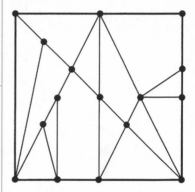

Figure 4.64 The long-lost Stomachion of Archimedes.

Möbius mirror. In theory, if the universe had characteristics similar to those of a Möbius strip, you could travel in space and return as your mirror image. August Möbius discovered the Möbius strip in 1840. You can create a Möbius strip by twisting a strip of paper 180 degrees and then taping the ends together. A Möbius strip has only one side. If that's hard to believe, build one and try to color one side red and the "other" side green.

By way of analogy, if a Flatlander lived in a Möbius world, he could be flipped easily by moving him along his universe without ever taking him out of the plane of his existence. If a Flatlander travels completely around the Möbius strip and returns, he will find that all of his organs are reversed (figure 4.65). A second trip around the Möbius cosmos would straighten him out again.

In the thirteenth century, Christian monks ripped the original manuscript apart, washed it, and covered it with religious text. Today, we cannot see the Stomachion with the naked eye, and ultraviolet light and computer-imaging techniques are needed to reveal the hidden mathematical gem. Scholars are uncertain whether Archimedes ever correctly solved the problem. (For further reading, see Gina Kolata, "In Archimedes' Puzzle, a New Eureka Moment," vol. CLIII, no. 52,697, *New York Times*, December 14, 2003, p. 1.)

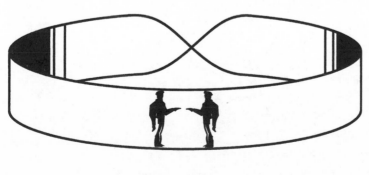

Figure 4.65 Möbius cosmos.

Platonic patent. Platonic solids (defined early in this chapter) are a big business these days, and U.S. patents exist for Platonic toys. One favorite example is U.S. patent 4,676,507, "Puzzles Forming Platonic Solids," issued to Bruce Patterson of Egg, Switzerland. The patent describes several 3-D puzzles formed from identical pieces. Figure 4.66 shows several views of Patterson's tetrahedron puzzle from his patent. Other figures in the patent show similar puzzles for other Platonic solids.

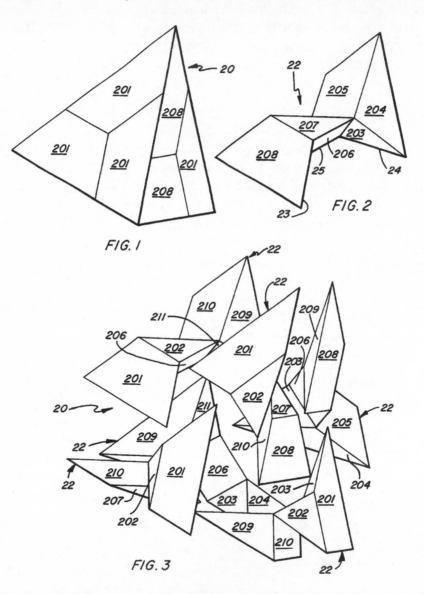

Figure 4.66   Platonic patent.

Tank charge. You are a tank, training in the Nevada desert. In the diagram, you can travel up, down, left, and right through electric fields. Only two of your many moves may be a diagonal move. If you travel through two minus (–) signs in a row, your battery is drained and you are stuck. If you travel through two plus (+) signs in a row, your battery overcharges and explodes. How can you travel from Start to End through every cell once and survive? (See Answer 4.83.)

S	+	–	+	–
–	+	–	–	+
–	+	+	+	–
+	–	+	–	+
E	+	–	+	–

- - - - - -

The tritar—a guitar made by genius mathematicians. In January 2004, the media were all abuzz about the tritar ("TREE-tar," sometimes spelled "tritare"), a Y-shaped guitar invented by two Canadian number theorists, Samuel Gaudet and Claude Gauthier, both at the University of Moncton. Some say that the tritar will revolutionize music.

Sangaku geometry. During Japan's period of isolation from the West (1639–1854), a tradition known as *sangaku*, or Japanese temple geometry, arose. Mathematicians, farmers, samurai, women, and children solved difficult geometry problems and inscribed the solutions on tablets. These colorful tablets were then hung under the roofs of temples. More than 800 tablets have survived, and many of them feature problems that concern tangent circles. As one example, consider figure 4.67, a late sangaku tablet from 1873 created by an eleven-year-old boy named Kinjiro Takasaka. Consider a fan, which is one-third of a complete circle. Given the diameter d_1 of the circle pointed to by the arrow, what is the diameter d_2 of the circle with the question mark? I believe that the answer is approximately

$$d_2 \approx d_1 (\sqrt{3072} + 62)/193.$$

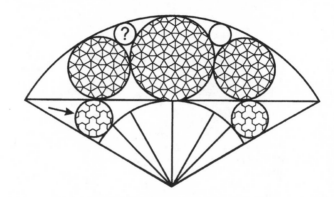

Figure 4.67   A Japanese sangaku from 1873.

In their research involving infinite sums called *p* series, the mathematicians invented a series of "hyperimaginary" numbers lying on a Y-shaped number line, and these numbers were useful for creating a musical instrument with Y-shaped strings. The tritar produces a range of extremely novel and beautiful sounds, from guitars to bells. If one string is plucked, it vibrates across all three of the fretboards. As I write this, the mathematicians have filed for a patent and are taking the tritar to trade shows, with hopes of commercializing the instrument.

Molecular madness.
Scientists visiting planet Nectar have discovered an amazing molecule with a fruity aroma. They call the molecule Nectarine. Each of the atoms shown here in this chemical diagram of Nectarine is connected to at least one other atom by one or more bonds, which we can draw as straight lines connecting the atoms (figure 4.68). The numbers in each atom tell the number of bonds that the atom makes to other atoms. The molecule has two double bonds and one triple bond, and the remaining bonds are single bonds. Black regions represent solvent molecules through which your bonds may not pass.

Can you solve the molecular structure by drawing the missing bonds as straight lines on this flat diagram? Your bonds must not cross one another or cross through the atoms. Is there more than one solution? (See Answer 4.84.)

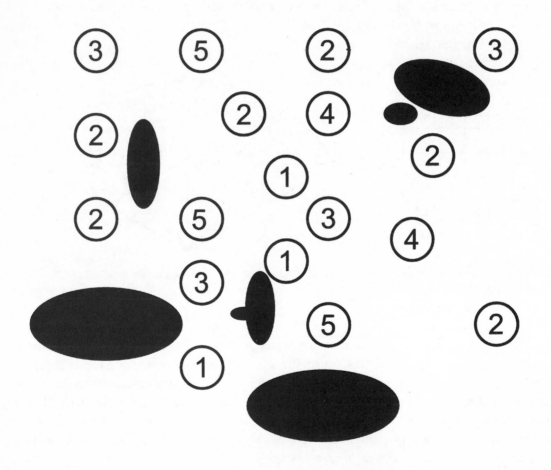

Figure 4.68   Alien molecule confounds scientists.

Drinking numbers. For centuries, arrays of numbers have been thought to confer magical powers. Consider the following square from India that a woman may use when searching for a husband. The numbers are drawn on a china plate with a crayon and then washed off the plate with water that the woman drinks:

24,762	24,768	24,771	25,320
24,770	24,758	24,763	25,341
24,759	24,773	24,766	25,325
24,767	24,761	24,760	25,344

Finding the Perfect Husband

Whenever possible, this number array should be written with a special ink known as *Ashat Gandh*. This is a mixture of several items, the most important of which is water from the Ganges River. I would be interested in hearing from readers who ascertain any significance in the use of these particular numbers.

- - - - - - - - - - - - - - - - - - - -

Clown with balls. A large intellectual dressed in a clown outfit approaches you with bright red pool (billiard) balls. His nickname is "Solenoid," and he has various springs, resistors, and capacitors in his hair.

Solenoid places the balls inside an equilateral triangle in preparation for playing a game (figure 4.69). Each ball has a radius of 1 unit length. What is the side length of the equilateral triangle? (See Answer 4.85.)

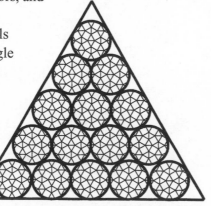

Figure 4.69   Ball puzzle.

Stained glass squares of Constantine. How many squares can you count in this window? (See figure 4.70.) I recently designed and created this stained glass window in gorgeous color, and my brainy colleagues always give different answers! Show this to friends, as no two friends will give you the same solution. (See Answer 4.86.)

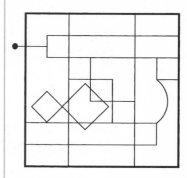

Figure 4.70   Stained glass squares of Constantine.

- - - - - -

Math luxury. "The science of calculation is indispensable as far as the extraction of the square and cube roots: Algebra as far as the quadratic equation and the use of logarithms are often of value in ordinary cases: but all beyond these is but a luxury; a delicious luxury indeed; but not to be indulged in by one who is to have a profession to

follow for his subsistence"
(Thomas Jefferson, in J.
Robert Oppenheimer's "The
Encouragement of Science,"
which appeared in Isabel

Gordon and Sophie Sorkin,
eds., *The Armchair Science
Reader*, 1959).

– – – – – –

Pythagorean Lute assembly. In an answer earlier in this chapter, we encountered the Lute of Pythagoras and all its golden ratios. Recall that the golden ratio is $\phi = (1 + \sqrt{5})/2 = 1.61803$ . . . Figure 4.71 shows an impressive assembly of multiple lutes created by Bruce A. Rawles, who is interested in geometry, art, and nature. Rawles is the author of *Sacred Geometry Design Sourcebook*, published by Elysian Publishing. You can learn more about his work at "Sacred Geometry Home Page," www.intent.com/sg/.

Figure 4.71   Pythagorean Lute assembly by Bruce Rawles.

Chocolate puzzle. Dr. Brain is a master chef and has his own personal chocolate bar problem for you. Your goal is to divide the 8×8 gridded chocolate square into rectangles or squares, such that each quadrilateral contains a number made of icing—and that number must designate the number of chocolate cells inside the quadrilateral. For example, I've drawn a rectangle around 4 cells that include the 4, just to show you what a "legal" quadrilateral could look like. Can you draw all the quadrilaterals so that one icing number is in each quadrilateral? What strategy did you use? How long did it take you to solve this? (See Answer 4.87.)

– – – – – –

Jerusalem crystal. You are digging in the ancient soil of Jerusalem and discover a mysterious crystal with Hebrew letters (figure 4.72). Beginning at the ש in the lower right, you must discover a secret path that will lead you to the ש in the upper left. How many solutions can you find, using the following rules?

1. You may move up, down, right, or left along each facet.

2. You may move only between adjacent facets that share an edge; you may not move between facets that touch only at their corners.

3. Your entire path must follow this repeating code: ש, ש, ש, B, B, B, ע, ע, ע, ש, ש, ש, B, B, B, ע, ע, ע . . . , and so on, where the "B" stands for a blank facet with no letter. That is, after moving through three ש facets, you must move through three blank facets; after the three blank facets, you must move through three ע facets, and so on, repeating the pattern. Good luck!

(See Answer 4.88.)

Figure 4.72
Jerusalem crystal.

Finish 👉

👉 Start

Schmidhuber circles. Figure 4.73 shows a fantastic collection of Schmidhuber circles, named after their discoverer, the German mathematician Jürgen Schmidhuber. To create this pattern, start by drawing a circle of arbitrary radius and center position. Randomly select a point on the first circle and use it as the center of a second circle with an equal radius. The first two circles are defined as "legal circles." The rules for generating additional legal circles are as follows:

- *Rule 1.* Wherever two legal circles of equal radius touch or intersect, draw another legal circle of equal radius with the intersection point as its center.

- *Rule 2.* Within every legal circle with center point p and radius r, draw another legal circle whose center point is also p but whose radius is $r/2$.

Figure 4.73 is the result of a recursive application of these rules (www.idsia.ch/~juergen/).

Figure 4.73 Schmidhuber circles, created by Jürgen Schmidhuber.

ⓘ Squiggle map coloring.

If you draw a map on a plane using a continuous line, without taking your pencil off the paper, and return to the starting point, you need only two colors to produce a map such that any regions with a common boundary line have different colors (figure 4.74). (In your coloring, two adjacent regions can share a common vertex and have the same color, but they can't share the same edge and have the same color.)

Figure 4.74   Squiggle map coloring.

Probability: Take Your Chances

I N WHICH WE EXPLORE CASINOS, LOGIC, GUESSING, DECISIONS, COMBINATIONS, permutations, competition, possibilities, games involving choice, monkeys typing *Hamlet*, Benford's law, combinatorics, alien gambits, Rubik's Cubes, card shuffles, marble mazes, nontransitive dice, dangerous movements of air molecules, the board game of the gods, and the tunnels of death and despair.

> Probability describes and quantifies uncertainty.
> It's the spinning roulette wheel in the casino of reality.

Gambling. Your friend Donald Trumpet is a gambling man. He approaches you and your friend on a street corner and wants to use a coin to determine which of you will get a brand new car. He will flip a coin to decide. Unfortunately, his coin is biased, which means it favors one side over the other and does not have a 50-50 chance of landing heads or tails. How can you instantly generate even odds from Donald's unfair coin? For example, if both of you know that Donald's coin is biased toward tails, how can you get the equivalent of a fair coin with several tosses of Donald's unfair coin? (See Answer 5.1.)

Mathematical garments. "The mathematician may be compared to a designer of garments who is utterly oblivious of the creatures whom his garments may fit. To be sure, his art originated in the necessity for clothing such creatures, but this was long ago; to this day a shape will occasionally appear which will fit into the garment as if the garment had been made for it. Then there is no end of surprise and delight!" (Tobias Dantzig, *The Bequest of the Greeks*, 1969).

Monkeys typing *Hamlet*. I put a single monkey in a room, typing randomly on a typewriter. What are the odds that the monkey will sit down and type out all of Shakespeare's *Hamlet* correctly? (See Answer 5.2.)

Many monkeys typing *Hamlet*. Let us assume for the previous problem that each day we *double* the number of monkeys that are trying to accomplish the task of typing *Hamlet*. How do the odds of at least one of the monkeys in the monkey collection typing *Hamlet* correctly change through time? (See Answer 5.3.)

Tic-tac-toe probability. In randomly played tic-tac-toe, the probability that the first mover wins is 737/1260 or 0.5849206.

Tic-tac-toe games. There are 255,168 possible games in tic-tac-toe, when you consider all possible games that end in 5, 6, 7, 8, and 9 moves.

The Book of Everything That Can Be Known. Moses, along with several angels, approaches you. In his hand is a massive book, leather bound with gold trim, titled *The Book of Everything That Can Be Known*. The book is 10,000 pages long. On page 314 is the statement "The universe is finite in spatial extent," and this statement is nowhere else in the book. On average, how many times would you have to flip open this book to be able to view this fact? (See Answer 5.4.)

Benford's bravura. Benford's law, also called the first digit law or the leading digit phenomenon, asserts that in numerous listings of numbers, the digit 1 tends to occur *in the leftmost position* with probability of roughly 30 percent, much greater than the expected 11.1 percent (i.e., one digit out of nine). Benford's law can be observed, for instance, in tables that list populations, death rates, baseball statistics, the areas of rivers or lakes, and so forth. Explanations for this phenomenon are very recent.

Benford's law is named for the late Dr. Frank Benford, a physicist at the General Electric Company. In 1938 he noticed that pages of logarithms that corresponded to numbers starting with the numeral 1 were dirtier and more worn than other pages, because the number 1 occurred as the first digit about 30 percent more often than any other did. In numerous kinds of data, he determined that the probability of any number n from 1 through 9 being the first digit is $\log_{10}(1 + 1/n)$. Even the Fibonacci sequence—1, 1, 2, 3, 5, 8, 13 . . . —follows Benford's law. Fibonacci numbers are far more likely to start with "1" than with any other digit. The next most popular digit is 2, and 9 is the least probable. It appears that Benford's law applies to any data that follows a "power law." For example, large lakes are rare, medium-sized lakes are more common, and small lakes are even more common. Similarly, eleven Fibonacci num-

bers exist in the range 1–100, but only one in the next three ranges of 100 (101–200, 201–300, 301–400). Fibonacci numbers become increasingly rare as we scan higher ranges of size 100.

Benford's law has often been used to detect fraud. For example, Dr. Mark J. Nigrini, an accounting consultant, used the law to detect fraudulent tax returns in which the occurrence of digits did not follow what would be expected by Benford's law. See Theodore Hill, "The First Digit Phenomenon," *American Scientist* 86 (1998): 358–63; Malcolm W. Browne, "Following Benford's Law, or Looking Out for No. 1," *New York Times*, August 4, 1998; L. C. Wash-

ington, "Benford's Law for Fibonacci and Lucas Numbers," *The Fibonacci Quarterly* 19 (1981): 175–77; and Ron Knott, "The Mathematical Magic of the Fibonacci Numbers," www.mcs.surrey.ac.uk/Personal/R.Knott/Fibonacci/fibmaths.html#msds. (Dr. Knott was formerly with the University of Surrey and now operates his own company for Internet resource provision, with an emphasis on mathematics and teaching.)

- - - - - -

Bookkeeper arrangements. In how many different ways can I arrange the letters in BOOKKEEPER? (See Answer 5.5.)

- - - - - -

Subway odds. You are riding the New York City subway when a huge, hulking, hairy man suddenly threatens you with a knife. You tell him to wait and then hand him a pen and nine slips of paper. "Write a different number on each paper," you say. "You can use any numbers, but don't use any consecutive numbers."

Your assailant writes down nine numbers. He places the slips of paper facedown so that you can't see his numbers. You boast, "I'll start turning over the slips of paper at random, and even though I don't know what you wrote, I'll stop turning on the highest number. If I'm right, you set me free. If I'm wrong, I give you all the money in my wallet."

What is your strategy? (See Answer 5.6.)

📖 Combinatorics defined.

Some of the problems in this section deal with "combinatorics"—the branch of mathematics that is concerned with the selection and the arrangement of objects. Combinatorics has a long history, and interest skyrocketed in the seventeenth and eighteenth centuries, when it was used to estimate gambling odds and assess the probabilities of game outcomes. The Italian scientist Galileo Galilei (1564–1642) used probability to study the outcomes of dice rolls.

One important feature of probability is the *multiplication principle*, which says that if m different selections can be made in succession, and the first selection can be made in n_1 ways, the second in n_2 ways, the third in n_3 ways, and so on, then the total number of selections is $n_1 \times n_2 \times n_3 \times \ldots n_m$. As an example, consider a pizza that can be with or without cheese, can be small or large, and can have pepperoni, peppers, or anchovies. The total number of pizza varieties is $2 \times 2 \times 3 = 12$, if no toppings are mixed. (I'd choose the small cheese pizza with anchovies.)

📖 Permutations defined.

Permutations are ordered arrangements of elements. Order is important. For example, two permutations of the same elements are considered different permutations if they are in different orders. Thus, *ABC* has six permutations: *ABC, ACB, BAC, BCA, CAB, CBA*. The number of permutations for n elements is n factorial. In this case, $3! = 3 \times 2 \times 1 = 6$.

Sometimes we may have a collection of objects from which we take several members. For example, starting with A, B, C, and D, we might ask how many permutations exist if we take only two letters from the batch. We can list all 12 permutations: *AB, AC, AD, BA, BD, BC, CA, CB, CD,* and *DA, DB,* and *DC*. More generally, if we take p elements from a set of n *different* elements and want to determine the number of permutations, we use the handy formula $_nP_p = n!/(n-p)!$ In our example, $n = 4$ and $p = 2$, so $_4P_2 = 4!/(4-2)! = 24/2 = 12$.

📖 Circular permutation defined.

A circular permutation is the number of different ways you can arrange items in a circle. A typical circular permutation involves people sitting around a table. As an example, in how many ways can 5 different people be seated around a circular table? If we asked the question for 5 people in a row, we have $5! = 120$ different possible arrangements. However, around a circular table, the 5! arrangements often produce the same order. For example, *ABCDE* is the same as *BCDEA* in a circle. The relative positions of people have not changed. In fact, *ABCDE* can be rotated 5 times, and still we have the same order. Thus we must reduce the 5! arrangements in a row by dividing by 5 to get the number of arrangements in a circle, which is the same as calculating $(n - 1)!$ arrangements, given n different objects arranged in a circle. For 5 people around a table, there are $4! = 24$ different arrangements or permutations.

- - - - - -

❗ Permutations with some identical elements.

Up to now, we have been discussing the number of permutations of objects that are different. If

The ancient paths of Dr. Livingstone. Dr. Livingstone is traveling from Town *A* through Town *B* to Town *C* in the deepest heart of Africa (figure 5.1). In how many ways can he travel, assuming that he does not travel a road more than once or through a town more than once? In other words, he doesn't backtrack. (See Answer 5.7.)

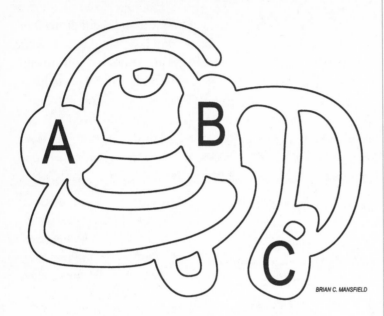

Figure 5.1   The ancient paths of Dr. Livingstone.

BRIAN C. MANSFIELD

we wish to determine the number of permutations of sets of objects that have some identical elements, the number of permutations is reduced. In particular, we can compute the number of permutations of *n* objects, *q* of one type, *r* of another, *s* of another, and so forth, using this formula: $n!/(q!r!s! \ldots)$. For example, the number of distinct permutations of the letters in BALLOON is $7!/(1!1!2!2!1!) = 5,040/4 = 1,260$.

- - - - - -

Randomness and uncertainty. "The popular image of mathematics as a collection of precise facts, linked together by well-defined logical paths, is revealed to be false. There is randomness and hence uncertainty in mathematics, just as there is in physics" (Paul Davies, *The Mind of God*, 1992).

- - - - - -

Organ arrangements. A stomach, a heart, a kidney, a brain, and a spleen are lined up in a neat row along a mad scientist's table. In how many different ways can they be arranged in the row? In how many ways can all possible sets of three organs from our collection of five organs be arranged? (See Answer 5.8.)

- - - - - -

Mombasa order. An archaeologist wishes to visit Jerusalem, Damascus, Cairo, Baghdad, Nairobi, Meru, and Mombasa. Before he and his wife leave for the trip, his wife insists that they visit Mombasa last. In how many ways can their order of visits vary? (See Answer 5.9.)

- - - - - -

Diamond arrangements. A movie star bearing a striking resemblance to Shannen Doherty has sixteen different-sized diamonds to string on a bracelet. In how many ways can this be done? (See Answer 5.10.)

Marble maze. You drop a marble into the maze in figure 5.2, and it is pulled down by gravity. At certain points, the marble has a choice of taking one of two forks in the tunnel. These choices are marked by dots. Assume that each time the marble has a valid choice of tunnels, it is equally likely to choose either route. You win a neural stimulator that gives you ecstatic pleasure if you emerge at the tunnel marked "Win." What is the likelihood that you will win the prize? (See Answer 5.11.)

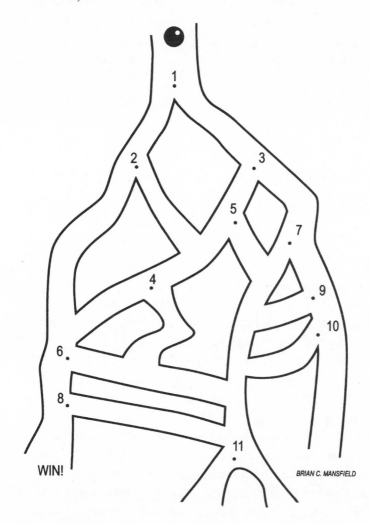

BRIAN C. MANSFIELD

Figure 5.2   Marble maze.

Sushi combinatorics. Sally is looking at a sushi menu. To make a single piece of sushi, she can choose tuna, squid, or shrimp, which can be either in the form of a roll or sliced. In addition, the piece may be either plain or coated with red, black, white, or green caviar. How many different pieces of sushi can Sally make? (See Answer 5.12.)

- - - - - -

Rationality. "What if our science rests on irrational impulses that we cannot measure? What if our mind is a ruler that cannot measure itself without always getting the same answer?" (George Zebrowski, "Is Science Rational?" 1994).

- - - - - -

Turnip permutations. You are buying vegetables in a local grocery store; you turn to a fellow shopper and ask, "How many different orderings or permutations can one make for the letters in the word TURNIP?" What should the shopper's response be? (See Answer 5.13.)

- - - - - -

Mathematics and freedom. "The essence of mathematics resides in its freedom" (Georg Cantor, "Über Unendliche, Lineare Punktmannischfaltigkeiten," *Mathematische Annalen*, 1883 [in this paper, Cantor extends the natural numbers to infinite ordinal numbers]).

The Africa gambit. Throw a dart at a map of the world. Assume that it lands randomly on the map. What is the chance that it lands on Africa? (See Answer 5.14.)

Rubik's Cube. An ordinary Rubik's Cube has 432,520, 032,274,489,856,000 different positions. The four-dimensional analog of a Rubik's Cube has 1,756,772, 880,709,135,843,168,526,079, 081,025,059,614,484,630, 149,557,651,477,156,021, 733,236,798,970,168,550, 600,274,887,650,082,354, 207,129,600,000,000,000,000 possible positions.

Sushi gambit. The dangerous Dr. Sushi places 5 pieces of tuna sushi and 5 pieces of octopus sushi in front of you. You have two dishes. You are about to place all of the pieces of sushi on the two dishes, distributing them however you like. After that, Dr. Sushi will choose one dish at random and eat a piece of sushi, chosen at random, from the dish. How should you distribute the sushi to maximize the chances that Dr. Sushi will eat octopus sushi? (See Answer 5.15.)

Cards of Minerva. Zeus and Minerva are playing with cards, each of which has one black side and one white side. Zeus has $n + 1$ cards, while Minerva has only n cards. Both players toss all of their cards simultaneously and randomly, and observe the number of their cards that come up black. What is the probability that Zeus obtains more black cards than Minerva? (See Answer 5.17.)

Alien gambit. An alien slaps a strip containing three buttons on your forehead. The buttons are labeled as follows:

The unit on your head contains a capsule of smelly, dangerous green dye that will open in 10 minutes. The dye is permanent and will color your face green for the rest of your life. You can choose whichever sequence of buttons you like to prevent the capsule's opening. However, before you press, consider that one of the buttons, if pressed, opens the capsule immediately—that's bad. Another button seals the capsule forever—that's good. The remaining button stimulates the pleasure center of your brain to such an extent that you don't care whether you turn green.

Are the chances greater that you will turn green today or that you will not turn green today? If you answer correctly, the alien will remove the capsule. You can't remove the buttons, but you can continue to press buttons to try to avoid turning green. (See Answer 5.16.)

Card shuffles. It is rumored that magicians who are excellent with their hands can shuffle a deck "perfectly" so that the two halves of the deck are exactly interleaved. Is it possible to say how many consecutive shuffles of this type will restore a deck to its original order? (See Answer 5.18.)

Piano probability. You are blindfolded and asked to select a piano key at random. What note are you most likely to play, out of the usual white keys, C, D, E, F, G, A, B, and the black keys, C#, D#, F#, G#, A#? In other words, if you're asked to select one of the 88 standard piano keys at random, with each key having an equal probability of being chosen, which step in the chromatic scale A, A#, B, C, C#, D, D#, E, F, F#, G, G# is your selected key most likely to represent? (See Answer 5.19.)

Martian shuffle. Consider a deck of three cards, one picturing Mars, another picturing the Martian moon of Phobos, and a third picturing the Martian moon of Deimos. They are in the order of Mars, Phobos, and Deimos. You randomly shuffle the deck. What are the chances that you will have the original order after you shuffle? (See Answer 5.20.)

Air molecules. It is possible, although extremely unlikely, for all of the air molecules in your room to suddenly go to one corner and suffocate you, based simply on random movements of the molecules. To better understand the chance of this, consider the following. How unlikely is it for just 10 molecules in your room to jump into a corner volume that is 10 percent the room's volume? (See Answer 5.21.)

Hannibal's organs. A crazy man named Dr. Hannibal motions to an opaque, formalin-filled jar. "The jar contains one organ, either a kidney or a brain," Hannibal truthfully says. You watch as Hannibal drops a brain into the jar, shakes the jar, and randomly withdraws an organ that proves to be a brain. One organ remains in the jar. "What is now the chance of removing a brain from the jar?" (See Answer 5.23.)

Probability and truth. "It has been pointed out already that no knowledge of probabilities, less in degree than certainty, helps us to know what conclusions are true, and that there is no direct relation between the truth of

Adventure. "He calmly rode on, leaving it to his horse's discretion to go which way it pleased, firmly believing that in this consisted the very essence of adventures" (Cervantes, *Don Quixote*, 1605).

Chess moves. You are sitting with a brilliant robot, Roberta, playing chess. You move first and have a choice of moves. Roberta then responds by making a move. How many different possible board configurations exist that consist of these first two moves? (See Answer 5.22.)

a proposition and its probability. Probability begins and ends with probability" (John Maynard Keynes, "The Application of Probability to Conduct," in James Newman, *World of Mathematics*, 1988).

🌀 Walrus and Carpenter.

Carpenter: You would agree that the probability of picking 1 from the first 10 numbers is 1/10?

Walrus: Of course.

C: And that the probability of picking 1 from the first 100 numbers is 1/100.

W: Of course.

C: And that as you increase the number to be selected from, the probability decreases to zero in the limit?

W: (after a pause) Yes, and this logic applies to any number.

C: Would you also agree that the probability of picking 1 or 2 from the first 10 numbers is 2/10? That the probability of picking one or another is the sum of the probabilities of picking either?

W: (pause) Yes, I would.

C: And, therefore, that the probability of picking 1 or 2 from the set of all integers is thus the sum of zero and zero—in other words, zero?

W: That must be so.

C: And that furthermore, the probability of picking any of a set of alternative integers from the set of all integers is zero?

W: Of course. The probability of picking any one of the integers from the set of all integers is also zero.

C: So pick an integer and tell it to me.

The Walrus is struck dumb.

(Graham Cleverley, personal communication, 2004).

❓ Alien pyramid.

An alien hands you ten colored pyramids numbered 11 through 20. The alien selects a pyramid at random, and it happens to be Pyramid 13. The alien takes a long breath. He now randomly picks another pyramid from the remaining collection. Is this pyramid more likely to be even than odd? (See Answer 5.24.)

- - - - - -

❓ Rabbit math.

Inside a cage is a brown rabbit and a polka-dot rabbit, and each rabbit species generally produces members that have an equal chance of being male or female. The rabbits listen to classical music all day as they enjoy their life of leisure. Today, we learn that at least one of the rabbits is male. What is the probability that both rabbits are male? (See Answer 5.25.)

- - - - - -

❓ Card deck cuts.

Danielle opens a brand new deck of playing cards containing 52 normal cards. She tells Craig, "I am not a good card shuffler. I will therefore cut the deck at random depths, several times, by taking a random number of cards off the top of the deck and placing them under the remaining deck. I want to be fair. Even though my hand will get tired, I will do this either ten times, a hundred times, or a thousand times to try to ensure that the deck is reasonably shuffled."

How many times does Craig want Danielle to cut the deck to ensure the greatest randomization? (See Answer 5.26.)

Russian American gambit. A woman and a man are sitting in metal chairs, facing one another. One person is American and the other is Russian, but we don't know which is which.

"I am an American," the woman says.

"I am Russian," the man says.

A robot in the room tells you, "At least one person is lying." Assume that the robot is telling the truth. My question: Which of the humans is lying? (See Answer 5.27.)

Quantum kings and the death of reality. Martin Gardner, a master of mathematical puzzles, once posed a variant

Car sequence. You go into Tony's spacious garage, located in an upper-class section of Perth Amboy, New Jersey. He has 10 cars, made in the years 1960, 1961, 1962, 1963, 1964, 1965, 1966, 1967, 1968, and 1969.

The cars are arranged neatly in a row and in random date order. What are the chances that the three cars nearest you have dates in descending order? (Here is one example of dates being in descending order: 1968, 1966, 1965.) (See Answer 5.28.)

of the following scenario to Marilyn vos Savant, who has an I.Q. score of 228—the highest ever recorded for an adult. Here's my variation:

Bill and Monica are playing bridge, and they always tell the truth. After a deal, a third person asks Monica, "Do you have a king in your hand?"

She nods. At this point, there is a certain probability that her hand holds at least one other king. Call this probability P_1.

After the next deal, the third person again asks Monica, "Do you have a king of spades?" Monica nods. Again, there is a certain probability that Monica's hand holds at least one other king. Call this probably P_2. Which

probability, P_1 or P_2, is greater? Or are they both the same? (See Answer 5.29.)

AIDS tests. You work for a computer company. Suppose that about 2 percent of the people in your company have AIDS. A nurse named Julia tests all of the people in your company for AIDS, using a test with the following characteristics. The test is 98 percent accurate, which we define as follows. If the individual has AIDS, the test will be positive 98 percent of the time, and if the person doesn't have AIDS, the test will be negative 98 percent of the time.

You are tested, and, sadly, the test turns out positive.

Woman in a black dress. A woman in a black dress approaches you with three cards; each card is red on one side and black on the other. The cards are randomly positioned. She places the three cards on the table. What is the probability that no two consecutive reds show? For her next puzzle, she places *n* cards on the table. What is the probability that no two consecutive reds show? (See Answer 5.31.)

You lose your health insurance. Would you conclude from this that you are highly likely to have AIDS? (See Answer 5.30.)

- - - - - -

Great discoveries. "Sixty percent of new great discoveries were made by an outsider to the field within the first year he was in the field" (Raymond Damadian, the inventor of the first human body scan with magnetic resonance imaging, "Scanscam," *New York Times Magazine*, 2003).

- - - - - -

Random chords. What is the probability that a randomly selected chord of a circle is shorter than the radius of that circle? If you were a gambling person, should you bet that a randomly selected chord is smaller than the circle's radius? (A chord is a line joining any two points on the circle. To create the ran-dom chord, select one point on a circle by choosing a random angle from the center and draw a line from this point to a second point on the circle that you chose by selecting another random angle.) (See Answer 5.32.)

- - - - - -

Bart's dilemma. Nelson is a mean, but very honest, bully. He captures Bart and tells him, "I will paint you red if you speak the truth or will shave your head if you lie." Obviously, Bart does not look forward to either of these odious choices. Bart may speak some words or remain silent. What would you advise Bart to do? (See Answer 5.33.)

- - - - - -

Multilegged creatures. Three robots named Mr. Eighty, Mr. Ninety, and Mr. Hundred are crawling in a dark laboratory. One robot has 80 legs; one robot has 90 legs; one robot has 100 legs. The robots are usually quite happy and enjoy the added mobility the multiple legs give them. Today, however, they are jealous because they hear that their creator might produce a robot with 1,000 legs.

"I think it is interesting," says Mr. Hundred, "that none of us has the same number of legs that our names would suggest."

"Who the heck cares?" replies the robot with 90 legs.

How many legs does Mr. Ninety have? Amazingly, it is possible to determine the answer, despite the little information given. (See Answer 5.34.)

- - - - - -

Mathematical reality. "We feel certain that the extraterrestrial message is a mathematical code of some kind. Probably a number code. Mathematics is the one language we might conceivably have in common with other forms of intelligent life in the universe. As I understand it, there is no reality more independent of our perception and more true to itself than mathematical reality" (Don DeLillo, *Ratner's Star*, 1976).

Three fishermen. Three Portuguese fishermen have three new buckets. One bucket holds just flounder, the other just mackerel, and the third both flounder and mackerel. The fishermen label their three buckets "flounder," "mackerel," and "flounder and mackerel."

Sadly, all three labels are incorrect.

The buckets are closed. How many fish does a fisherman have to look at to correctly label the buckets? (See Answer 5.35.)

- - - - - -

Mummy madness. An Egyptologist is examining the mummies of an ancient cat, a hyena, and a mouse. He turns to you. "You can't see the mummies clearly,

but I tell you this: At least one of the mummies is missing a tail. What is the probability that the cat is missing a tail?"

You think about this for a moment, staring at the shrouded figures. The Egyptologist says to you, "If you guess correctly, I will reward you by giving you the golden staff of King Tut." Can you solve this? (See Answer 5.36.)

- - - - - -

Heart attack. A surgeon in a dimly lit operating room removes five pulsating hearts from five patients. She places the hearts back in the bodies at random. What are the chances that only four hearts are returned to their correct bodies? (See Answer 5.37.)

Floating boat game. The sun is shining on the Mississippi River, and your three friends are playing with paper boats. Once a person puts a boat in the water, there is a 50 percent chance that it will float and a 50 percent chance that it will immediately sink. Don, Melissa, and Carl place their boats on the river in succession. First, Don puts his boat on the water. Next, Melissa. Finally, Carl. The winner will be the first to float (i.e., not sink). What are their respective chances of winning? (Assume that the game is over after Carl's turn, and the game stops when someone wins.) (See Answer 5.39.)

- - - - - -

Clown's dreams. A friendly clown with purple hair places two aquaria on your kitchen table. One contains grape juice, the other vinegar. Both aquaria contain exactly the same volume of liquid. The clown takes a glass of vinegar from the vinegar-filled aquarium and mixes it into the juice-containing aquarium, then he takes a glass from the juice aquarium and mixes it with vinegar. Both aquaria are now "contaminated." If

Sushi play. Pam and Nick each have ten pieces of sushi, branded with numbers 1 through 10.

🐟 🐟 🐟 🐟 🐟 🐟 🐟 🐟 🐟 🐟 (Pam)

🐟 🐟 🐟 🐟 🐟 🐟 🐟 🐟 🐟 🐟 (Nick)

Pam gives sushi piece 2 to Nick as a gift. Nick now has 11 pieces of sushi and withdraws a piece at random from his collection and places it in a bright red box. He turns to Pam. "I can only tell you this. The sushi piece in the box does not have an odd number on it." What are the chances that the sushi piece has a 2 on it? (See Answer 5.38.)

you can determine which aquarium is more contaminated, the clown will leave your home and stop asking this silly question. Does the vinegar now contain more juice than the juice does vinegar, or the other way around? (See Answer 5.40.)

Lottery. Over the last ten years, you play the same lottery number and never win. Would you have a better chance of winning if you played different numbers instead of always playing the same number? (See Answer 5.41.)

Fossil lock. You want to send a valuable fossil to your friend Homer. You have a padded sphere that can be fitted with multiple locks to lock the sphere's door, and you have several locks and their corresponding keys. However, Homer owns some locks, but he does not have any keys to your locks, and if you send a key in an unlocked sphere, the key could be copied en route. How can you send the fossil securely? Assume that a key is required to both lock and unlock a lock. (See Answer 5.42.)

Nontransitive dice. Here is an impressive game to try on your friends, based on a recent discovery made by the statistician Bradley Efron of Stanford University. Magic dice are displayed in figure 5.3 so that you can clearly see their six faces. The dice demonstrate a probability that, at first, seems to defy logic and can be used by the unscrupulous to make millions of dollars.

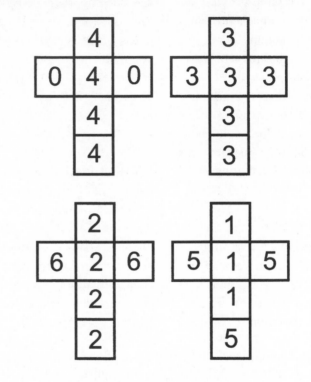

Figure 5.3   Nontransitive dice.

Let's assume that Bill and Monica are playing a game with these dice. Bill asks Monica to select any one of the four dice. Bill selects another. They throw the dice at the same time for a predetermined number of times. On each toss, the person with the highest number wins the throw. Bill can almost always beat Monica in a game of ten throws. After each game, Monica is free to choose another die, and Bill then chooses a die. Whichever die Monica selects, Bill always wins. How can that be? (See Answer 5.43.)

Sublime math. "There was a blithe certainty that came from first comprehending the full Einstein field equations, arabesques of Greek letters clinging tenuously to the page, a gossamer web. They seemed insubstantial when you first saw them, a string of squiggles. Yet to follow the delicate tensors as they contracted, as the superscripts paired with subscripts, collapsing mathematically into concrete classical entities—potential; mass; forces vectoring in a curved geometry—that was a sublime experience. The iron fist of the real, inside the velvet glove of airy mathematics" (Gregory Benford, *Timescape*, 1980).

Farmer McDonald's moose. Farmer McDonald's male moose brings him 100 pairs of red gloves and 100 pairs of blue gloves and accidentally drops them into a stinking sewer. "Darn!" McDonald says. In complete darkness, how many gloves must he take from the sewer in order to be sure to get a pair that match? (See Answer 5.45.)

Tic-tac-toe. Despite the apparent simplicity of the game of tic-tac-toe, sometimes the best strategy from different configurations can be difficult to see. Many players will seek a draw against a weaker player, instead of playing to maximize their chances of benefiting from a weak move. For example, assume that you are X in the following unusual opening. What is your best move after O has moved into position 8? (See Answer 5.44.)

1	X	3
4	5	6
7	O	9

Black versus red. Brad is staring at two 1-dollar bills. One is painted black on both sides. The other is painted black on one side and left alone on the other. His friend Jennifer randomly selects a bill and shows it to Brad. He sees a black bill. What's the probability that the other side is black? If Brad were a gambler, should he bet that the other side is black? (See Answer 5.46.)

The language of math. "The Universe is a grand book which cannot be read until one first learns to comprehend the language and become familiar with the characters in which it is composed. It is written in the language of mathematics" (Galileo, *Opere Il Saggiatore*, 1623).

Brunhilde's moose. Brunhilde throws a dart twice at the word *moose*, which hangs on her living room wall. Assume that Brunhilde's dart always lands on some letter, that each letter is equally likely to be hit, and that having a dart on a letter doesn't affect whether the next dart will land on the letter. What is the chance that Brunhilde will hit the letter *O* followed by another *O*? (See Answer 5.47.)

Board game of the gods. Consider the following simple, yet diabolic, board game, played by Aphrodite and Apollo in a magnificent palace on the heights of Mount Olympus (figure 5.4).

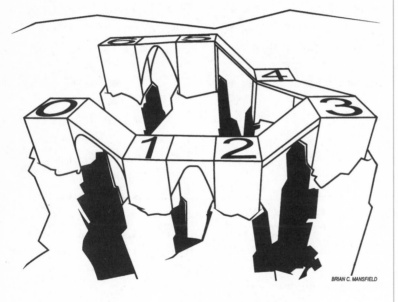

BRIAN C. MANSFIELD

Figure 5.4   Board game of the gods.

Players move from 1 to 2 to 3 to 4 to 5 and finally to 6. All movements are on numbered squares only, and unnumbered connecting ramps in the diagram are ignored. The toss of a coin determines the movements on the board. A tail means that one moves by advancing one square. A head means to advance two squares. Play starts with the player's piece on square zero, prior to the first toss. On average, how many coin tosses will it take for a player to win, that is, to reach or go beyond the finishing 6 square? (See Answer 5.48.)

Chimps and gibbons. Three primates approach you. Their names are Mr. Chimp Gibbon, Mr. Chimp Chimp, and Ms. Gorilla Lemur. You are blindfolded. One of the primates, chosen at random, truthfully says, "My name is Chimp," but you don't know whether he is referring to his first or last name. (Assume that a primate is equally likely to call out his first or last name.) Are the odds ½, ⅓, or some other value that the primate's full name is Chimp Gibbon? (See Answer 5.49.)

- - - - - -

Cars and monkeys. Andrea, Barbara, and Claire are driving red sportscars with leather seats. Each person is driving the car of one friend and has the pet monkey of another. The woman who has Claire's monkey is driving Barbara's car. From this amazingly scant information, can we determine who is driving Andrea's car? (See Answer 5.50.)

- - - - - -

Guessing numbers. You are with your friends Bill and Julie, gazing at a large chunk of petrified wood. "Earth is roughly 5 billion years old," you say. "Julie, randomly write down a year between 0 and 5 billion." You turn to

Bill. "I bet that my guess of Julie's year will be closer to the year than your guess. You'll have one guess, while I have two, but to make up for my slight advantage, not only can you guess first, but I'll buy you pizza every day for the next month if I lose. If I win, you'll buy me pizza."

What strategy do you use to win? Would you use the same strategy if it meant that you would lose or gain a great deal of money? (See Answer 5.51.)

- - - - - -

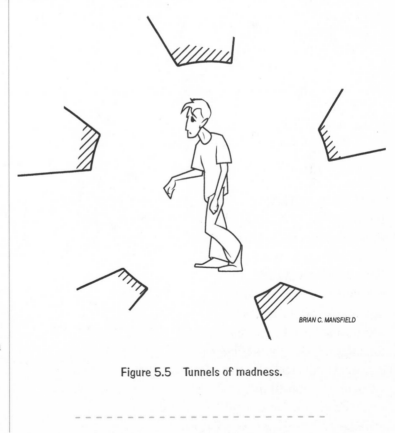

Flying numbers. "Arithmetic is where numbers fly like pigeons in and out of your head" (Carl Sandburg, "Arithmetic," 1993).

- - - - - -

Martian identities. In a large crater on Mars, each Martian carries an identification card. Each Martian's card is different and contains just one letter and one 1-digit number, including zero. How many Martians at most live in the crater? (Assume that the Martian number system and alphabet are the same as ours.) (See Answer 5.52.)

Robotic ants. A large robot ant named "Punisher" places you in a central chamber in his ant colony, miles beneath Manhattan. Five tunnels radiate away from the chamber, which is filled with eggs and larvae (figure 5.5). Four of the five tunnels lead to deadly warrior ants. Only one leads upward to Fifth Avenue and freedom. Punisher tells you to pick a tunnel. You randomly pick Tunnel 1 to travel through. But wait! Punisher, who knows where each tunnel leads, truthfully tells you that Tunnel 5 terminates in a warrior ant. He gives you the opportunity to change your mind and choose a tunnel other than Tunnel 1. You can assume that whenever Punisher takes prisoners, he always points to a tunnel that terminates in a warrior ant. Should you forget your choice of Tunnel 1 and choose another tunnel? (See Answer 5.53.)

BRIAN C. MANSFIELD

Figure 5.5   Tunnels of madness.

- -

Venusian insects. On beautiful Venus, 85 percent of the Venusians eat beetles, 80 percent eat wasps, 42 percent eat ants, 97 percent eat crabs, and 96 percent eat oysters. None of the Venusians eat all five items, but they all eat four of the five foods. What percentage of the Venusians eat insects? (See Answer 5.54.)

- - - - - -

Domain names. In one of my novels, a character makes money by buying every five-letter Internet domain name, like doggy.com or lions.com, and then selling them to people a year later. Just how many five-letter domain names are there? Will we run out of domain names in the future? (See Answer 5.55.)

- - - - - -

The physics of reality. "I think that modern physics has definitely decided in favor of Plato. In fact the smallest units of matter are not physical objects in the ordinary sense; they are forms and ideas which can be expressed unambiguously only in mathematical language" (Werner Heisenberg, quoted in "Physics for the Fatherland," 1992).

Where are the sea horses? Consider a strange aquarium with five tanks connected by tubes (figure 5.6). You add thousands of sea horses to Chamber A, indicated by the hand. Assume that the sea horses are swimming randomly through the structure, like mindless molecules of gas filling a balloon. Also assume that the sea horses are much smaller than those pictured schematically in the diagram. Where do the most sea horses reside after a long time? How would your answer change if there were an additional tunnel connecting Chamber C to D? (See Answer 5.56.)

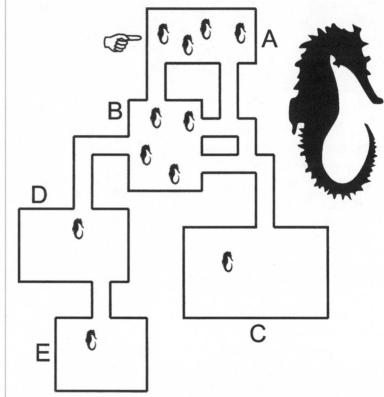

Figure 5.6   Where are the sea horses?

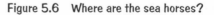

Lotteries and superstition. Bill and Monica are participating in a lottery in which six numbers in the range of 13 to 16 are drawn at random. (In other words, the numbers drawn can be 13, 14, 15, and 16.) You have to pick all six numbers in the correct order to win a large bag of extraordinary blue sapphires from Sri Lanka.

You hear thunder in the distance. Monica gasps, "I'm superstitious and for the next year will avoid the number 13." Bill says, "If you do that, you'll decrease your chances of winning. You're eliminating lots of potential choices."

Is Bill right or wrong? Will Monica's superstition put her at a disadvantage? Justify your answer. Assume that after a number is drawn, it is replaced and has a chance of being drawn again. In other words, each number drawn can be 13, 14, 15, or 16. (See Answer 5.57.)

- - - - - -

Robot jaws. You are trying to open a robot's mouth that has two push buttons on the upper lip. They can each toggle between ten positions. The proper positions will open this top lock. A lock on the bottom of the jaw has two buttons of the same kind. This second lock has its own combination, which may or may not be the same as the top lock's combination. You need to get both combinations right to open the mouth. Assume that you will randomly try positions.

What are the odds that you will find the right combination on the first try to open the mouth? (See Answer 5.59.)

- - - - - -

Educated guessing. "Consider a dramatic moment at the casino in Monte Carlo on August 18, 1913. A roulette wheel, when spun, has an equal chance of showing a red or black number. On this evening, black came up 26 times in a row. . . . A frantic rush to bet on red enriched

Coin flips and the paranormal. Your friend approaches with a coin. He flips it several times. We expect 50 percent heads and 50 percent tails. How many heads in a row would cause an ordinary person to say, "This is a biased coin," or that something was amiss? How many heads in a row would cause a *mathematician* to say, "This is a biased coin"?

Do people who believe in the paranormal respond differently than skeptics do in experiments of this nature? (See Answer 5.58.)

the casino considerably" (Samuel Kotz and Donna F. Stroup, *Educated Guessing*, 1983).

- - - - - -

Flies in a cube. I place 1,000 large flies in a cube that is one mile on a side. How would you attempt to compute the average distance traveled by the flies between collisions with one another, assuming that the flies fly randomly? (Also assume that each fly can be modeled as a sphere with a half-inch diameter.) (See Answer 5.60.)

- - - - - -

Proportion. "There is no excellent beauty that hath not some strangeness in the proportion" (Francis Bacon, *Of Beauty*, c. 1601).

- - - - - -

❓ Boomerang madness. I toss two boomerangs at the same time toward twenty maple trees. What's the probability of both boomerangs getting caught in any one tree together? (Assume that my toss is random, that the boomerangs are equally likely to end up in any tree, and that the boomerangs always end up in some trees.) (See Answer 5.61.)

❓ Apes in a barrel. A man places thirty apes in a large barrel. Your ape is named Tiffany. The apes are released, one at a time, and the last ape to emerge is the winner. Thus, if Tiffany emerges last, you win. What are the chances of you winning? (Assume that the apes emerge in a random order.) (See Answer 5.62.)

❓ Napoleon and Churchill. Napoleon and Churchill are sitting in an underground bunker, three miles south of London. Today, they are tossing a coin in the air. They are betting the fate of the world on a particular triplet that might result from three successive fair tosses of the coin. Napoleon picks HHH and Churchill picks THH. They keep tossing coins until either Napoleon's or Churchill's triplet has appeared. The triplets may appear in any three consecutive tosses; for example, the THH triplet might start in position one, two, or three in the long sequence of attempts. The winner is the player whose triplet appears first. Whom do you pick as the winner? (See Answer 5.64.)

- - - - - -

❓ Dark hallway and vapor men. You are traveling down a long hallway. Strange scratching sounds come from beneath the floor, and scarlet beetles climb the walls. Along the hallway are ten shadowy men spaced roughly at equal lengths along the hallway (figure 5.7).

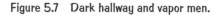

In ... Out

Figure 5.7   Dark hallway and vapor men.

Assume that each time you pass a man, he touches your hand. Each times he touches your hand, you have a 50 percent chance of being vaporized.

The world is offered total peace and prosperity if you can correctly guess which man will vaporize you as you walk down the hallway. If you were a gambling person, which person do you think will vaporize you? How many attempts would you predict it would require for a person to walk from one end of the hallway to the other without being vaporized (assuming that the person is reconstituted each time so that he or she can try again and again until safely making it to the end)? (See Answer 5.63.)

❀ The mine of math. "Mathematics may be likened to a large rock whose interior composition we wish to examine. The older mathematicians appear as persevering stone cutters slowly attempting to demolish the rock from the outside with hammer and chisel. The later mathematicians resemble expert miners who seek vulnerable veins, drill into these strategic places, and then blast the rock apart with well placed internal charges" (Howard Eves, *Mathematical Circles*, 1969).

- - - - - -

Triangles and spiders. You are visiting a moon of Uranus, where a spiderlike alien randomly throws a meatball at a wall on which is drawn a large triangle with vertices at (0,0), (0,4), and (5,0) (figure 5.8). The meatball sticks to the triangular target. What is the probability that the *x*-coordinate of the meatball is less than the *y*-coordinate? (See Answer 5.65.)

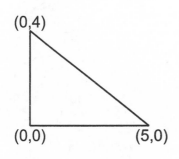

Figure 5.8   Triangles and spiders.

- - - - - -

Stylometry. What is stylometry, and who wrote *The Royal Book of Oz*, the fifteenth book in the Oz series? (See Answer 5.66.)

Factors. A wise man wearing a long flowing robe chooses 3 integers at random from 1 to 10^{100}. What is the probability that no common factor will divide them all? If you can't tell me this exactly, what is your estimate? (A divisor of a number is a number that divides *n*—also called a factor.) (See Answer 5.67.)

- - - - - -

Mathematical existence. "One cannot escape the feeling that these mathematical formulae have an independent existence and an intelligence of their own, that they are wiser than we are, wiser even than their discoverers, that we get more out of them than we originally put in to them" (Heinrich Hertz [commenting on Maxwell's equations], quoted in Eric Bell, *Men of Mathematics*, 1937).

Card arrangements. The 52 cards in a playing card deck can be arranged in 52! different ways, or

$$80658175170943878571660636856403766975289505440883277824000000000000$$

or about 8.07×10^{67}. This is much larger than the number of atoms that make up Earth.

- -

Alien Patterns. How many patterns of *n* creatures in a row exist if every alien is next to at least one other alien? We are to choose from an auditorium filled with aliens and humans. For example, with four creatures, there are seven patterns:

In general, if $P(n)$ is the number of patterns of *n* creatures, then the ratio $P(n + 1)/P(n)$ approaches 1.754877666247 . . . , a solution to the equation $x^3 = 2x^2 - x + 1$ (in John Conway and Richard Guy, *The Book of Numbers*, 1996, p. 205).

Plato's choice. Sitting beneath the shade of the Parthenon in ancient Greece were five mortals and five gods who found five emeralds (figure 5.9). They soon began to fight about who should get the jewels.

Zeus bellowed, "We'll arrange ourselves in a circle, and count out individuals by a fixed interval, and give an emerald to an individual as he leaves the circle." Zeus was clever and arranged the circle so that by counting a certain mortal as "one," he could count out all the gods first.

The arrangement of gods and mortals is shown in figure 5.9, which follows a sequence: M M M G G G M G G M.

The count starts with the mortal at the top and continuously goes clockwise. Each individual who is "counted out" steps out of the circle, gets an emerald, and is not included in the count thereafter. For example, if the interval of "decimation" is 5, the first three individuals to get emeralds would be: god 5, mortal 10, and god 6. So far, pretty good for the gods.

It turned out, to Zeus's dismay, that the mortal at the top who was counted as 1 was none other than the all-wise Plato.

"Wait!" Plato cried. "I insist on my right to choose the interval of decimation."

Plato's astute choice counted out all of the mortals first. What interval will count out the five gods first, and what will count out the five mortals first? (See Answer 5.68.)

Figure 5.9 Plato's choice.

BRIAN C. MANSFIELD

Kravitz arrays. Sidney Kravitz once asked in the *Journal of Recreational Mathematics* (31, no. 4 [2003]: 304): In how many ways can we place the digits 1 to 9 into a 3-by-3 array so that each digit is smaller than the digit immediately to its right and smaller than the digit immediately below it? There are 42 such arrangements.

Big Numbers and Infinity

I N WHICH WE EXPLORE VERY LARGE NUMBERS, THE EDGES OF COMPREHENSION, infinity, the Funnel of Zeus, the infinite gift, the omega crystal, the Skewes number, the Monster group, Göbel's number, the awesome quattuordecillion, the Erdös-Moser number, Archimedes' famous "Cattle Problem," classic paradoxes, Gauss's "measurable infinity," and Knuth's arrow notation.

> Infinity is an endless train weaving
> its way through the landscape of
> reality. But who made the train
> tracks, and where is the conductor?

Infinity and God. "The study of the infinite is much more than a dry, academic game. The intellectual pursuit of the Absolute Infinite is a form of the soul's quest for God. Whether or not the goal is ever reached, an awareness of the process brings enlightenment" (Rudy Rucker, *Infinity and the Mind*, 1982).

- - - - - -

Counting. Assuming that you count a number a second, how long would it take you to count to a billion (1,000,000,000)? (See Answer 6.1.)

- - - - - -

Large numbers. "Large numbers have a distinct appeal, a majesty if you will. In a sense, they lie at the lim-its of the human imagination, which is why they have long proved elusive, difficult to define, and harder still to manipulate. Modern computers now possess enough memory and speed to handle quite impressive figures. For instance, it is possible to multiply together million-digit numbers in a mere fraction of a second. As a result we can now characterize numbers about which earlier mathematicians could only dream" (Richard E. Crandall, "The Challenge of Large Numbers," 1997).

- - - - - -

Wonders. "He who wonders discovers that this in itself is a wonder" (M. C. Escher, *The Graphic Work*, 1992).

Disembodied intelligence. "A formal manipulator in mathematics often experiences the discomforting feeling that his pencil surpasses him in intelligence" (Howard W. Eves, *Mathematical Circles*, 1969).

- - - - - -

Large law. What is the "law of truly large numbers"? (See Answer 6.2.)

- - - - - -

The ultrahex. The following is an extremely large number:

$$6^{6^{6^{6^{6^{6}}}}}$$

Although humanity may never be able to compute this and write down all its digits, what can we say, if anything, about its last digit? (See Answer 6.3.)

- - - - - -

Life experiences. The math fanatic Robert Munafo says that $10^{10^{18}}$ is an estimate of the number of possible life experiences a person can have, based on a sensory bandwidth of 10^{10} bits per second.

- - - - - -

Decillion. One decillion is 1 followed by 33 zeros:

100000000000000000000000000000000

or 10^{33}. It's the largest power of 10 known to humans that can be represented as the product of two numbers that themselves contain no zero digits:

$$10^{33} = 2^{33} \times 5^{33} =$$

$$(8,589,934,592) \times (116,415,321,826,934,814,453,125)$$

I do not think that humanity will ever be able to find a larger power of 10 that can be represented as the product of two numbers that themselves contain no zero digits.

The wonderful quattuordecillion. A few favorite number names: 10^{21} is a sextillion; a decillion is 10^{33}; 10^{45} is a *quattuordecillion*; 10^{60} is a novemdecillion; and 10^{63} is a vigintillion.

The ubiquitous 7. How many numbers have a 7 in them, if we consider all the numbers between 1 and googolplex? (See Answer 6.4.)

Knuth notation for ultra-big numbers. In 1976, the mathematician and computer scientist Donald Knuth published an amazing arrow notation to represent huge numbers. Thus $3 \uparrow 3$ is simply 3 cubed or 3^3; $3 \uparrow\uparrow 3$ means $3 \uparrow (3 \uparrow 3)$ or 3^{3^3}. The arrows always "associate to the right," so that $a \uparrow b \uparrow c \uparrow d \ldots$ means $a \uparrow (b \uparrow (c \uparrow d)) \ldots$. Here's another example: $2 \uparrow\uparrow\uparrow 3 = 2 \uparrow\uparrow 2 \uparrow\uparrow 2 = 2 \uparrow\uparrow 4 = 65{,}536$. Using this logic for making huge numbers compact, do you think that $3 \uparrow\uparrow 4$ represents a number that is smaller than the number of atoms in the visible universe? (See Answer 6.5.)

Definition of googolplex. A "googolplex" is equal to 10^{googol} or $10^{10^{100}}$; that is, 1 followed by a googol of zeroes. Like googol, the term *googolplex* was coined by the mathematician Edward Kasner in conversations with others.

The software engineer Frank Pilhofer has tried to put a googolplex in perspective when he writes, "[Compared to a googolplex], a googol is nothing special; the total number of elementary particles in the known universe is about 10^{80}. If this space was packed solid with neutrons, so there was no empty space anywhere, there would be about 10^{128} particles in it. This is quite a lot more than a googol. But you simply cannot express the kind of *googolplex*'s numerical dimension with terms other than '10 to the power of something.' . . . If one takes a moderately large fraction of the mass of our Local Group of galaxies, puts it into a black hole, and asks how many states there are with a similar macroscopic appearance, one would get a googolplex" (Frank Pilhofer, "Googolplex," www.fpx.de/fp/Fun/Googolplex/, with material from Don Page, CIAR Cosmology Programme, Theoretical Physics Institute, University of Alberta, 2002).

Big number. "I believe there are

$$15{,}747{,}724{,}136{,}275{,}002{,}577{,}605{,}653{,}961{,}181{,}555{,}468{,}044{,}$$
$$717{,}914{,}527{,}116{,}709{,}366{,}231{,}425{,}076{,}185{,}631{,}031{,}296$$

protons in the universe and the same number of electrons" (Sir Arthur Eddington, *The Philosophy of Physical Science*, 1939).

The magnificent 7. As you may have gleaned, I love the number 7. Consider the following scenario. You are alone on an infinite plane, examining a list of integers in numerical order, starting at 1, 2, 3, 4, 5, 6, . . . and ending with some huge number like googolplex. You get into a spaceship and journey far away, then shoot a dart randomly at the integer list. What is the chance that you will hit a number with a 7 in it? (See Answer 6.6)

Computers and big numbers. What is the largest integer number that a future computer could in principle work with or yield as a result of a computation? (The numbers are to be stored in binary format as a string of 0s and 1s.) (See Answer 6.7.)

- - - - - -

Eternity. "That's the problem with eternity, there's no telling when it will end" (Tom Stoppard, *Rosencrantz and Guildenstern Are Dead*, 1967).

- - - - - -

Three digits. The number $\Theta = 9^{9^9}$ is the largest number that can be written using only three digits. The number Θ contains 369,693,100 digits. If typed on paper, it would require about 2,000 miles of paper strip. Since the early 1900s, scientists have tried to determine some of the digits of this number. Fred Gruenberger recently calculated the last 2,000 digits and the first 1,200 digits.

- - - - - -

Counting every rational number. Show that it is theoretically possible to count all the rational numbers by

Four digits. The German mathematician Carl Friedrich Gauss (1777–1855) called

$$\mathfrak{I} = 9^{9^{9^9}}$$

"a measurable infinity." The number has $10^{369693100}$ digits, a number far larger than the number of atoms in the visible universe. If typed on paper, $\mathfrak{I}$ would require $10^{369693094}$ miles of paper strip, according to Joseph Madachy. If the ink used in printing $\mathfrak{I}$ was a one-atom-thick layer, there would not be enough total matter in millions of our universes to print the number. Shockingly, the last 10 digits of $\mathfrak{I}$ have been computed. They are 1,045,865,289.

- - - - - - - - - - - - - - - -

1597 problem. Consider the formula $y = \sqrt{1597 x^2 + 1}$. Is y ever an integer for any integer x greater than 0? (Hint: Don't bother to solve this unless you use a computer.) (See Answer 6.9.)

means of an infinite list, whereas it is not possible to count all the real numbers by means of such a list. (See Answer 6.8.)

- - - - - -

Continuum hypothesis. As we discussed in chapter 2, the total number of possible integers or possible rational numbers corresponds to the smallest level of infinity, usually denoted as $\aleph_0$. The number of possible real numbers is $2^{\aleph_0}$, which is larger than $\aleph_0$. (See Cantor's diagonalization process in answer to the previous factoid.) The

continuum hypothesis suggests that there are no sets intermediate in size between $\aleph_0$ and $2^{\aleph_0}$. In other words, the hypothesis states that there is no set whose size is strictly between that of the integers and that of the real numbers. We can map the algebraic numbers onto the naturals, and the cardinality of the algebraic numbers will be the same as for naturals. Here are some cool rules involving $\aleph_0$:

$$\aleph_0 = 1 + \aleph_0$$
$$\aleph_0 = \aleph_0 + \aleph_0$$
$$\aleph_0 = \aleph_0 \times \aleph_0$$

Mathematical theology. *Manifold*—the only journal ever published entirely on the subject of mathematical theology—began as a publication created by several graduate students at the University of Warwick in England. The journal started in 1968 and ran for twenty issues, until 1980. Tim Poston, a researcher at the Institute of Systems Science at the National University of Singapore, led me to several interesting articles regarding God and mathematics that appeared in *Manifold*. For example, in *Manifold* 6 (1970, pp. 48–49), Vox Fisher published "Ontology Revisited," a fascinating paper on a mathematical proof of God's existence. The paper starts,

Theorem: The axiom of choice is equivalent to the existence of a unique God (St. Anselm, Aquinas, and others).

Proof: Partially order the set of subsets of the set of all properties of objects by inclusion. This set has maximal elements. God is by definition (due to Anselm) a maximal element set:

God $\subseteq$ God $\cup$ {existence}, so God = God $\cup$ {existence}
$\therefore$ God exists.

Uniqueness: If God and God′ are two gods, then God $\cup$ God′ $\supseteq$ God (due to Aquinas), and $\therefore$ God $\cup$ God′ = God, $\therefore$ God $\subseteq$ God′ and similarly God′ $\subseteq$ God, $\therefore$ God = God′.

Given a set $\{A_\alpha\}_{\alpha \in R}$ of sets, let the unique God pick $x_\alpha \in A_\alpha$ for each $\alpha \in A$. (He can do so by omnipotence.) Then

$$\{x_\alpha\}_{\alpha \in R} \in \prod_{\alpha \in R} A_\alpha$$

as required.

Infinite awe. "Infinity commonly inspires feelings of awe, futility, and fear" (Rudy Rucker, *Infinity and the Mind*, 1982).

Infinite points. The number of points on a line, a plane, or a 3-D space are all the same and are equal to $C = 2^{\aleph_0}$.

Infinity and divinity. "Cantor was careful to stress that despite the actual infinite nature of the universe, and the reasonableness of his conjecture that corporeal and aetherical monads were related to each other as powers equivalent to transfinite cardinals $\aleph_0$ and $\aleph_1$, this did not mean that God necessarily had to create worlds in this way" (Joseph Dauben, *Georg Cantor*, 1990).

- - - - - - -

Mathematics and God. As I said in chapter 1, in many ways, the mathematical quest to understand infinity parallels mystical attempts to understand God. Of course, there are also many *differences* between mathematics and religion. For example, while various religions *differ* in their beliefs, there is remarkable *agreement* among mathematicians.

Philip Davis and Reuben Hersh, in *The Mathematical Experience*, suggest that "all religions are equal because all are incapable of verification or justification." Similarly, certain valid branches of mathematics seem to yield contradictory or different results, and it seems that there is not always a "right" answer.

Feeding the spirit. "The single most compelling reason to explore the world of mathematics is that it is beautiful, and pondering its intriguing ideas is great fun. I'm constantly perplexed by how many people do not believe this, yet over 50,000 professional mathematicians in America practice their trade with enthusiasm and fervor. . . . To study the deep truths of number relationships feeds the spirit as surely as any of the other higher human activities of art, music, or literature" (Calvin Clawson, *Mathematical Mysteries*, 1996).

- - - - - -

Infinite surface. The Greek God Zeus hands you a gallon of red paint and asks how you would completely paint an infinite surface with this gallon of paint. What would be an excellent answer to win Zeus's admiration? What surface would you choose? (Hint: I am thinking of a shape that resembles a horn or a funnel.) (See Answer 6.10.)

- - - - - -

I like infinity. "I like infinity. I believe that infinity is just another name for mother

nature. Nature provides infinite possibilities all the time. But because we have suffered through this world of wars and woes, we sometimes fail to get this. We see the world as a little stingy at times" (Fred Wolf, *Parallel Universes*, 1990).

- - - - - -

Infinity keyboard. Imagine that an extremely agile (supernatural) being alternately presses the J key and the H key on your PC keyboard. The being presses the J key for ½ of a second, then it presses the H key for ¼ of a second, then it presses the J key for ⅛ second, the H key for ¹⁄₁₆ of a second, and so on. It turns out that this infinite series ($\frac{1}{2} + \frac{1}{4} + \frac{1}{8} + \dots$) adds up to 1. Therefore, at the end of 1 second, the keys have switched an infinite number of times because

1 second = ½ second + ¼ second + ⅛ second . . .

Note that the being has stopped typing after 1 second. His eyes glow a bright green. He is finished and smiling.

What is the last letter that was typed at the end of 1 second? You seem to have all the necessary information to determine whether the last letter is either a J or an H. What is your answer? (See Answer 6.11.)

- - - - - -

Numerical evidence. It was once conjectured that $313 \times (x^3 + y^3) = z^3$ has no positive integer solutions; however, it was discovered that the smallest counterexample has numbers with more than 1,000 digits!

- - - - - -

Lengthening horizons. "The heavens call to you, and circle about you, displaying to you their eternal splendors, and your eye gazes only to earth" (Dante, *Purgatorio*, 1319).

- -

Long journeys. "We live on a placid island of ignorance in the midst of black seas of infinity, and it is not meant that we should voyage far" (H. P. Lovecraft, *The Call of Cthulhu*, 1928).

The classic barber paradox. Suppose there is a village with only one barber. Every day, this barber shaves all men, and only these men, who do not shave themselves. But who shaves the barber? If he does shave himself, then he shouldn't shave himself. This is the paradox.

The barber paradox, attributed to Bertrand Russell, is a paradox with importance to mathematical logic and set theory. It is a paradox because such a barber can neither shave himself nor not shave himself, yet at the same time must shave himself if he doesn't.

- - - - - -

Computers and the infinite. High-speed computers have allowed investigators to check the validity of many mathematical conjectures, and much mathematical evidence now involves unimaginably large numbers. However, numerical evidence should be viewed with caution and is sometimes inadequate. For example, consider the fact that

$$\int_{2}^{x} 1/\ln t\, dt - \pi(x)$$

is positive for all values of $x \leq 10^{12}$ and probably far beyond. Here, $\pi(x)$ is the number of prime numbers less than or equal to x. This massive amount of "numerical" evidence previously led researchers astray; however, it is now known that a sign change *does* occur, but all that is currently known is that the first sign change occurs below 1.65×10^{1165}.

- - - - - -

Mathematics and sex. "What is the origin of the urge, the fascination that drives physicists, mathematicians, and presumably other scientists as well? Psychoanalysis suggests that it is sexual curiosity. You start by asking where little babies come from, one thing leads to another, and you find yourself preparing nitroglycerine or solving differential equations. This explanation is somewhat irritating, and therefore probably basically correct" (From David Ruelle, *Chance and Chaos*, 1993).

- - - - - -

The infinite window. The White Queen was admiring her reflection in a window that consisted of alternating pink and green panes, stuck together with no frames. No matter how hard Alice peered, she could see no end to the window, not to the right, not to the left, not upward.

She remarked as much.

The Queen replied, "Yes, isn't it marvelous? Every pane has another pane to its right, to its left, and above it, extending to infinity."

Alice said, "It must really be valuable. But is it not in danger?"

The Queen said, "Valuable, yes. But what danger do you mean?"

Number definition paradox. Let A be the set of all positive integers that can be defined in under 100 words. Since there are only finitely many of these, there must be a smallest positive integer n that does not belong to A. But haven't I just defined n in under 100 words? (Wording by Timothy Gowers.)

- -

Large perfect number. In 2001, $4.27764198 \times 10^{8107891} = 2^{13466916}(2^{13466917} - 1)$ was the largest known perfect number.

"Well," said Alice, "suppose some boys were playing cricket and mishit the ball, and it flew towards the window?"

"It would not matter," said the Queen, pointing to a particular pane. "Consider how likely would the ball be to hit this particular pane?"

"Not very," said Alice. "In fact, it wouldn't be possible at all. There are so many other panes."

"Very good," said the Queen. "And the same would be true of all the panes, would it not?"

Alice considered and said, reluctantly, "I suppose so."

"So," said the Queen, "if it couldn't hit any of the panes, then the magnificent structure is in no danger."

Alice nodded, threw a ball, and destroyed the wonderful structure.

(Graham Cleverley, personal communication, 2004).

- - - - - -

Strange paper title. In 1975, George Englebretsen published "Sommers' Proof That Something Exists" in the prestigious *Notre Dame Journal of Formal Logic* (16: 298–300).

The Skewes number. S_1 is one of the largest numbers that has occurred in a mathematical proof:

$$S_1 = 10^{10^{10^{34}}}$$

This number was given by Skewes in 1933 and decreased, in the 1980s, to

$$S_1 = e^{e^{27/4}} \approx 8.185 \times 10^{370}$$

A related number, called the "second Skewes number," is even larger:

$$S_2 = 10^{10^{10^{1000}}}$$

The mathematician G. H. Hardy called the Skewes number "the largest number which has ever served any definite mathematical purpose in mathematics."

Infinity in a nutshell. "I could be bounded in a nutshell and count myself a king of infinite space" (Shakespeare, *Hamlet*, 1603).

- - - - - -

Universe game. G. H. Hardy determined that if one "played chess" with all the particles in the universe (which he estimated to be 10^{87}), where a move meant simply interchanging any two particles, then the number of possible games was roughly equal to Skewes's original number:

$$10^{10^{10000000000000000000000000000000000000}}$$

- - - - - -

Göbel's sequence is defined by $x_n = \frac{1 + x_0^2 + x_1^2 + \ldots + x_{n-1}^2}{n}$ with $x_0 = 1$. The first few values of x_n are 1, 2, 3, 5, 10, and 28. This baby grows large very quickly. Mathematicians are still exploring the various intriguing properties of this recursive sequence. One thing we do know is that the first non-integer value of x_n occurs at x_{43}, which is about equal to

$$x_{43} = 5.4093 \times 10^{178485291567}$$

This "Göbel number," x_{43}, is so large that humanity will *never* be able to compute all of its digits. If this little bit of information has intrigued you, you can learn more by reading R. K. Guy, "A Recursion of Göbel," in *Unsolved Problems in Number Theory*, 2nd ed. (New York: Springer-Verlag, 1994), pp. 214–15.

Mystery of patterns. "It was all a pattern, as surely as a spiderweb is a pattern, but a pattern does not imply a purpose. Patterns exist everywhere, and purpose is at its safest when it is spontaneous and short lived" (Anne Rice, *The Witching Hour*, 1993).

- - - - - -

Kinky number. The author Calvin Clawson, in his book *Mathematical Mysteries* (p. 37), reports that the number of kinks in the core of an "embedded tower" is roughly

$$E = 10^{10^{10^{10^{10^7}}}}$$

- - - - - -

Twain math. "I had been to school most of the time, and could spell, and read, and write just a little, and could say the multiplication table up to six times seven is thirty-five, and I don't reckon I could ever get any further than that if I was to live forever. I don't take no stock in mathematics, anyway" (Huck Finn, in Mark Twain, *The Adventures of Huckleberry Finn*, 1912).

- - - - - -

Infinite gift. You are visiting your neighbor down the street, who bears a striking resemblance to the actress Julia Roberts. It is your birthday, and she has a special gift for you. On her couch is an attractive assembly of attached boxes with ever-diminishing sides. The largest box is wrapped with a sparkling red ribbon (figure 6.1).

The smallest boxes are so tiny that you would need a microscope to see them. The sides of the boxes diminish in an interesting sequence:

$$1, \frac{1}{\sqrt{2}}, \frac{1}{\sqrt{3}}, \frac{1}{\sqrt{4}}, \dots \frac{1}{\sqrt{n}} \dots$$

She smiles as she touches the largest box at the top of her stack, which has an edge that is 1 foot in length. The next box has an edge length of 1 over the square root of 2, and the next box has an edge length of 1 over the square root of 3, and so forth. This series diverges, or gets bigger and bigger, which means that your friend's gift is a structure of infinite length! If you wanted to paint the faces of the gift, you would need an infinite supply of paint.

Remarkably, even though the length is infinite, the volume of the gift is finite! What is the volume? (See Answer 6.12.)

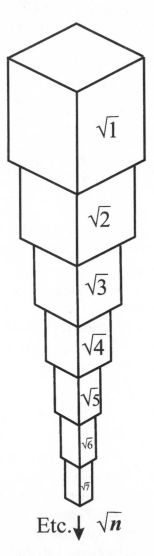

Figure 6.1   Infinite gift.

Mystery of infinity. "Infinity is where things happen that don't" (S. Knight).

- - - - - -

Large number contest. Construct the expression for the largest number possible, using only the following eight symbols:

$$1 \ 2 \ 3 \ 4 \ (\ ) \ . \ -$$

Each digit can be used only once. The "." is a decimal point. The "–" is a minus sign. If you like, you can raise numbers to powers of one another. For example 31^{42} has an amazing 63 digits! Can you do better than this?

For the second part of the contest, construct the expression for the largest number possible, using only the following six symbols:

$$1 \ 0 \ (\ ) \ . \ -$$

Each digit can be used five times, at most. (See Answer 6.13.)

- - - - - -

Leviathan number. The first six digits of the Leviathan number $\aleph = (10^{666})!$ are 134,072. (The ! is the symbol for factorial.) Alas, the number is so large that humanity will never be able to compute all the digits of the Leviathan number.

Monster number. One of the largest individual numbers that occurs naturally in a theorem is

8080 17424 79451 28758 86459 90496
17107 57005 75436 80000 0000

This is the order of the so-called *Monster simple group*. An example of a finite group is a collection of integers from 1 to 12 under the operation of "clock arithmetic," so that, for instance, $9 + 6 = 3$. The concept sounds simple, but it gives rise to a mathematical jungle. For decades, mathematicians have tried to classify all of the finite groups. One of the strangest groups discovered is the "monster group," which has over 10^{53} elements and a little-understood structure. Robert Griess constructed this beast in 1982 as a group of rotations in 196,883-dimensional space.

Eternity. "High up in the North in the land called Svithjod, there stands a rock. It is a hundred miles high and a hundred miles wide. Once every thousand years a little bird comes to the rock to sharpen its beak. When the rock has thus been worn away, then a single day of eternity will have gone by" (Hendrik Willem Van Loon, *The Story of Mankind*, 1922).

- - - - - -

Distance. "I looked round the trees. The thin net of reality. These trees, this sun. I was infinitely far from home. The profoundest distances are never geographical" (John Fowles, *The Magus*, 1965).

- - - - - -

Buddhist asankhyeya. One of the highest numbers for which there exists a name is the Buddhist *asankhyeya*, which is equal to 10^{140}— much bigger than a googol. This number is mentioned in Jainist writings of the first century B.C. India's ancient writings often contain references to large numbers with names; however, it is sometimes difficult to assign a precise value to these humongous numbers because of conflicting or ambiguous uses.

— — — — — —

🛈 **Great aeon.** According to Bhikkhu Bodhi's *A Comprehensive Manual of Abhidhamma*, Buddhist philosophy identifies three kinds of aeons—an interim aeon, an incalculable aeon, and a great aeon. An *interim aeon* (*antarakappa*) is a period of time required for the life span of humans to rise from ten years to the maximum of many thousands of years, and then fall back to ten years. Twenty such interim aeons equal a single *incalculable aeon* (*asankhyeyakappa*), and four incalculable aeons constitute a single *great aeon* (*mahakappa*). Buddha says that the length of a great aeon is longer than the time it would take for a man to wear away a mountain of solid granite that is one *yojana* (about 7 miles) high and wide by stroking it once every hundred years with a silk cloth.

In Daniel Boorstin's *The Creators*, Gautama reached full enlightenment in the course of three incalculable aeons. In the first, he did not know whether he would become a Buddha. In the second aeon, he knew it but did not reveal it. In the third, he declared it. One Buddhist scholar has calculated the length of one incalculable aeon as 1 followed by 352 septillion kilometers of zeros, with each zero being 0.001 mm in width. A septillion is 10 to the power of 24.

🛈 **Finite and infinite.** "There was from the very beginning no need for a struggle between the finite and infinite. The peace we are so eagerly seeking has been there all the time" (Daisetz Teitaro Suzuki, *Introduction to Zen Buddhism*, 1957).

— — — — — —

🛈 **The game of Go.** There are approximately 4.63×10^{170} possible positions in the Asian game of Go, which is played with black and white stones on a 19×19 board. (M. Beeler, et al., Item 96, in M. Beeler, R. W. Gosper, and R. Schroeppel, "HAKMEM," Cambridge, Mass.: MIT Artificial Intelligence Laboratory, Memo AIM-239, p. 35, February 1972.) Opponents play Go by alternately placing stones on the board to surround as much territory as possible. For comparison, there are about 10^{12} positions in checkers.

— — — — — —

🛈 **A slave to limit.** "The will is infinite and the execution confined. The desire is boundless and the act a slave to limit" (Shakespeare, *Troilus and Cressida*, 1602).

🛈 **An infinity of worlds.** "It is known that there is an infinite number of worlds, simply because there is an infinite amount of space for them to be in. However, not every one of them is inhabited. Therefore, there must be a finite number of inhabited worlds. Any finite number divided by infinity is as near to nothing as makes no odds, so the average population of all planets in the Universe can be said to be zero. From this it follows that the population of the whole Universe is also zero, and that any people you may meet from time to time are merely the products of a deranged imagination" (Douglas Adams, *The Restaurant at the End of the Universe*, 1982).

❶ Archimedes' "Cattle Problem." The solution to the restricted version of Archimedes' famous "Cattle Problem" is 7.760271406 486818269530232833213 . . . $\times 10^{202544}$. The problem can be stated as follows:

O stranger, compute the number of cattle of the Sun, who once upon a time grazed on the fields of the Thrinacian isle of Sicily, divided into four herds of different colors—one milk white, another glossy black, the third yellow, and the fourth dappled. The number of white bulls was equal to ($\frac{1}{2} + \frac{1}{3}$) the number of black bulls plus the total number of yellow bulls. The number of black bulls was ($\frac{1}{4} + \frac{1}{5}$) the number of dappled bulls plus the total number of yellow bulls. The number of spotted bulls was ($\frac{1}{6} + \frac{1}{7}$) the number of white bulls, plus the total number of yellow bulls. The number of white cows was ($\frac{1}{3} + \frac{1}{4}$) the total number of the black herd. The number of black cows was ($\frac{1}{4} + \frac{1}{5}$) the total number of the dappled herd. The number of dappled cows was ($\frac{1}{5} + \frac{1}{6}$) the total number of the yellow herd. The number of yellow cows was ($\frac{1}{6} + \frac{1}{7}$) the total number of the white herd.

If you can accurately tell, O stranger, the total number of cattle of the Sun, including the number of cows and bulls in each color, you would not be called unskilled or ignorant of numbers, but not yet shalt thou be numbered among the wise. But understand also these conditions: The white bulls could graze together with the black bulls in rows, such that the number of cattle in each row was equal and that number was equal to the total number of rows, thus forming a perfect square. And the yellow bulls could graze together with the dappled bulls, with a single bull in the first row, two in the second row, and continuing steadily to complete a perfect triangle. If you are able, o stranger, to find out all these things and gather them together in your mind, giving all the relations, you shall depart crowned with glory and knowing that you have been adjudged perfect in this species of wisdom.

If we attempt to solve the first part of the problem, the smallest solution for the total number of cattle is 50,389,082. But if you add the additional two constraints contained in the second part, the solution is much higher— about 7.76×10^{202544}. It took until 1880 to find this approximate answer. The actual number was first calculated in 1965 by Williams, German, and Zarnke, using an IBM 7040.

⟳ Walnuts and infinity. "We are told that the shortest line segment contains an infinity of points. Then even the shell of a walnut can embrace a spatial infinity as imponderable as intergalactic space" (William Poundstone, *Labyrinths of Reason*, 1988).

⟳ Having fun. Graffiti in the math department's restroom:

I am happy because I know

$$\aleph_0 + \aleph_0 = \aleph_0$$

🄐 **A special Rubik's number:**

$$1010973622236244622911804223695320000000$$

$$= 54! / ((9!)^6) \sim 1.0109 \times 10^{38}$$

According to Robert Munafo, this is the number of combinations of the $3 \times 3 \times 3$ Rubik's Cube if you are allowed to take the stickers off and replace them in places that are different from their original locations.

- -

🄐 **Jainism.** The number of years in the longest time period in the Jainist religion is $1.0130653244 \ldots \times 10^{177} = 2^{588}$.

🄖 **Watching Lila.** "He watched her for a long time and she knew that he was watching her and he knew that she knew he was watching her, and he knew that she knew that he knew; in a kind of regression of images that you get when two mirrors face each other and the images go on and on and on in some kind of infinity" (Robert Pirsig, *Lila*, 1991).

- - - - - -

🄐 **Big number in fiction.** The largest known number described in a work of fiction is $2.74858523 \ldots \times 10^{80588} = 2^{267709}$. Douglas Adams, in *The Hitchhiker's Guide to the Galaxy*, uses this number when stating the odds against his characters being rescued by a passing spaceship. According to Robert Munafo, this number is mentioned in the radio program version and was changed to 2^{260199} for the book.

- - - - - -

🄖 **Brain chimeras.** "The knowledge we have of mathematical truths is not only certain, but *real* knowledge; and not the bare empty vision of vain, insignificant chimeras of the brain" (John Locke, *An Essay Concerning Human Understanding*, 1690).

- - - - - -

🄐 **Erdös-Moser.** The *Erdös-Moser number* is huge, a real treat, and sure to excite your friends at your lunch table. Paul Erdös conjectured that there is no solution (other than $1^1 + 2^1 = 3^1$) to the following equation sum involving integers:

$$\sum_{j=1}^{m-1} j^n = m^n$$

Mathematicians still do not know whether this conjecture is true or not. All we know is that there is no solution for

$$m < 1.485 \times 10^{9321155}$$

I call this value for m the *Erdös-Moser number*, after early researchers of this problem, although William Butske and colleagues actually determined this number in 1999.

You can learn more about this large number here: William Butske, Lynda M. Jaje, and Daniel R. Mayernik, "On the Equation $\Sigma_{p|N} \, 1/p + 1/N = 1$, Pseudoperfect Numbers, and Partially Weighted Graphs," *Mathematics of Computation* 69 (2000): 407–20; and Leo Moser, "On the Diophantine Equation $1^n + 2^n + 3^n + \ldots + (m - 1)^n = m^n$," *Scripta Mathematica* 19 (1953): 84–88.

- - - - - -

Tegmark number. According to the astrophysicist Max Tegmark, the number of meters you must travel to find an exact copy of yourself, assuming that the universe is homogeneous and infinite, is $10^{10^{29}}$.

Chess. Are any of you fans of chess? If so, $1.15 \ldots \times 10^{42} = 64! / (32! \times 8!^2 \times 2!^4 \times 2^4)$ is the number of possible chess positions, based on a 1950 article by Shannon (Claude Shannon, "Programming a Computer for Playing Chess," *Philosophical Magazine* 41 [1950]: 256–75). Theoretically speaking, you can arrange all 32 pieces in any position whatsoever (giving 64!/32!). However, this number is reduced somewhat because pawns of a given color are equivalent (8! for each color). Other interchanges also serve to reduce this number. Note that this number is only an estimate because it does not take into account the fact that a pawn cannot switch columns or move past the opposing pawn in its column unless it captures that pawn. Also, the number does not take possible pawn promotion into consideration.

G. H. Hardy has estimated that the number of chess games is much higher than the number of chess positions:

$$\Lambda = 10^{10^{50}}$$

Superfactorial. In 1987, A. Berezin published "Super Super Large Numbers" in the *Journal of Recreational Math* (19, no. 2: 142–43). This paper discusses the mathematical and philosophical implications of the "superfactorial" function defined by the symbol \$, where $N\$ = N!^{N!^{N!}} \ldots$ The term $N!$ is repeated $N!$ times. In other words, there are $N!$ repetitions of $N!$ on the right-hand side. The number grows rather quickly. The first two values are 1 and 4, but subsequently grow so rapidly that 3\$ already has a very large number of digits. In particular, the third term, 3 superfactorial, is

$$3\$ = 3!^{3!^{3!^{3!^{3!^{3!}}}}} = 6^{6^{6^{6^{6^{6}}}}} = 10^{10^{10^{2.0691973765 \times 10^{36305}}}}$$

It has way too many digits to compute exactly.

- - - - - - - - - - - - - - - - - - -

Zeroless powers. According to Michael Beeler and William Gosper, there is at least one zero in the decimal expression of each power of 2 between

$$2^{86} = 77,371,252,455,336,267,181,195,264 \text{ and } 2^{30739014}$$

Notice how 2^{86} itself is zeroless. Their computer program was stopped at $2^{30739014}$. If digits of such powers were random, $1/10^{411816}$ is the approximate probability that the next number, $2^{30739015}$, has no zeros. According to Schroeppel, a power of 2 can have arbitrarily many nonzero digits.

10^{500} environments. "On the theoretical side, an outgrowth of inflationary theory called eternal inflation is demanding that the world be a megaverse full of pocket universes that have bubbled up out of inflating space like bubbles in an uncorked bottle of Champagne. At the same time string theory, our best hope for a unified theory, is producing a landscape of enormous proportions. The best estimates of theorists are that 10^{500} distinct kinds of environments are possible" (Leonard Susskind, "The Landscape," an interview with John Brockman at Edge.org, 2003).

– – – – – –

🌀 **Transionic ϑ.** Transions are transfinite numbers with probabilistic values. For example, if there is some probability that space is quantized as integer multiples of a certain minimum distance and that our universe is infinite, we may represent the cardinality of points in such a universe as $\aleph_0$; however, if space is not quantized and infinitely divisible, the cardinality of universal points might be represented as $C = 2^{\aleph_0}$. As my colleague Graham Cleverley notes, if we allocate a 95 percent probability to the first hypothesis and a 5 percent probability to the second, then we obtain a transionic cardinality for the set of universal points. Thus,

the universe's shimmering transionic value can be denoted $\vartheta_5^{95} = \aleph_0 \,|\, 2^{\aleph_0}$.

Note that the idea that space, and even time, comes in discrete lumps is not far-fetched. Loop quantum gravity predicts that space comes in tiny fragments, the smallest of which is about a cubic Planck length, or 10^{-99} cubic centimeter. Time flows in discrete ticks of about a Planck time, or 10^{-43} second.

Of course, the notion of transions need not be confined to probabilities in cosmology or quantum mechanics but has general application when probabilities can be assigned to the cardinality of a set or the elements that compose a set.

🌀 **Dream of mathematics.** "Science in its everyday practice is much closer to art than to philosophy. When I look at Gödel's proof of his undecidability theorem, I do not see a philosophical argument. The proof is a soaring piece of architecture, as unique and as lovely as a Chartres cathedral. The proof destroyed Hilbert's dream of reducing all mathematics to a few equations, and replaced it with a greater dream of mathematics as an endlessly growing realm of ideas" (Freeman Dyson, in the introduction to *Nature's Imagination*, 1995).

– – – – – –

🌀 **The Paradox of Tristram Shandy.** Bertrand Russell once formulated a paradox in which Tristram Shandy requires a year to write about the first day of his life and laments that at this rate, material would accumulate faster than he could deal with it, so that he could never come to an end of his biography. However, Russell notes that if Shandy had lived forever and had not wearied of his task, then no part of Shandy's biography would have remained unwritten because there is a one-to-one correspondence between each year that Shandy writes in and each day he writes about. Therefore, no matter what day of his life you care to consider, there will eventually come a year in which he will be able to write about it. There is no part of his life that can never be written down. Nevertheless, Shandy gets further and further behind in his task!

More mathematics and reality. "Our brains evolved so that we could survive out there in the jungle. Why in the world should a brain develop for the purpose of being at all good at grasping the true underlying nature of reality?" (Brian Greene, in Susan Kruglinski's "When Even Mathematicians Don't Understand the Math," *New York Times*, May 25, 2004).

- - - - - -

Bubbles in a black sea. "Consider the true picture. Think of myriads of tiny bubbles, very sparsely scattered, rising through a vast black sea. We rule some of the bubbles. Of the waters we know nothing . . . " (Larry Niven and Jerry Pournelle, *The Mote in God's Eye*, 1974).

Mathematics
and Beauty

IN WHICH WE EXPLORE ARTISTIC FORMS GENERATED FROM MATHEMATICS—
delicate fungi, twenty-second-century cityscapes, fractal necklaces and seed
pods, alien devices, and a rich panoply of patterns that exhibit a cascade of
detail with increasing magnifications.

> Beauty is eternity gazing at itself in a mirror.
> —Kahlil Gibran, *The Prophet*, 1923

The line between science, mathematics, and art is a fuzzy one; the three are fraternal philosophies formalized by ancient Greeks such as Eratosthenes and Ictinus. Computer graphics helps reunite these philosophies by providing convenient ways to represent natural, mathematical, and artistic objects. In the short art gallery that follows, simple mathematical formulas or sets of rules are used to generate a surprising variety of beautiful and unpredictable patterns.

Mathematicians and scientists have begun to enjoy and present bizarre mathematical patterns in new ways—ways motivated as much by a sense of aesthetics as by the needs of logic. Moreover, computer graphics allows nonmathematicians to experience a little of the delight that mathematicians take in their work and to better appreciate the complicated behavior of simple formulas.

Most of the figures that follow are fractals. Fractals usually exhibit self-similarity, which means that various copies of an object can be found in the original object at smaller-sized scales. The detail continues for many magnifications like an endless nesting of Russian dolls. Some of the shapes remind me of intricate flowers, futuristic cityscapes, fractal jewelry, or alien devices that will forever remain beyond our understanding. As Francis Bacon said, "The job of the artist is always to deepen the mystery."

Figure 7.1 Fractal object by Abram Hindle is formed from a simple set of mathematical rules. (See chriscoyne.com/cfdg/ for details.)

Figure 7.2   Pop-art flower produced by Chris Coyne using a simple set of mathematical rules. Artists and computer programmers can create a magnificent landscape of new shapes and forms using what Chris Coyne calls "context-free design grammar." The grammar includes rules for growth and various weighting factors. (See chriscoyne.com/cfdg/ for details.)

Figure 7.3   Magnification of a region of the previous figure, highlighting a potentially
endless cascade of detail. (See chriscoyne.com/cfdg/ for details.)

Figure 7.4   Tendrilous, botanical artwork produced by Chris Coyne using a simple
set of mathematical rules. (See chriscoyne.com/cfdg/ for details.)

Figure 7.5   Magnification of a portion of the previous figure. Image processing methods were applied to create a slight three-dimensional effect. (See chriscoyne.com/cfdg/ for details.)

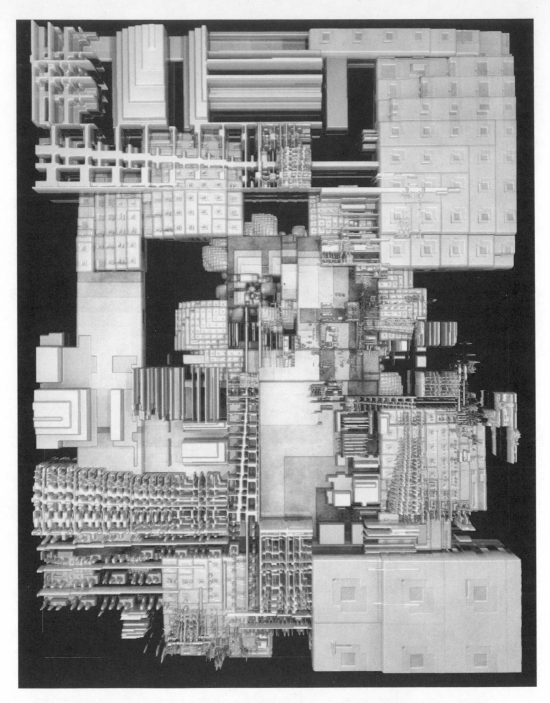

Figure 7.6   Futurist fractal device and city by Jock Cooper, who writes, "I use mathematics to create objects that inspire the imagination. I want the viewer to wonder what the results represent—some alien piece of art, or technology from another dimension." (See fractal-recursions.com for details.)

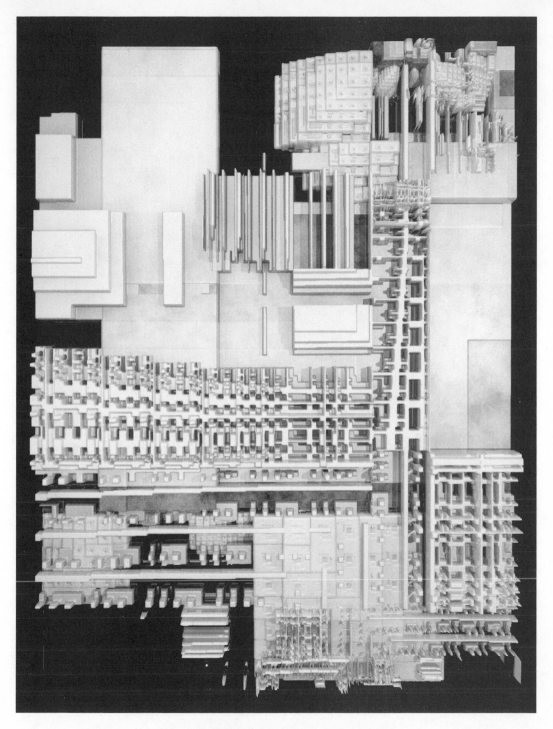

Figure 7.7   Magnification of a region of figure 7.6.
(See fractal-recursions.com for details.)

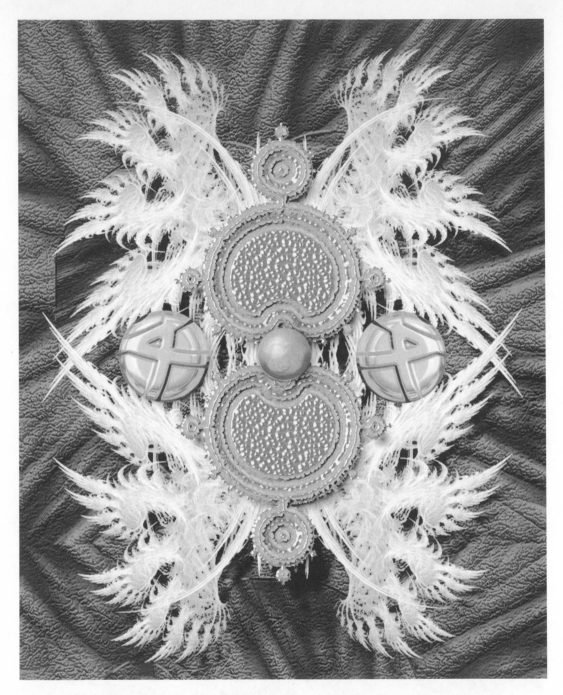

Figure 7.8   Fractal collage by Linda Bucklin, an expert at using various fractal-generating and image-processing tools to produce her works. She assembles her fractal collages using software that includes Ultra Fractal, KPT FraxFlame, and Apophysis. Adobe Photoshop, Corel Photopaint, and other tools are used to assemble the final work. (See lindabucklin.com for details.)

Figure 7.9   "Infocity" by Sally Hunter, who writes, "As I created the object using mathematics and various software tools, I imagined an information storage device and contemplated the fractal nature of information—the more you find out, the more questions you have." (See sallyhunter.co.uk for details.)

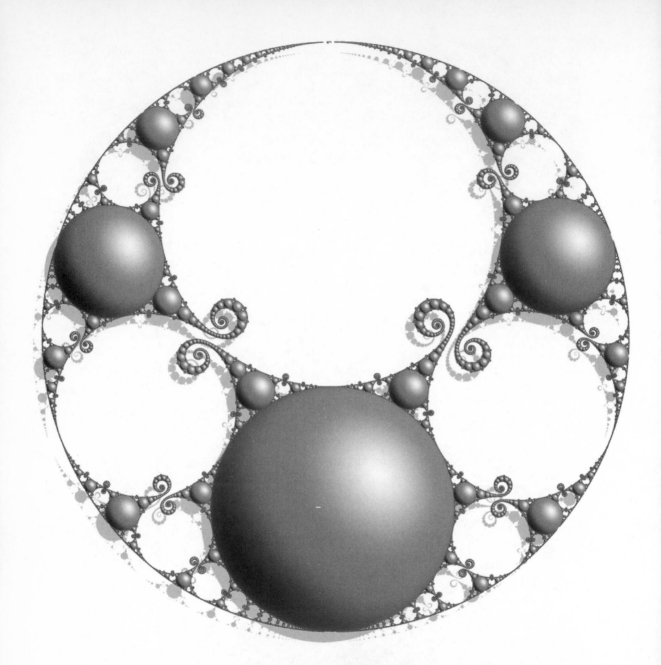

Figure 7.10   A Kleinian group image, a limit set generated by Möbius transformations of the form $z \rightarrow (ab + z)/(cz + d)$. More particularly, this fractal image by Jos Leys was generated with two Möbius transformations and their inverse transformations. This iterative process will repeatedly displace an initial point in the complex plane. The resultant set of points forms the limit set, represented graphically in this figure. No matter how often and in what order the displacements are repeated, the new points fall somewhere on the figure's curved shapes. Möbius transformations will transform circles to circles, and this property yields the spherelike objects in the image. (See josleys.com for details.)

Figure 7.11   Magnification of a region of figure 7.10. (See josleys.com for details.)

Figure 7.12   "Martian Missile Defense," a fractal created by Roger A. Johnston using the Apophysis software tool. (See community.webshots.com/user/rajahh for details.)

Figure 7.13   Halley map created by the author reveals the intricate behavior of a root-finding method. The problem of finding zeros of a function by iterative methods occurs frequently in science and engineering. These approximation techniques start with a guess and successively improve upon it with repetitions of similar steps. The resultant graphics give an indication of how well one of these iterative methods, Halley's method, works—where it can be relied upon and where it behaves strangely. (See pickover.com for details.)

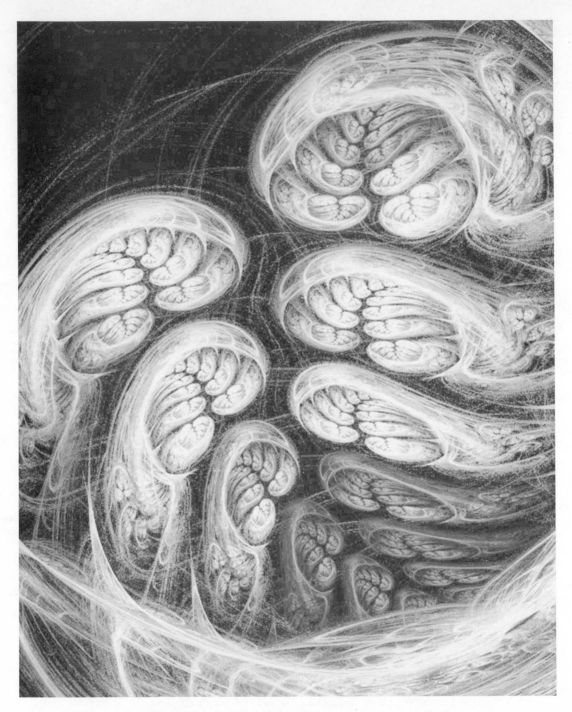

Figure 7.14 Fractal "Seed Pods," created by Cory and Catska Ench using an "iterated functions system" that consists of affine transformations involving rotations, scalings by a constant ratio, and translations. (See enchgallery.com for details.)

Answers

I. Numbers, History, Society, and People

I.I Ancient counting. Among the oldest direct evidence of human counting is a baboon's thigh bone marked with 29 notches. The bone is 35,000 years old and was discovered in the Lebombo Mountains of Africa.

To put this in perspective, the oldest fossils of modern humans are nearly 150,000 years old. The Lebombo bone resembles the calendar sticks still used today by Bushmen clans in Namibia.

I.2 Mathematics and reality. In my mind, we don't invent mathematics and numbers, but rather we discover them. In chapter 2, when we discuss exotic kinds of numbers like transcendental numbers, I have the feeling that they are out there in the realm of eternal ideas. They have an independent existence from us. These ideas are controversial, and there are certainly other points of view. However, to me, mathematics and numbers transcend us and our physical reality. The statement "3 + 1 = 8" is either true or false. It's false. Was the statement false before the discovery of integers? I believe it was. Numbers and mathematics exist whether humans know about them or not. In *Are Universes Thicker Than Blackberries?* Martin Gardner stated this as, "If two dinosaurs joined two other dinosaurs in a clearing, there would be four there, even though no humans were around to observe it, and the beasts were too stupid to know it." G. H. Hardy, in his famous *Apology*, wrote, "I believe that mathematical reality lies outside us, that our function is to discover and *observe* it, and that the theorems which we prove, and which we describe grandiloquently as our 'creations,' are simply our notes of our observations." The nineteenth-century mathematician Leopold Kronecker said, "God created the integers—all else is human invention!"

I think that mathematics is a process of discovery. Mathematicians are like archaeologists. The physicist Roger Penrose felt the same way about fractal geometry. In his book *The Emperor's New Mind*, he says that fractals (for example, intricate patterns such as the Julia set or the Mandelbrot set) are out there waiting to be found:

> It would seem that the Mandelbrot set is not just part of our minds, but it has a reality of its own. . . . The computer is being used in essentially the same way that an experimental physicist uses a piece of

experimental apparatus to explore the structure of the physical world. The Mandelbrot set is not an invention of the human mind: it was a discovery. Like Mount Everest, the Mandelbrot set is just there.

I think we are uncovering truths and ideas independently of the computer or mathematical tools we've invented.

Penrose went a step further about fractals in *The Emperor's New Mind*: "When one sees a mathematical truth, one's consciousness breaks through into this world of ideas. . . . One may take the view that in such cases the mathematicians have stumbled upon works of God."

Anthony Tromba, the coauthor of *Vector Calculus*, said in a July 2003 University of California press release, "When you discover mathematical structures that you believe correspond to the world around you, you feel you are seeing something mystical, something profound. You are communicating with the universe, seeing beautiful and deep structures and patterns that no one without your training can see. The mathematics is there, it's leading you, and you are discovering it" (www.ucsc. edu/news_events/press_releases).

James Gleick, in *Chaos: Making a New Science*, writes, "The Mandelbrot set . . . exists. It existed before . . . Hubbard and Daudady understood its mathematical essence, even before Mandelbrot discovered it. It existed as soon as science created a context—a framework of complex numbers and a notion of iterated functions. Then it waited to be unveiled."

Other mathematicians disagree with my philosophy and believe that mathematics is a marvelous invention of the human mind. One reviewer of my book *The Zen of Magic Squares* used poetry as an analogy when "objecting" to my philosophy. He wrote,

> Did Shakespeare "discover" his sonnets? Surely all finite sequences of English words "exist," and Shakespeare simply chose a few that he liked. I think most people would find the argument incorrect and hold that Shakespeare *created* his sonnets. In the same way, mathematicians create their concepts, theorems, and proofs. Just as not all sequences of words are sonnets, not all grammatical sentences are theorems. But theorems are human creations no less than sonnets.

Similarly, the molecular neurobiologist Jean-Pierre Changeux believes that mathematics is invented: "For me [mathematical axioms] are expressions of cognitive facilities, which themselves are a function of certain facilities connected with human language."

Also, I should point out that the development of "higher" math skills is not inevitable as a culture matures or evolves. In fact, higher math, unlike counting and adding, is extremely rare. John Barrow, in *Pi in the Sky: Counting, Thinking, and Being*, suggests,

> Having a notion of quantity is a long way from the intricate abstract reasoning that today goes by the name of mathematics. Thousands of years passed in the ancient world with comparatively little progress in mathematics. . . . It is not good enough to possess the notion of quantity. One must develop an efficient method of recording numbers . . . more crucially still, the adoption of a place value system with

a symbol for zero was a watershed. A good notation permits an efficient extension to the ideas of fractions and the operations of multiplication and division. . . . Again, we find these discoveries are deep and difficult; almost no one made them (pp. 103–104).

My colleague Chuck G. comes down on the side of mathematics as a creation, not a discovery:

A block of stone contains every possible statue that can be carved from it. When a sculptor selects one of these statues, it's said to be an act of creativity. If mathematics already exists, then so do all possible mathematics, including an infinity of incorrect, worthless, boring, irrelevant, useless, ridiculous, and incomprehensible mathematics. Shouldn't the finding of worthwhile mathematics be given the same consideration as finding a work of art in a stone and be called creative? If not, then isn't every human endeavor just blundering around finding things?

My colleague Graham C. suggests that the statement $3 + 1 = 8$ is neither true nor false until we define what we mean by it:

Is $a + b = c$ true or false? Obviously neither. It is only given meaning in context, and it is a human mind that gives it context. Philosopher Ludwig Wittgenstein (1889–1951) would have said that you need to know the rules of the game before you play it. And someone has to define the rules. Aristotle said that a statement had to be either true or false. Much as I admire Aristotle, he was

wrong. A statement can be meaningless, and therefore neither true nor false. (Popper, Carnap, or Ayer would have held that the statement "God exists" is neither true nor false.) I interpret Martin Gardner to be saying that cardinality, not numbers, existed without a conscious mind. That is, if two dinosaurs joined two other dinosaurs in a clearing, and two chickens joined two other chickens in the same clearing, each dinosaur would have been able to eat one chicken. There was a one-to-one mapping. So, I think cardinality existed and was discovered. Numbers, on the other hand, were invented as a method for describing and manipulating cardinality. Cardinality is in the real world, number in the mind. This is why some tribes can get by with only a few words for numbers. They observe the cardinality, but cannot describe it accurately. In fact, pigeons can observe the cardinality but cannot describe it at all because they haven't invented number. As an analogy, mass exists out there in the real world. Before the rise of humans, the mass of the Earth was much the same as it is now. However, we couldn't measure it until someone invented units of mass. Without units of mass, we can tell if one thing is more massive than another, but we cannot, for instance, say how much more massive.

Finally, my colleague Pete B. responds, "The words 'elephants have trunks' do not create elephants with swinging noses; likewise, the words 'one plus one equals two' do not create the mathematical relationship. However, in both instances, nature has created what we use language to describe."

I.3 Mathematics and murder. André Bloch (1893–1948). After a family meal, he murdered his brother, his uncle, and his aunt. He was confined for life in a psychiatric hospital, where he wrote breakthrough papers on a large range of topics: function theory, geometry, number theory, algebraic equations, and kinematics. The Académie des Sciences awarded him its distinguished Becquerel Prize just before his death.

When asked why he'd committed the gruesome murders, he replied, "There had been mental illness in my family." He saw it as his duty to eliminate the madness.

I.4 Leaving mathematics and approaching God. The French geometer Blaise Pascal (1623–1662). Pascal and Pierre de Fermat independently founded probability theory. Pascal also invented the first calculating machine, studied conic sections, and produced important theorems in projective geometry. His father, a mathematician, was responsible for his education.

Pascal was not allowed to begin learning a subject until his father thought he could easily master it. As a result, the eleven-year-old boy worked out for himself, in secret, the first twenty-three propositions of Euclid. At sixteen, he published essays on conics that Descartes refused to believe were the handiwork of a teenager.

In 1654, Blaise Pascal decided that religion was more worthy of his intense dedication than was mathematics. You can read more about him later in this chapter.

I.5 Counting and the mind. Seven. In 1949, Kaufman, Lord, Reese, and Volkmann flashed random patterns of dots on a screen. When subjects looked at patterns containing up to five or six dots, the subjects made no errors. The performance on these small numbers of dots was so different from the subjects' performance with more dots that the observation methods were given special names. Below seven, the subjects were said to *subitize*; above seven, they were said to *estimate*. For more information, see E. L. Kaufman, M. W. Lord, T. W. Reese, and J. Volkmann, "The Discrimination of Visual Number," *American Journal of Psychology* 62 (1949): 498–525. Also see George Miller, "The Magical Number Seven, Plus or Minus Two: Some Limits on Our Capacity for Processing Information," *The Psychological Review* 63 (1956): 81–97.

I.6 Circles. Sometime around 2400 B.C., the ancient Sumerians noticed the apparent circular track of the Sun's annual path across the sky and knew that it took about 360 days to complete the journey. Thus, it was reasonable for them to divide the circular path into 360 degrees to track the Sun's daily movement. This eventually led to our modern 360-degree circle.

I wonder whether modern scientists, with their metric systems, have considered replacing the ancient 360-degree circle with a 100-degree circle. In some sense, 360 degrees may be more useful than 100 degrees, simply because 360 has so many factors that provide a larger number of easily definable units: 2, 3, 4, 5, 6, 8, 9, 10, 12, 15, 18, 20, 24, 30, 36, 40, 45, 60, 72, 90, 120, and 180. Of course, for real metric aficionados, there's always the *grad*, which is defined such that there are 100 grads in a right angle. Thus, 1 degree equals 100/90 grads, and 400 grads correspond to a complete revolution around the circle. In the 1800s, the grad unit was introduced in France, where it is called the grade.

1.7 The world's most forgettable license plate?

My friend, the physicist Paul Moskowitz, just bought a new car and chose a particularly difficult license plate to memorize, illustrated in figure A1.1.

Figure A1.1 The physicist Paul Moskowitz's difficult-to-remember car license plate.

If a bank robber were attempting to generate a license for his car, he might consider only numbers that are easily confused with one another. For instance, an 8 can be mistaken for a 0 or a 3 at a distance, and some people might confuse 2 and 5. If he can use letters, this approach works even better because other similar-looking symbols can be used, such as 0 and D, and 8 and B. Thus, a plate like 103BI8I1 may be hard to remember and to recognize. Also, any sequence of letters that would be hard to pronounce would be useful.

Some scientists I consulted suggested that they would use the concept of Shannon entropy for assessing disorder in a license plate sequence, although I don't think most bank robbers would use this approach. Shannon entropy is beyond the scope of this book, but it quantifies the ease with which we can compress (or simply represent) a string of characters. Shannon entropy is an important measure that is used to evaluate structures and patterns in our data. The lower the entropy, the more structure that exists in the string.

Other colleagues suggested that anything one could do to break the symmetry of the digit string would help to make the plate less memorizable. For example, a symmetrical plate such as 11000011 would be too easy to memorize.

My favorite mathematical friends have approached the problem of selecting difficult-to-remember plates by focusing on plates that consist of only 1s and 0s. Obviously, humans are good at recognizing patterns in numbers. Plates with 11111111 and 00000000 would stand out. Likewise 10000000 and 00000001 would draw attention.

My brainy colleague Dr. Joseph Pe has come up with a mathematical measure of the memorability $M(n)$ of a binary license plate. General readers may wish to skip reading the following mathematical exposition because it involves some terms that may be unfamiliar. For those brave souls who remain—consider a number n. $M(n)$ should be large if the information $I(n)$ conveyed by n is large; that is, the greater the information, the greater the memorableness. For example, "11111111" is more memorable than "11010010," and it contains more information by our definition. (If you think of 1s as heads and 0s as tails, it is harder to get 8 heads in 8 tosses than 4 heads and 4 tails in 8 tosses.) So, if $p(n)$ denotes the probability of getting the same number of heads and number of tails as n, then, using the binomial distribution, we find that $p(11111111) \leq p(11010010)$.

Joseph Pe asks us to consider that $I(n) = -\log p(n)$, where $p(n)$ is the probability of obtaining a string with the same number of 0s and the same number of 1s as n. So $p(11111111) = 1/2^8$, and $p(11010010) = C(8,4)(1/2^8)$. However, $M(n)$ should be large if $K(n)$ (the sum of the lengths of the substrings of n consisting of more than one consecutive 0 or more than one consecutive 1) is large. The idea is that a substring of consecutive 0s or 1s is easier to remember than a substring with mixed 0s and 1s. For example, $K(11110000) = 4 + 4 = 8$, $K(11010010) = 2 + 2 = 4$, and $K(11111111) = 8$. Let's consider the measure $M(n) = I(n) + K(n)$ and run through an actual computation. For example, consider the probability of getting four 1s and four 0s, which is $p(11110000) = p(11010010)$. This equals $C(8,4) \times 1/2^8 = 35/128$. Also, $p(11111111) = 1/2^8 = 1/256$. Hence, $M(11111111) = I(11111111) + K(11111111) = -\log(1/256) + 8 = 8 + \log(256)$; $M(11110000) = I(11110000) + K(11110000) = -\log(35/128) + 8 = 8 + \log(128/35)$; $M(11010010) = I(11010010) + K(11010010) = -\log(35/128) + 4 = 4 + \log(128/35)$. Therefore, $M(11111111) > M(11110000) > M(11010010)$, which meets the test of common sense.

To sum up, Joseph Pe proposes a final measure, $M(n) = I(n) + K(n)$. The problem now reduces to minimizing this expression over all length-8 binary strings, which can be done by computer. I'd think the answer is not unique.

My friend Daniel Dockery points out that a binary plate is rather limiting as to the number of different possible plates. For example, there are only $2^8 = 256$ different possible plates that have a mixture of two possible val-ues. On the other hand, there are 100,000,000 possible 8-digit plates using any of the possible digits, and 2,821,109,907,456 possibilities if we use A–Z and 0–9. With only 256 possibilities, it might be simple for law enforcement to identify the exact plate and car, particularly if other details like the make and the model of the vehicle are available. Dennis Gordon notes that a license plate consisting of zeros and Os would be quite difficult to remember.

I.8 Calculating π.

William Shanks (1812–1882) spent a good part of his life calculating the value of π to 707 decimal places. In fact, this feat took Shanks over fifteen years—in other words, he averaged only about one decimal digit per week! He died a happy man, thinking that he had left behind a major contribution to mathematics. To compute π, he had used the formula

$$\pi/4 = 4 \tan^{-1}(\tfrac{1}{5}) - \tan^{-1}(\tfrac{1}{239}).$$

Unfortunately for Shanks, in 1944 D. F. Ferguson calculated π and found that Shanks had made an error in the 528th place, and the digits thereafter were also incorrect.

I.9 A mathematical nomad.

Paul Erdös (1913–1996). During the last year of his life, at age eighty-three, he continued to churn out theorems and deliver lectures, defying conventional wisdom that mathematics was a young person's sport. On this subject, Erdös once said, "The first sign of senility is when a man forgets his theorems. The second sign is when he forgets to zip up. The third sign is when he forgets to zip down."

Paul Hoffman, the author of *The Man Who*

Loved Only Numbers, notes that "Erdös thought about more problems than any other mathematician in history and could recite the details of some 1,500 papers he had written. Fortified by coffee, Erdös did mathematics 19 hours a day, and when friends urged him to slow down, he always had the same response: 'There'll be plenty of time to rest in the grave.'" Erdös traveled constantly and lived out of a plastic bag, focusing totally on mathematics at the expense of companionship, sex, and food.

Erdös made an early mark on mathematics at age eighteen, when he discovered an elegant proof of the theorem that for each integer n greater than 1, there is always a prime number between n and double the number, $2n$. For example, the prime number 3 lies between 2 and 4.

1.10 Mirror phobia.

The British mathematician G. H. Hardy (1877–1947). He hated mirrors to such an extent that he covered mirrors in any hotel rooms that he entered. He also intensely disliked having his photograph taken and being touched by others.

Hardy's *Course of Pure Mathematics*, published in 1908, was one of the first rigorous English treatises on numbers, functions, and limits, and it transformed college mathematics education. He is most famous for his collaborations with the mathematicians Littlewood and Ramanujan, and he contributed much to the fields of Diophantine analysis, summation of series, and the distribution of primes.

Hardy believed that the most profound and beautiful mathematics was actually the least useful. In his view, mathematical usefulness detracted from the beauty of mathematics.

Like many other great male mathematicians, Hardy did not seem to be very interested in women. Calvin Clawson writes in *Mathematical Mysteries*, "Whether Hardy, himself, was a homosexual is not known, but he had no meaningful relationships with women during his life except with his mother, and sister Gertrude."

1.11 Animal math.

The meaning of "counting" by animals is a highly contentious issue among animal behavior experts. However, it seems clear that animals have some sense of number. H. Kalmus, writing in *Nature* ("Animals as Mathematicians," 202 [June 20, 1964]: 1156), notes that

> There is now little doubt that some animals such as squirrels or parrots can be trained to count. . . . Counting faculties have been reported in squirrels, rats, and for pollinating insects. Some of these animals and others can distinguish numbers in otherwise similar visual patterns, while others can be trained to recognize and even to reproduce sequences of acoustic signals. A few can even be trained to tap out the numbers of elements (dots) in a visual pattern. . . . The lack of the spoken numeral and the written symbol makes many people reluctant to accept animals as mathematicians.

Rats have been shown to "count" by performing an activity the correct number of times in exchange for a reward. Chimpanzees can press numbers on a computer that match numbers of bananas in a box. Tetsuro Matsuzawa, of the Primate Research Institute at Kyoto University in Japan, taught a chimpanzee to identify numbers from 1 to 6 by pressing the appropriate key when she was shown a certain number of objects on the computer screen.

Michael Beran, at Georgia State University, trained chimps to use a computer screen and a joystick. The screen flashed a numeral and then a series of dots, and the chimps had to match the two. One chimp learned numerals 1 to 7, while another managed to count to 6. When the chimps were tested again after a gap of three years, both chimps were able to match numbers, but with double the error rate. For more information, see "Chimps Remember How to Play the Numbers Games for Years," *New Scientist* 180, no. 2421 (2003): 16.

1.12 Mystery mathematician.
His name was Diophantus, often known as the "father of algebra," who died at age 84 in the third century A.D. The famous puzzle is said to be Diophantus's epitaph, and it commemorates his work on algebra, including the study of Diophantine equations. Most of the details of Diophantus's life (including details that may be fictitious) come from the Greek *Anthology*, compiled by Metrodorus around A.D. 500. This particular puzzle in *Anthology* is said to have been written on Diophantus's tombstone. We can solve the problem as follows. Let x be the number of years he lived. Thus we have,

$$(1/6)x + (1/12)x + (1/7)x + 5 + (1/2)x + 4 = x$$

which simplifies to

$$(25/28)x + 9 = x$$
$$(25/28)x - x = -9$$
$$-(3/28)x = -9$$
$$x = 84 \text{ years}$$

We can visualize the great one's life on the number line in figure A1.2. Thus Diophantus lived 84 years. His son's age at death was $(1/2)x$ or 42 years old.

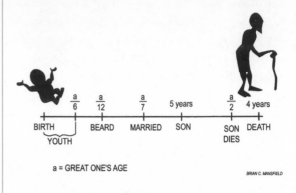

Figure A1.2   The life of Diophantus.

1.13 Mathematician starves.
Kurt Gödel (1906–1978), an eminent Austrian mathematician and perhaps the most brilliant logician of this century. The implications of his "incompleteness theorem" are vast, not only applying to mathematics, but also touching on areas such as computer science, economics, and nature. One of his closest friends was Albert Einstein when Gödel was at Princeton.

When his wife, Adele, was not with him to coax him to eat—because she was in a hospital recovering from surgery—Gödel stopped eating. He was paranoid and felt that people were trying to poison him. On December 19, 1977, he was hospitalized but refused food. He died on January 14, 1978. During his life, he had also suffered from nervous breakdowns and hypochondria.

Gödel is perhaps the most interesting example of a mathematician studying cosmic questions. Sometime in 1970, Gödel's mathematical proof of the existence of God began to circulate among his colleagues. The proof was less than a page long and caused quite a stir. You can see the proof later in this chapter.

Gödel's academic credits were impressive. For example, he was a respected mathematician and a member of the faculty of the University of Vienna starting in 1930. He also was a member of the Institute of Advanced Study in Princeton, New Jersey. He emigrated to the United States in 1940.

Gödel is most famous for his theorem, published in 1931, demonstrating that there must be true formulas in mathematics and logic that are neither provable nor disprovable, thus making mathematics essentially incomplete. Gödel's theorem had quite a sobering effect upon logicians and philosophers because it implies that within any rigidly logical mathematical system, there are propositions or questions that cannot be proved or disproved on the basis of axioms within that system, and, therefore, it is possible for basic axioms of arithmetic to give rise to contradictions. The repercussions of this fact continue to be felt and debated. Moreover, Gödel's theorem put an end to a centuries-long attempt to establish axioms that would provide a rigorous basis for all of mathematics.

Over the span of his life, Gödel kept numerous notes on his mathematical ideas. Some of his work is so complex that mathematicians believe that many more decades will be required to decipher all of it. The author Hao Wang writes on this very subject in his book *Reflections on Kurt Gödel* (1987): "The impact of Gödel's scientific ideas and philosophical speculations has been increasing, and the value of their potential implications may continue to increase. It may take *hundreds of years* for the appearance of more definite confirmations or refutations of some of his larger conjectures."

Gödel himself spoke of the need for a physical organ in our bodies to handle abstract theories. He also suggested that philosophy will evolve into an exact theory "within the next hundred years or even sooner." He even believed that humans will eventually disprove propositions such as "there is no mind separate from matter."

I.14 **Mathematician murdered.** Hypatia of Alexandria (A.D. 370–415) was martyred by being torn into shreds by a Christian mob— partly because she did not adhere to strict Christian principles. She considered herself a neo-Platonist, a pagan, and a follower of Pythagorean ideas. Interestingly, Hypatia is the first woman mathematician in the history of humanity of whom we have reasonably secure and detailed knowledge. She was said to be physically attractive and determinedly celibate. When asked why she was obsessed with mathematics and would not marry, she replied that she was wedded to the truth.

In one of her mathematical problems for her students, she asked them for the integer solution of the pair of simultaneous equations $x - y = a$, $x^2 - y^2 = (x - y) + b$, where a and b are known. Can you find any integer values for x, y, a, and b that make both of these formulas true?

The Christians were her strongest philosophical rivals, and they officially discouraged her teachings, which were Pythagorean in nature, with a religious dimension. On a warm March day in A.D. 414, a crowd of Christian zealots seized her, stripped her, and proceeded to scrape her flesh from her bones using sharp shells. They then cut up her body and burned the pieces. Like many victims of terrorism today, she may have been seized merely because she was a famous person on the other side of the religious divide. It was not until after the Renaissance that another woman, Maria Agnesi, made her name as a famous mathematician.

1.15 Going to the movies. *Billion Dollar Brain*, 1967, starring Michael Caine as secret agent Harry Palmer. An older movie that also qualifies is *The Billion Dollar Scandal*, 1932, in which three ex-cons become involved with a financier who plans to swindle the government.

1.16 Movie title. "Fifty Million Frenchmen Can't Be Wrong," sung by Sophie Tucker and performed with the Ted Lewis Band. The song was recorded in 1927. Interestingly, the population of Frenchmen was nowhere near 50 million in 1927!

Lesser known and shorter movies include *Genesis: Four Billion Years in the Making*, starring Paul Novros, 1999 (run time: 32 minutes). If one considers "infinity" a number, another classic movie is *Slave Girls from Beyond Infinity*, 2002, in which a pair of bikini-clad babes escape from a prison in outer space and land on a strange jungle planet. Also, Matthew Broderick and Patricia Arquette starred in the 2002 movie *Infinity*—a true-life romantic drama based on the life of the famous A-bomb physicist Richard Feynman.

1.17 Math and madness. Georg Cantor (1845–1918), a German mathematician who is best known as the creator of modern set theory. He is recognized by mathematicians for having extended set theory to the concept of transfinite numbers, including the cardinal and the ordinal number classes. *Cardinal numbers* are used to describe how many objects are in a collection, that is, elements in a set. *Ordinal numbers* are used to find the proper order in a set. A number in a street address is an example of an ordinal number. Whenever Cantor suffered from periods of depression, he tended to turn away from mathematics and toward philosophy.

To understand "transfinite numbers," consider a set that has a finite number of elements. For example, such a set may contain 8 turtles (elements). The number of elements in this set is called a cardinal number and is finite in this case. Cardinal numbers are numbers used to denote the size of a set. If the set is infinite, the corresponding cardinal number is called a transfinite (or infinite) cardinal number. The first transfinite number is denoted $\aleph_0$. (Aleph, or $\aleph$, is the first letter of the Hebrew alphabet.) For example, $\aleph_0$ is the cardinal number of the set of all integers. Mathematicians deal with "higher" levels of infinity, denoted by $\aleph_1$, $\aleph_2$, and so forth, which will be clarified in chapter 2.

Cantor was also fascinating because of his religious interests. Interestingly, the Jesuits used his theories of transfinite numbers to "prove" the existence of God and the Holy Trinity. Although Cantor, who was also an excellent theologian, distanced himself from such "proofs," he did equate his concept of the Absolute Infinite with God. He once wrote, "I entertain no doubts as to the truths of the transfinites, which I recognized with

God's help and which, in their diversity, I have studied for more than twenty years; every year, and almost every day brings me further in this science." He frequently suffered from bouts of depression, which forced him to become hospitalized.

Cantor believed that God ensured the existence of these transfinite numbers. Cantor regarded the transfinite numbers as leading directly to the Absolute, to the one "true infinity" that was incomprehensible within the bounds of man's understanding. Constantin Gutberlet, one of Cantor's contemporaries, worried that Cantor's work with mathematical infinity challenged the unique, "absolute infinity" of God's existence. However, Cantor assured Gutberlet that instead of diminishing the extent of God's dominion, the transfinite numbers actually made it greater. After talking to Gutberlet, Cantor became even more interested in the theological aspects of his own theory on transfinite numbers.

Gutberlet subsequently made use of Cantor's ideas, including that God ensured the existence of Cantor's transfinite numbers. In addition, God ensured the ideal existence of infinite decimals, the irrational numbers, and the exact value of pi. Gutberlet also believed that God was capable of resolving various paradoxes that seem to arise in mathematics.

Cantor's own religiosity grew as a result of his contact with various Catholic theologians. In 1884, Cantor wrote to the Swedish mathematician Gösta Mittag-Leffler, explaining that he was not the creator of his new work but merely a reporter. God had provided the inspiration, leaving Cantor responsible only for the way in which his papers were written, for the organization and style, but not for their content. Cantor claimed and believed in the absolute truth of his "theories" because they had been *revealed* to him. Thus, Cantor saw himself as God's messenger, and he desired to use mathematics to serve the Christian faith. Cantor said that he knew the transfinites were real because "God had told me so," and it would have diminished God's power had God only created *finite* numbers.

1.18 Numerical religion. Pythagoras (569 B.C.–475 B.C.). This Greek philosopher was responsible for important developments in mathematics, astronomy, and the theory of music. He and his followers are attributed with discovering that the sum of the angles of a triangle is equal to two right angles and determining that for a right-angled triangle, the square of the hypotenuse is equal to the sum of the squares of the other two sides. Pythagoras and his followers were vegetarians and wore white clothes. They believed that numbers were divine ideas that created and maintained the universe. Studying arithmetic was the way to perfection and led to an understanding of the divine plan for the universe.

1.19 Modern mathematical murderer. Theodore Kaczynski (b. 1942) is certainly one of the most murderous mathematicians. Ted Kaczynski, also known as the Unabomber, was a mathematician who rose swiftly to academic heights even as he became an emotional cripple, a loner, and a murderer. Kaczynski's twenty-five-year self-imposed exile in the Montana woods was particularly appropriate for this man who had always been alone.

Before he had become a hermit, Kaczynski wrote several notable papers on the mathematical properties of functions in circles and boundary functions. Although his IQ was measured as 170, he exhibited many odd characteristics: excessive (pathological) shyness, a fascination with body sounds, a metronomic habit of rocking, and frequent concerns about germs, infections, and other health matters. His room at school stank of rotting food and was piled high with trash.

After teaching for two years and publishing mathematical papers that impressed his peers and put him on a tenure track at one of the nation's most prestigious universities, he suddenly quit. He spent nearly half his life in the woods, killed three strangers, and injured twenty-two others using bombs sent in the mail. Kaczynski's killings grew from his deluded attempts to fight against what he perceived as the evils of technological progress. He was finally arrested in 1996 at his remote cabin in Montana. One of his math papers is T. J. Kaczynski, "Boundary Functions for Bounded Harmonic Functions," *Transactions of the American Mathematics Society* 137 (1969): 203–09.

I.20 The ∞ symbol.
The English mathematician John Wallis (1616–1703) introduced the mathematical symbol for infinity (∞) in 1655 in his *Arithmetica Infinitorum*. I don't think it appeared very often in print again until Jakob Bernoulli (1654–1705) published *Ars Conjectandi* (posthumous publication in 1713).

I.21 Mathematics of tic-tac-toe.
As simple as the playing board is, players can place their Xs and Os on the tic-tac-toe board in $9! =$ 362,880 ways. As James Curran notes at noveltytheory.com, we may arrive at this number by observing that the first player (X) has a choice of making 9 opening moves. Next, the O player can move to any of 8 vacant cells. Hence, there could be 72 (9×8) possible boards with one X and one O. On the third move, the X player can choose any of the 7 remaining spots, and then O has 6 cells to choose from. If we continue this logic, we find that there are 362,880 ($9!$ or $9 \times 8 \times 7 \times 6 \times 5 \times 4 \times 3 \times 2 \times 1$) different complete games. In real life, there are fewer games, because many real games finish in a few moves.

Although this is the number of different move sequences for a "game" (not considering that a game could end before the board is filled with symbols), it is not the number of different patterns or configurations that the board can exhibit. For example, note that many different sequences of moves can give rise to the same pattern of Xs and Os on a board. Consider that there are 9 cells that can exist in one of three states (X, O, or empty). This implies that there can be only 19,683 (3^9 or $3 \times 3 \times 3 \times 3 \times 3 \times 3 \times 3 \times 3 \times 3$) different board configurations or images. This number is reduced further by our noting that many of these configurations will never be achieved in normal tic-tac-toe play. For example, a board filled with nine Os will never occur.

When we consider that various possible games are either a reflection or a rotation of others, there are 48 possible rational outcomes of tic-tac-toe, all of them draws.

I.22 Chickens and tic-tac-toe.
I can only guess at how the chicken challenge works. If it's like any of the tic-tac-toe-playing chicken

machines in the past, the chicken does not know how to play tic-tac-toe. The board is placed in such a way that the human player can't really see where the chicken is pecking. It's true that the chicken pecks at the board and gets some food as reward, but a computer decides where the actual move is placed. The player sees the chicken peck and an X or an O light up on the board. The human beings are so distracted that they never think about where the chicken actually pecks.

I.23 First female doctorate. Sofia Kovalevskaya (1850–1891). Kovalevskaya made valuable contributions to the theory of differential equations and was the first woman to receive a doctorate in mathematics.

Like most other mathematical geniuses, Sofia fell in love with mathematics at a very young age. She wrote in her autobiography: "The meaning of these concepts I naturally could not yet grasp, but they acted on my imagination, instilling in me a reverence for mathematics as an exalted and mysterious science which opens up to its initiates a new world of wonders, inaccessible to ordinary mortals." When Sofia was eleven years old, the walls of her bedroom were papered with the mathematician Mikhail Ostrogradski's lecture notes on differential and integral analysis. In 1874 Kovalevskaya received her doctorate, *summa cum laude*, from Göttingen University. However, despite this doctorate and enthusiastic letters of recommendation from Weierstrass, Kovalevskaya was unable to obtain an academic position because she was a woman.

I.24 Mathematics and homosexuality. Alan Turing, a computer theorist, whose code-breaking work helped to shorten World War II. For this contribution, he was awarded the Order of the British Empire. When he called the police to investigate a burglary at his home, a homophobic police officer suspected that Turing was homosexual. Turing was forced to make a decision. He could either go to jail for a year or take experimental drug therapy. His death two years after the therapy, in 1954, at age forty-two, was a shock to his friends and family. Turing was found in bed. The autopsy indicated cyanide poisoning. Perhaps he had committed suicide, but to this day we are not certain.

I.25 Mathematician cooks. Jean Baptiste Fourier (1768–1830) was a French mathematician and a physicist who is best known for initiating the investigation of Fourier series—a representation of a periodic function as a sum of trigonometric functions. He is also famous for explaining the propagation of heat. Perhaps his interest in heat led him to believe that heat was healthful. Toward the end of his life, he covered his body with layers of clothing and kept his room so hot that it was unbearable to others. He left his home less and less. According to Theoni Pappas, the author of *Mathematical Scandals*, the heat may have exacerbated a heart condition that led to his death.

I.26 Mathematics and money. I asked a number of colleagues this question, thinking that surely an increased salary for mathematics teachers would improve mathematics education. However, most of my colleagues said that a disparity in income between mathematics teachers and teachers in other fields would be harmful both to mathematics education

and to education as a whole. Some worried that this sudden surge in salary would cause fewer teachers to be hired because either private or public funds would have to pay for the increase. In addition, many teachers outside of mathematics would suddenly become unhappy and would attempt to become mathematics teachers, which would not actually improve mathematics education. Also, this diaspora of teachers from other areas would decrease quality in those fields.

Many respondents wondered where the money would come from. If the amount of money allocated to education was constant, the money would have to come from other education areas, which would be harmful. Over time, however, the mathematical educational levels would rise, as new people chose careers as mathematics teachers, due to the higher salaries. Competition would increase, and only the best candidates would be successful in getting a position. Nonetheless, shortages might occur in other areas, decreasing educational quality overall.

Some colleagues said that increasing the number of math teachers would be useful, but they thought it would be more useful to show the current math teachers *how* to teach math to make it "twice as interesting and something students consider worthwhile learning." Others suggested that the greatest improvements would come from allowing math teachers to teach advanced classes that are arbitrarily small, yet still pay them for working. A common sentiment was expressed in one respondent's statement: "Calculus in my high school was available with a class size of 25, but number theory would never fly with a class size of 5." Some suggested that educational resources were directed to the best- and poorest-performing students, but the average student was sometimes overlooked.

Another respondent said that the main effect of doubling math teachers' salaries would be to "get all the non-math teachers very angry." He wrote, "A problem as important as that of teachers' salaries is student motivation. Today, there is much less interest in learning and understanding anything technical or rigorous than there has been in the past. I see this in my own kid and also in the decreasing number of students selecting technical majors in college and other ways. Does it matter? I think so. What to do about it? I don't know. I think there's a certain kind of anti-intellectualism which is fostered by mass media, by computer games, and by parents. This is difficult to fix, even after you've convinced the world that it's worth fixing."

Because my respondents painted such a bleak picture with respect to increasing the salary of math teachers, I would be interested in hearing from those of you with opposing views.

I.27 Mad mom tortures mathematician daughter.
Ada Lovelace, the daughter of Lord Byron (the poet) and the first computer programmer. She analyzed and expanded upon Charles Babbage's plans for difference and analytical engines, and she explained how the machines could tackle problems in astronomy and mathematics. While married to William King, she fell in love with the mathematician John Crosse and became obsessed with gambling.

During the last year of her life, Ada's cervical cancer progressed slowly, and her mother took charge of her care. When Ada confessed her affair with Crosse, her mother promptly discarded all of Ada's morphine and opium—the only thing holding the horrific pain at bay—so that Ada's soul would be redeemed. Ada's last days were spent in agony, as her mother watched but did nothing.

I.28 Mathematician pretends. Alhazen

(965–1039). Al-Hakim, the ruler of Egypt, was angry with Alhazen when Alhazen made gross errors in his ability to predict and control the Nile's flooding. To save himself from execution, Alhazen pretended to be insane and was placed under house arrest. When he was not feigning insanity, Alhazen made important discoveries in optics, describing various aspects of light reflection, magnification, and the workings of the eye.

I.29 Mathematician believes in angels. Sir Isaac

Newton (1642–1727), a brilliant English mathematician, a physicist, and an astronomer. We discussed Newton earlier in this chapter. He and Gottfried Leibniz invented calculus independently. Isaac Newton was so influential that some extra background into his odd life may appeal to you. Newton's father died before his son was born on Christmas Day, 1642. In his early twenties, Newton invented calculus, proved that white light was a mixture of colors, explained the rainbow, built the first reflecting telescope, discovered the binomial theorem, introduced polar coordinates, and showed that the force causing apples to fall (gravity) is the same as the force that drives planetary motions and produces tides.

Many of you probably don't realize that Newton was also a biblical fundamentalist,
believing in the reality of angels, demons, and Satan. He believed in a literal interpretation of Genesis and believed Earth to be only a few thousand years old. In fact, Newton spent much of his life trying to prove that the Old Testament is accurate history.

One wonders how many more problems in physics Newton would have solved if he had spent less time on his biblical studies. Newton appears not to have had the slightest interest in sex; he never married and almost never laughed (although he sometimes smiled). In 1675, Newton suffered a massive mental breakdown, and some have conjectured that throughout his life, he was a manic depressive with alternating moods of melancholy and happy activity. Today we would classify this as bipolar disorder.

I.30 History's most prolific mathematician. Leon-

hard Euler (1707–1783), whom we mentioned before as also being a religious mathematician. He was the Swiss mathematician who, when he was completely blind, made great contributions to modern analytic geometry, trigonometry, calculus, and number theory. Euler published over 8,000 books and papers, almost all in Latin, on every aspect of pure and applied mathematics, physics, and astronomy. In analysis, he studied infinite series and differential equations, introduced many new functions (e.g., the gamma function and elliptic integrals), and created the calculus of variations. His notations, such as e and π, are still used today. In mechanics, he studied the motion of rigid bodies in three dimensions, the construction and the control of ships, and celestial mechanics.

Leonhard Euler was so prolific that his

papers were still being published for the first time two centuries after his death. His collected works have been printed bit by bit since 1910 and will eventually occupy more than seventy-five large books.

1.31 Suicidal mathematician. Évariste Galois

(1811–1832) was famous for his contributions to group theory and for formulating a method of determining when a general equation could be solved by radicals.

Although he obviously knew more than enough mathematics to pass the Lycée's examinations, Galois's solutions were often so innovative that his university examiners failed to appreciate them. Also, Galois would perform so many calculations in his head that he did not bother to clearly outline his arguments on paper. For these reasons, in addition to his temper and rashness, he was denied admission to the Ecole Polytechnique.

When he was taunted into a duel, he accepted, knowing he would die. The circumstances that led to Galois's death have never been fully explained. It has been variously suggested that it resulted from a quarrel over a woman, that he was challenged by royalists who detested his republican views, or that an agent provocateur of the police was involved. In any case, when preparing for the end, Galois spent the entire night feverishly writing his mathematical ideas and discoveries in as complete a form as he could.

The next day, Galois was shot in the stomach. He lay helpless on the ground. There was no physician to help him, and the victor casually walked away, leaving Galois to writhe in agony until he died. Not until 1846 had group theory advanced sufficiently that his discoveries could be appreciated. His legacy has greatly impacted twentieth-century mathematics, and his mathematical reputation rests on fewer than one hundred pages of posthumously published work of original genius.

1.32 Marry a mathematician? I asked dozens of

colleagues if they would rather marry the best mathematician in the world or the best chess player. Not a single person chose the chess player.

Jon A. replied, "I choose the best mathematician. I would tire quickly of playing chess with someone who can beat me in less than twenty moves. I could probably have a lot more stimulating conversation with a mathematician."

David J. assumed that the best chess player would have a very rigorous touring schedule, leaving many nights where he could not be with his wife. Thus, he preferred marrying the mathematician. However, he wondered what it would mean to be "the best mathematician in the world."

Edith R. says she'd prefer the mathematician. However, she says she's met good chess players and good mathematicians, and "both species are often prone to soaring egos and boring speech making. However, while mathematicians are obsessive, they are generally open, at least occasionally, to conversations on other subjects. Chess nuts tend to be single-minded."

David P. said that there isn't much to talk about in chess except for move sequences and historical games.

1.33 Mathematical corpse. I posed this question

to dozens of colleagues. A majority of friends

feared the corpse with the fancier equation because they felt that it implied a genius murderer on par with the infamous, brilliant Hannibal Lecter, portrayed by Anthony Hopkins in the movie *The Silence of the Lambs*.

I.34 Blind date.

I posed this question to brilliant mathematical colleagues. They all said that they would not go on the date because they recognized the tattoos as representing the famous "casualty equation" that F. W. Lanchester created in 1914. In particular, the Lanchester equation describes enemy casualty rates in war! R and B represent the numbers, at the time t, of opposing Red and Blue armies, and k_B and k_R are the killing rates of the Blue and the Red armies. My friend Graham Cleverley responded, "The fact that the blind date has these for tattoos suggests a somewhat gloomy preoccupation." He also would not go on a date with a woman who had these tattoos:

$$\frac{dx}{dt} = Ax - Bxy$$

$$\frac{dy}{dt} = -Cy + Dxy$$

These are the Lotka-Volterra equations that describe predator-prey relationships, such as the number of foxes and hens as a function of time. A is the growth rate of prey; B is the rate at which predators destroy prey; C is the death rate of predators, and D is the rate at which predators increase by consuming prey. Graham considers this tattoo indicative of a possible "fatal attraction." Of course, it might

not be clear whether the prospective dating partner saw herself as predator or as prey.

I.35 Fundamental Anagram of Calculus.

Isaac Newton and Gottfried Wilhelm Leibniz had a feud over who *really* discovered calculus. One incident from 1677 is particularly interesting and relevant. At this time Newton was answering several of Leibniz's questions about infinite series. In his letter, Newton came close to revealing his "fluxional method"—that is, Newton's own version of calculus. However, instead of revealing the method, Newton concealed it in the form of an anagram. Perhaps Newton used the anagram because he didn't want Leibniz to scoop him and also because Newton wanted a way to demonstrate that he had actually known about calculus should he have to prove this at a later date. Whatever the reason, Newton was not ready to give a full explanation. After Newton described his methods of tangents and finding maxima and minima, he wrote to Leibniz:

> The foundations of these operations are evident enough, in fact; but because I cannot proceed with the explanation of it now, I have preferred to conceal it thus
>
> 6accdae13eff7i3l9n4o4qrr4s8t12ux
>
> On this foundation I have also tried to simplify the theories which concern the squaring of curves, and I have arrived at certain general theorems. (Richard Westfall, *Never at Rest* [New York: Cambridge University Press, 1980], 265. Also see www.seanet. com/~ksbrown/kmath414.htm.)

The anagram expresses, in Newton's terminology and in Latin, the fundamental theorem of calculus: "*Data aequatione quot-cunque fluentes quantitates involvente, flux-iones invenire; et vice versa,*" which means "Given an equation involving any number of fluent quantities to find the fluxions, and vice versa." (*Fluxion* was Newton's term for "derivative.") The numbers in the anagram count the number of letters in the sentence. For example, "6a" corresponds to the 6 occurrences of the letter *a* in the sentence. There are 13 occurrences of the letter *e*, and so forth.

Interestingly, neither Leibniz nor Newton had published papers on calculus at the time this letter was exchanged, although both probably were well versed in the subject. Thus, if Newton had avoided using the anagram and had sent Leibniz an explicit statement about his calculus, Newton might have established his superior knowledge beyond doubt. Instead, Newton's secret anagrams caused him to lose his possible claim to calculus and led to heated arguments that plagued his and Leibniz's lives for years to come.

I.36 Parallel universes and mathematics. This question would be impossible to answer unless you are familiar with the work of Max Tegmark. For the moment, let's talk about the multiverse, God, and the *physics of polytheism*. In 1998, Max Tegmark, a physicist formerly at the Institute for Advanced Study at Princeton, New Jersey, used a mathematical argument to bolster his own theory of the existence of multiple universes that "dance to the tune of entirely different sets of equations of physics." The idea that there is a vast "ensemble" of universes (a multiverse) is not new—the idea occurs in the many-worlds interpretation of quantum mechanics and the branch of inflation theory, suggesting that our universe is just a tiny bubble in a tremendously bigger universe. In Marcus Chown's "Anything Goes," appearing in the June 1998 issue of *New Scientist*, Tegmark suggests that there is actually greater simplicity (e.g., less information) in the notion of a multiverse than in an individual universe.

To illustrate this argument, Tegmark gives the example of the rational numbers. A useful definition of something's complexity is the length of a computer program that is needed to generate it. Consider how difficult it could be to generate an arbitrarily chosen number between 0 and 1, the arbitrary number specified by an infinite number of digits. Expressing the number would require an infinitely long computer program. On the other hand, if you were told to write a program that produced *all* the rational numbers, the instructions would be easy:

```
K := 2;
while K > 1 do begin
    if K mod 2 = 0 then
        for J:= 1 to K - 1 do output
        (J,'/',K - J)
    else for J:= K - 1 down to 1 do
        output(J,'/',K - J)
    K := K + 1;
    end;
```

This produces all the rational numbers, starting with 1/1, 2/1, 1/2, 1/3, 2/2, 3/1, 4/1, 3/2, 2/3, 1/4, ... (The mod statement in the program lists the rationals in the order used by Cantor to produce his proof of their countability, which you will see in chapter 6.) This program would be easy to write, which means that creating all possibilities may, in some cases, be much simpler than creating one very specific possibility. Tegmark extrapolates this idea to suggest that the existence of infinitely many universes $\gamma_1, \gamma_2, \gamma_3, \ldots \gamma_\infty$ is simpler, less wasteful, and more likely than just a single universe γ.

Could one extend Tegmark's reasoning to God and argue that the existence of infinitely many gods, $\gamma_1, \gamma_2, \gamma_3, \ldots \gamma_\infty$, is more likely than just a single God, γ? Obviously, this is a controversial application of Tegmark's theories and a wild stretch of the imagination, but the Old Testament gives numerous examples that may suggest the existence of multiple gods, γ_n, $n > 1$; for example, "Among the gods there is none like you, O Lord; no deeds can compare with yours" (Psalm 86:8); "I will bring judgment on all the gods of Egypt" (Exodus 12:12); or, "You shall have no other gods before me" (Exodus 20:3).

The physicist Dan Platt points out that the notion of a "multiverse" can be traced back to the mathematician Gottfried Wilhelm Leibniz (1646–1716), who explored one aspect of theodicy—why is there evil in the world if God is all-powerful and good? Leibniz suggested that our universe, γ, was the best of all possible universes, $\gamma_1, \gamma_2, \gamma_3, \ldots \gamma_\infty$, and the evil in it was unfortunate—but at least our universe exhibited the smallest amount of evil one could hope for.

I.37 One-page proof of God's existence. Some-
time in 1970, Kurt Gödel's mathematical proof of the existence of God began to circulate among his colleagues. The proof was less than a page long and caused quite a stir:

Gödel's Mathematical Proof of God's Existence

Axiom 1. (Dichotomy) A property is positive if and only if its negation is negative.

Axiom 2. (Closure) A property is positive if it necessarily contains a positive property.

Theorem 1. A positive property is logically consistent (i.e., possibly it has some instance).

Definition. Something is Godlike if and only if it possesses all positive properties.

Axiom 3. Being Godlike is a positive property.

Axiom 4. Being a positive property is (logical, hence) necessary.

Definition. A property P is the essence of x if and only if x has P and P is necessarily minimal.

Theorem 2. If x is Godlike, then being Godlike is the essence of x.

Definition. $NE(x)$: x necessarily exists if it has an essential property.

Axiom 5. Being NE is Godlike.

Theorem 3. Necessarily there is some x such that x is Godlike.

I obtained this proof from Hao Wang, *Reflections on Kurt Gödel* (Cambridge, Mass.: MIT Press, 1987), page 195. How shall we judge such an abstract proof? How many people on Earth can really understand it? Is the proof a result of profound contemplation or the raving of a lunatic? Recall that Gödel's academic credits were impressive. For example, he was a respected mathematician and a member of the faculty of the University of Vienna starting in 1930. He emigrated to the United States in 1940 and became a member of the Institute of Advanced Study in Princeton, New Jersey.

1.38 Song lyrics.
According to Robert Munafo, the answer is 100 billion, a number mentioned in "Message in a Bottle" by the Police. The precise lyric is, "A hundred billion bottles washed up on the shore."

1.39 Jews, π, the movie.
In the 1998 movie π, the Hasidic Jews interested in Kabbalah are searching for a sequence of 216 Hebrew letters, or a 216-digit number equivalent. This discovery may hold the key to unlocking God's true name, which was destroyed with the Temple by the Romans.

The movie is notable in that it features a mathematical genius named Max who is obsessed with numbers. In fact, he rarely leaves his apartment and sees numbers and number patterns in everything around him. When his computer displays the mystical 216-digit number, it crashes and strangeness ensues. Some believe that the number actually makes the computer conscious. Max decides that the 216-digit number is responsible for his own ill health, and he drills a hole in his own skull as a cure.

The computer guru Michael Egan points out that the movie does not always display the same 216 digits when referring to the mystical number, and sometimes it shows a 218-digit number as the God number:

94143243431512653932105487239048668285129134748760276719592346023582958304725016523252592969257276553643634627271840120126431475463294501278472648410756223478962672859285829534750277226264645621761398482951947541239850l

Why does the movie usually focus on 216 digits? Notice that if we multiply the digits in the number of the beast, we get: $6 \times 6 \times 6 = 216$. Also, Kabbalists suggest that a mysterious name for God exists that has 72 syllable, or parts, each composed of three Hebrew letters. Thus, this name for God has 216 letters. This name is composed by scrambling letters in the words of Exodus 14:19–21, each of these verses having 72 letters. For other possible reasons for using 216, see the factoid *Anamnesis and the number 216* in this chapter.

1.40 Mathematics and romance.
It's My Turn (1980) stars Jill Clayburgh, who plays a mathematics professor. In the opening scene, she proves the famous "snake lemma" of homological algebra to an obnoxious graduate student. You can learn more about the snake lemma here: Mathworld, "SnakeLemma," mathworld.wolfram.com/SnakeLemma.html.

1.41 Greek death.
According to Adrian Room's *The Guinness Book of Numbers*, if we multiply 1 through 8 by 8 and then add the resultant digits until one digit remains, the 8 in the last column gradually "dies" to 1. The Greeks and other cultures considered this significant.

Multiply	Add Digits	Add Digits Again
1 × 8 = 8	8	8
2 × 8 = 16	7	7
3 × 8 = 24	6	6
4 × 8 = 32	5	5
5 × 8 = 40	4	4
6 × 8 = 48	12	3
7 × 8 = 56	11	2
8 × 8 = 64	10	1

1.42 The *Matrix*. "IS 5416." In Isaiah 54:16, we find: "Behold, I have created the smith that bloweth the coals in the fire, and that bringeth forth an instrument for his work; and I have created the waster to destroy." Agent Smith was a destroyer and an adversary to *The Matrix*'s main character, Neo.

1.43 At the movies. *20,000 Leagues under the Sea*, by Jules Verne (1870). In the movie, a professor (played by Paul Lukas) seeks the truth about a legendary sea monster in the years just after the Civil War. Later the professor, his aide (Peter Lorre), and a harpoon master (Kirk Douglas) discover that the monster is actually a submarine run by Captain Nemo (James Mason).

Also, note that there was a popular movie, *The Beast from Twenty Thousand Fathoms* (1953), that depicts how an atom bomb awakens a prehistoric monster. However, the title for the short story version of *Beast* was merely "The Foghorn." In Ray Bradbury's "The Foghorn," a sea monster mistakes a foghorn for the mating cry of a female.

1.44 First mathematician. Thales of Miletus (634–548 B.C.), who established a school of mathematics and philosophy. He lived in Miletus, Asia Minor (now Turkey).

1.45 Game show. *The $64,000 Question* has its origin in the 1940 U.S. radio show *Take It or Leave It*, which offered prizes ranging from $1 to $64. The original colloquialism was "the $64 question," which signified an "important" question. The phrase was so popular that the name of the radio show changed to *The $64 Question* in 1950. The number 64 is useful because it is a power of 2 and thus fits well with games that involve doubling of monetary awards, as in the sequence: 1, 2, 4, 16, 32, 64.

1.46 Famous epitaphs. The variable x is Isaac Newton, and y is Ludwig Boltzmann. "$S = k \ln W$" describes the entropy of a system. Of this equation, Adrian Cho writes in the August 23, 2002, issue of *Science* ("A Fresh Take on Disorder"): "No less important than Einstein's $E = mc^2$, the equation provides the mathematical definition of entropy, a measure of disorder that every physical system strives to maximize. The equation serves as the cornerstone of 'statistical mechanics,' and it has helped scientists decipher phenomena ranging from the various states of matter to the behavior of black holes to the chemistry of life."

1.47 Secret mathematician. A mathematician named "N. Bourbaki" never existed. In the 1930s, a group of mostly French mathematicians referred to themselves as "N. Bourbaki." Their primary goal was to write a very rigorous unified account of all mathematics, and the group produced the following volumes: I *Set Theory*, II *Algebra*, III *Topology*, IV *Functions of One Real Variable*, V *Topological Vector Spaces*, VI *Integration*, and, later, VII *Commutative Algebra*, and VIII *Lie Groups*. Members of Bourbaki were very secretive and had to resign by age fifty.

I.48 How much? Of course, there is no easy answer to this question. In my book *The Loom of God*, I note how difficult it would be for a chimpanzee to understand the significance of prime numbers, yet the chimpanzee's genetic makeup differs from ours by only a few percentage points. These minuscule genetic differences in turn produce differences in our brains. Additional alterations of our brains would admit a variety of profound concepts to which we are now totally closed. What mathematics is lurking out there that we can never understand? How do our brains affect our ability to contemplate God? What new aspects of reality could we absorb with extra cerebrum tissue? And what exotic formulas could swim within the additional folds? Philosophers of the past have admitted that the human mind is unable to find answers to some of the most important questions, but these same philosophers rarely thought that our lack of knowledge was due to an organic deficiency shielding our psyches from higher knowledge.

If the Yucca moth, with only a few ganglia for its brain, can recognize the geometry of the yucca flower from birth, how much of our mathematical capacity is hardwired into our convolutions of cortex? Obviously, specific higher mathematics is not inborn, because acquired knowledge is not inherited, but our mathematical capacity is a function of our brain. There is an organic limit to our mathematical depth.

How much mathematics can we know? The body of mathematics has generally increased from ancient times, although this has not always been true. Mathematicians in Europe during the 1500s knew less than Grecian mathematicians did at the time of Archimedes. However, since the 1500s, humans have made tremendous excursions along the vast tapestry of mathematics. Today several hundred thousand mathematical theorems are proved each year.

In the early 1900s, a great mathematician was expected to comprehend the whole of known mathematics. Mathematics was a shallow pool. Today the mathematical waters have grown so deep that a great mathematician can know only about 5 percent of the entire corpus. What will the future of mathematics be like, as specialized mathematicians know more and more about less and less until they know everything about nothing?

I.49 Nobel Prize. Nobel prizes were instituted by the will of Alfred Nobel, a Swede who was a chemist, an industrialist, and the inventor of dynamite. Since 1901, prizes have been awarded for achievements in Physics, Chemistry, Physiology or Medicine, Literature, and Peace—but not mathematics. Legend says that Nobel ignored mathematics because his mistress or wife rejected him for Gösta Mittag-Leffler, a mathematician. However, little or no historical evidence exists for this story. Rather, Nobel probably excluded mathematics simply because he did not consider it a "practical" science that had much impact on society. Also, other prizes for mathematics existed during Nobel's life.

1.50 Roman numerals. In the Middle Ages, most of Europe switched from Roman to "Arabic numerals," similar to those we use today. This isn't to say that Roman numerals disappeared entirely in the Middle Ages. Many accountants still used them because addition and subtraction can sometimes be easy with Roman numerals. For example, if you want to subtract 15 from 67, in the Arabic system you subtract 5 from 7, and 1 from 6. But in the Roman system, you'd simply erase an X and a V from LXVII to get LII. In many instances, it's subtraction by erasing.

However, Arabic numerals are superior to Roman numerals because Arabic numerals have a "place" system, in which the value of a numeral is determined by its position. A "1" can mean 1, 10, 100, or 1,000, depending on its position in a numerical string. This is one reason it's so much easier to write 1998 than MCMXCVIII—1,000 (M) plus 100 less than a 1,000 (CM) plus 10 less than a 100 (XC) plus 5 (V) plus 1 plus 1 plus 1 (III). Try doing arithmetic with this Roman monstrosity. Positional notation greatly simplifies all forms of written numerical calculation, and I suspect that many of our most abstract theories in physics and math could not have been contemplated if we still used Roman numerals.

My guess is that during the Middle Ages, the calculational demands of capitalism eliminated any resistance to the "infidel symbol" of zero and ensured that by the early seventeenth century, Arabic numerals reigned supreme.

Even during Roman times, Roman numerals were used more to *record* numbers, while most calculations were done by using the abacus and piling up stones.

Note that the number of characters in the Roman numerals for 1, 2, 3, 4, 5, 6, 7, 8, 9, 10, . . . (or I, II, III, IV, V, VI, VII, VIII, IX, X, . . .) is 1, 2, 3, 2, 1, 2, 3, 4, 2, 1, 2, 3, 4, ... When this sequence is plotted for different ranges—like the range 1 to 20, 1 to 200, and 1 to 2,000—a scale-invariant, fractal-like stairstep pattern emerges.

1.51 Mathematical universe. First, let me give some background to this question. Marilyn vos Savant is listed in the *Guinness Book of World Records* as having the highest IQ in the world—an awe-inspiring 228. She is the author of several delightful books and is the wife of Robert Jarvik, M.D., the inventor of the Jarvik 7 artificial heart. Her column in *Parade* magazine is read by 70 million people every week. One of her readers once asked her, "Why does matter behave in a way that is describable by mathematics?" She replied, "The classical Greeks were convinced that nature is mathematically designed, but judging from the burgeoning of mathematical applications, I'm beginning to think simply that mathematics can be invented to describe anything, and matter is no exception."

Marilyn vos Savant's response is certainly one with which many people would agree. However, the fact that reality can be described or approximated by *simple* mathematical expressions suggests to me that nature has mathematics at its core. Formulas like $E = mc^2$, $\vec{F} = m\vec{a}$, $1 + e^{i\pi} = 0$, and $\lambda = h/mv$ all boggle the mind with their compactness and profundity. $E = mc^2$ is Einstein's equation relating energy and mass. $\vec{F} = m\vec{a}$ is Newton's second law: force acting on a body is proportional to its mass and its acceleration. $1 + e^{i\pi} = 0$ is Euler's formula relating three fundamental mathematical terms: e, π, and i. The last equation, $\lambda = h/mv$, is De Broglie's wave equation, indicating that matter has both wave and particle characteristics. Here the Greek letter lambda, λ, is the wavelength of the wave-particle, and m is its mass. These examples are not meant to suggest that *all* phenomena, including subatomic phenomena, are described by simple-looking formulas; however, as scientists gain more fundamental understanding, they hope to simplify many of the more unwieldy formulas. I see no reason why aliens will not discover the same truths.

I side with both Martin Gardner and Rudolf Carnap, whom I interpret as saying, Nature is almost always describable by simple formulas, not because we have invented mathematics to do so but because of some hidden mathematical aspect of nature itself. For example, Martin Gardner writes in his classic 1950 essay "Order and Surprise":

> If the cosmos were suddenly frozen, and all movement ceased, a survey of its structure would not reveal a random distribution of parts. Simple geometrical patterns, for example, would be found in profusion— from the spirals of galaxies to the hexagonal shapes of snow crystals. Set the clockwork going, and its parts move rhythmically to laws that often can be expressed by equations of surprising simplicity. And there is no logical or a priori reason why these things should be so.

Here Gardner suggests that simple mathematics governs nature from the molecular to galactic scales. Rudolf Carnap, an important twentieth-century philosopher of science, profoundly asserts, "It is indeed a surprising and fortunate fact that nature can be expressed by relatively low-order mathematical functions."

To best understand Carnap's idea, consider the first great question of physics: "How do things move?" Imagine a universe called JUMBLE, where Kepler looks up into the heavens and finds that most planetary orbits cannot be approximated by ellipses but rather by bizarre geometrical shapes that defy mathematical description. Imagine Newton dropping an apple whose path requires a 100-term equation to describe. Luckily for us, we do not live in JUMBLE. Newton's apple is a symbol of both nature and simple arithmetic from which reality naturally evolves.

1.52 A universe of blind mathematicians.

Countless examples exist of brilliant mathematicians who are totally blind. When the blind mathematician Bernard Morin (b. 1931) was asked how he computed the sign of a complicated calculation, he said he did so "by feeling the weight of the thing, by pondering it." The blind English mathematician Nicholas Saunderson (1682–1739) used abacuslike devices to aid calculations. Some mathematicians have even speculated that blindness may aid in certain areas of mathematics, where sight can prejudice or bias mathematicians working in particular fields of geometry. Famous blind mathematicians also include Lev Semenovich Pontryagin

(1908–1988), Louis Antoine (1888–1971), and the modern mathematicians Emmanuel Giroux, Lawrence Baggett, Norberto Salinas, John Gardner (also a physicist), and Zachary J. Battles. Salinas and Gardner have developed a new form of Braille that has eight dots instead of the usual six. The two additional dots are reserved for providing a wealth of mathematical notation. For further reading, see Allyn Jackson, "The World of Blind Mathematicians," *Notices of the AMS* 49, no. 10 (November 2002): 1246–51.

I.53 Close to reality. As with any deep topic on the edge of mathematics and philosophy, opinions are divided on this subject. In order to save space, I present my favorite opinion and urge readers to consult essays such as Martin Gardner's "How Not to Talk about Mathematics" (1981) for a range of viewpoints. The British mathematician G. H. Hardy speaks on this topic in *A Mathematician's Apology* (1940):

> There is probably less difference between the positions of a mathematician and of a physicist than is generally supposed, and the most important seems to me to be this, that the mathematician is in much more direct contact with reality. This may seem a paradox, since it is the physicist who deals with the subject-matter usually described as "real"; but a very little reflection is enough to show that the physicist's reality, whatever it may be, has few or none of the attributes which common sense ascribes instinctively to reality. . . . A chair or a star

is not in the least like what it seems to be; the more we think of it, the fuzzier its outlines become in the haze of sensations which surrounds it; but "2" or "317" has nothing to do with sensation, and its properties stand out the more clearly the more closely we scrutinize it.

For Hardy, a chair may be a collection of whirling electrons or an idea in the mind of God. The job of the physicist is to try to correlate the "incoherent body of crude facts confronting him with some definite and orderly scheme of abstract relations, the kind of scheme which he can borrow only from mathematics." Hardy concludes that 317 is a prime, not because we think so, or because our minds are shaped in one way rather than another, but because it *is* so.

The philosopher and the psychologist William James, in his book *The Meaning of Truth*, argues that truths of science and mathematics, not yet verified, are "sleeping truths." The quadrillionth decimal digit of pi, for example, "sleeps" in "the world of geometrical realities," even though "no one may ever try to compute it."

Martin Gardner's essays on mathematical reality, which include "How Not to Talk about Mathematics" and "Order and Surprise," are reprinted in *Order and Surprise* (Buffalo, N.Y.: Prometheus Books, 1983).

I.54 Why learn mathematics? Of course, in some technical fields like engineering, people do make use of mathematics beyond what is taught in fourth grade. But is this small minority of people sufficient to require that *all* students learn algebra, geometry, and related forms of higher-level math? A 2004 survey conducted by the Organization for Economic Cooperation and Development ranked the United States 28th out of 40 countries in mathematics, for example, far below Finland and South Korea. The survey tested mostly practical skills like estimating the size of Antarctica. But does America's poor showing really matter? After all, "America may lose math literacy surveys, but it dominates number-crunching in every sphere from corporate profits to supercomputers to Nobel Prizes," according to Donald G. McNeil Jr., writing in "The Last Time You Used Algebra Was . . . " (*New York Times*, Week in Review, Section 4, Sunday, December 12, 2004, p. 3). However, it's important to note that a third of the Americans who won Nobel Prizes were born abroad; thus, many American Nobel Prize winners may have received early training outside America.

McNeil gives several possible reasons why we should study math. One is that potential employers and higher educational institutions use mathematics as a "filter for the lazy or stupid, as passing freshman physics is for premed students." Perhaps a better answer comes from Miss Collins, McNeil's daughter's math teacher: "Kids don't study poetry just because they're going to grow up and be poets. It's about a habit of mind. Your mind doesn't think abstractly unless it's asked to—and it needs to be asked to from a relatively young age. The rigor and logic that goes into math is a good way for your brain to be trained."

2. Cool Numbers

— — — — —

2.I Pi. Symbolized by the Greek letter π, pi is the ratio of a circle's circumference to its diameter and is approximately 3.14159 . . . We don't know who was the first to recognize that a circumference of a circle was about 3 times the diameter. Perhaps it was not long after this that wheels were commonly used. Cecil Adams, the author of *The Straight Dope*, suggests that around 3500 B.C., a person may have noticed that for every revolution of a wheel, a cart moves forward about three times the diameter of the wheel. An ancient Babylonian tablet states that the ratio of the circumference of a circle to the perimeter of an inscribed hexagon is 1:0.96, implying a value of pi of 3.125. The Greek mathematician Archimedes (c. 250 B.C.) was the first to give us a firm, mathematically rigorous range for pi as lying between 223/71 and 22/7. The latter is called the "Archimedean value" of pi, but this approximation was in use long before his time.

William Jones (1675–1749) introduced the symbol π in 1706, most likely after the Greek word for periphery ($\pi\varepsilon\rho\iota\varphi\varepsilon\rho\varepsilon\iota\alpha$). Leonhard Euler (1707–1783) used the symbol π in his famous 1736 textbook *Mechanica*, and we've used the symbol ever since.

Note that π is the most famous ratio in mathematics, both on Earth and probably for any advanced civilization in the universe. The number π, like other fundamental constants of mathematics such as $e = 2.718$ . . . , is a transcendental number, a term defined later in this chapter. The digits of π and e never end, nor has anyone detected an orderly pattern in their arrangement.

2.2 An avalanche of digits. First, I'm not sure I would consider the mathematicians and the computer scientists obsessed, but they are "driven." Note that although π cannot be expressed by a fraction, it can be expressed as an infinite *series* of fractions. For example, in 1671, the mathematician Gottfried Leibniz discovered that the series $1 - \frac{1}{3} + \frac{1}{5} - \frac{1}{7} + \ldots$ slowly converges to $\pi/4$. Today, mathematicians use increasingly sophisticated series that converge more quickly to π.

The speed with which a computer can compute π is an interesting measure of a computer's computational ability. Today, we know pi's digits to over a trillion digits, due to fast computers, better algorithms, and increased understanding of how to mathematically manipulate large numbers that are over a billion digits long.

2.3 Evenness. Yes. An even number leaves no remainder when divided by 2. So, 0/2 = 0 and has no remainder. Also, an integer n is called "even" if there exists an integer m such that $n = 2m$, and n is called "odd" if $n + 1$ is even. Thus, 0 is even by this criterion as well.

2.4 Billion. Either way, a billion is big. If you started counting today, saying a number a second, you wouldn't reach a billion (American) until about 30 years later. Your friend in England would have to count for 32,000 years!

The word *billion* originally meant a million million, and in England and Germany it still does, at least in common talk. The prefix "bi" in billion implies two "million" written side by side

1,000,000 (6 zeros) and
1,000,000 (6 zeros)

1,000,000,000,000
(a British billion, 12 zeros)

Americans had a pressing need for a simple word for a number with a mere 9 zeros and simply took the word *billion* for this purpose. I believe that the American meaning of a billion has permeated most of the world these days, especially in scientific and mathematical literature.

The word *billion* is relatively recent; it was not in common use until the sixteenth century. One of the earliest uses in an American book occurred by a man named Greenwood in 1729. Greenwood gave the billion as 1 with 9 zeros.

2.5 Pick an integer, any integer. Because there are an infinite number of integers, this task would seem to be impossible—only "small" integers are accessible for selection. My colleague James "Jaymz" Salter writes to me, "Selection implies physical specification. The ability to physically specify an integer N diminishes as N approaches infinity. Certainly, integers of "infinite" size are beyond specification. So any integer specified would be necessarily drawn from a pool of accessible integers, and therefore biased towards smaller integers."

What do we mean when we say "selected at random" in this question? Usually, random numbers are required to be independent, so that there are no correlations between successive numbers. If no other information is given, the word *random* usually means "random with a uniform distribution."

Does my initial question imply "physical specification," or can we mean some other kind of random specification? Salter writes to me, "'Selection of a number' requires at least an infrastructure to represent the number by some physical means, under some 'system of meaning' (or counting convention) that is absolutely dependent on the existence of some physical substance, whether it is brain cells, transistors, or beads."

Regardless of all the ingenious shorthands (e.g., superfactorial, discussed in chapter 6), the ability to represent arbitrarily large numbers requires matter. It takes more matter to represent or specify larger numbers (e.g., more paper to write down long numbers, more brain cells to imagine large numbers, more transistors to hold more bits, etc.). Numbers that are inconceivably large are unfairly disadvantaged in the selection process.

Could we in fact hypothetically produce a random integer by tossing a dart at the number line from 0 to 1 and then taking the reciprocal of whatever the dart lands on? The answer would appear to be a resounding no because the method is biased toward small integers, even if the dart is thrown randomly on the interval. Half the time, the dart would hit a value between 0.5 and 1, and the reciprocal would be rounded to 1 if we are seeking to produce an integer.

2.6 Superthin dart. As you learned in previous paragraphs, your superthin dart would "always" land on a transcendental number. The infinity of transcendental numbers is so much larger than the infinity of algebraic numbers that your dart is virtually guaranteed to land on a transcendental number.

2.7 Bombs on magic squares. Here is one solution, and I am aware that other solutions exist.

0	62	2	60	11	53	9	55
15	49	13	51	4	58	6	56
16	46	18	44	27	37	25	39
31	33	29	35	20	42	22	40
52	10	54	8	63	1	61	3
59	5	57	7	48	14	50	12
36	26	38	24	47	17	45	19
43	21	41	23	32	30	34	28

2.8 Liouville constant. Joseph Liouville showed that his unusual number was transcendental, thus making this number among the first to be proved transcendental. Notice that the constant has 1 in each decimal place, corresponding to a factorial, and zeros elsewhere. This means that the 1s occur in the 1st, 2nd, 6th, 24th, 120th, 720th, and so on, places, and zeros are elsewhere. More generally, we can write

$$\sum_{k=1}^{\infty} a_k r^{-k!}$$

where $a \leq a_k \leq r$. The numbers a_k are integers. The resulting number is a Liouville number of base r. If the values for a_k are all 1, and $r = 10$, we get $1/10 + 1/10^{1 \times 2} + 1/10^{1 \times 2 \times 3} + \ldots = 0.110001000000000000000001000 \ldots$

Aside from his mathematical pursuits, Liouville was interested in politics and was elected to the French Constituting Assembly in 1848. After a later election defeat, Liouville became depressed. His mathematical ramblings were interspersed with poetical quotes (much like this present book!). Nonetheless, during the course of his life, Liouville wrote over 400 mathematical papers.

2.9 The paradox of pepperonis. The fallacy is that when you count seven pepperonis, you are mixing up addition and subtraction in the middle of a counting process. This is the same fallacy that occurs in much more complicated classic problems of gentlemen paying for hotel rooms. Also, by the twisted logic, you could count backward, "6, 5, 4, 3, 2, 1" and say that there is only one pepperoni in the set.

2.10 Taxicab numbers. The renowned British mathematician G. H. Hardy (1877–1947) was visiting Srinivasa Ramanujan (1887–1920), the self-taught yet brilliant mathematician from India whom we have already discussed. Hardy mentioned that the number of the taxicab that had brought him was 1729, which Hardy thought was "rather a dull" number. Ramanujan smiled and replied instantly, "No, it is a very interesting number; it is the smallest number expressible as a sum of two positive cubes in two different ways."

Ramanujan was thinking of $1{,}729 = 1^3 + 12^3$ and $1729 = 9^3 + 10^3$. Ramanujan was so quick with numbers that it was as if he were intimately familiar with every number! Indeed, numbers were his friends.

Today, we know that there exist an infinite number of "taxicab numbers" with integer solutions of the form $i^3 + j^3 = k^3 + l^3$. Several modern mathematicians enjoy searching for higher-order taxicab numbers, such as triple-pair solutions to $i^3 + j^3 = k^3 + l^3 = m^3 + n^3$, where all the numbers are integers.

In 1957, John Leech (1926–1992) discovered the smallest number expressible as the sum of two positive cubes in *three* different ways: $87{,}539{,}319 = 167^3 + 436^3 = 228^3 + 423^3 = 255^3 + 414^3$.

2.11 Transcendence. Mathematicians don't always know whether sums and products of transcendental numbers are transcendental. For example, all that mathematicians know about $(\pi + e)$ and $(\pi \times e)$ is that at *least one* of these two numbers is transcendental, but transcendence has not been proved for either number on its own! All that we know with certainty is that they both can't be algebraic.

If one of these numbers is transcendental and one is algebraic, we can't say for sure which is which. John dePillis and others prove some of our statements with the following argument. Consider the roots of the polynomial

(1) $z^2 - az + c = 0$, where $a = \pi + e$ and
$$c = \pi \times e$$

This is the same as saying

(2) $(z - \pi) \times (z - e) = 0$

The roots of Polynomial (2) are $z = \pi$ and $z = e$. The first step of the argument is to realize that polynomials with algebraic number coefficients must have roots that are algebraic. Let us assume for the moment that a and c in Polynomial (1) are both algebraic (i.e., not transcendental). We can show that this is impossible because this implies that the roots $z = \pi$ and $z = e$ must be algebraic. But we know that π and e are transcendental (not algebraic). This contradiction means that at least one of the coefficients ($\pi + e$) or ($\pi \times e$) is transcendental.

Note that e^π was proved to be transcendental in 1929 and $2^{\sqrt{2}}$ in 1930. I believe that no one knows whether π^π is transcendental.

2.12 Special class of numbers.
For many years, mathematicians have studied this cool class of numbers. Here's how to understand them. If a number is less than the sum of its proper divisors, it is called *abundant*. (A positive proper divisor is a positive divisor of a number n, excluding n itself.) As an example, the proper divisors of 12 are 1, 2, 3, 4, and 6. And these proper divisors add up to 16. The number 12 is less than 16, so 12 is abundant. The first few abundant numbers are 12, 18, 20, 24, 30, 36, . . . The first odd abundant number is 945. (Its prime factorization is $945 = 3^3 \times 5 \times 7$, and the sum of its factors is 975.)

2.13 Champernowne's number.
If you concatenate positive integers, 1, 2, 3, 4, . . . , and lead with a decimal point, you get Champernowne's number, 0.12345678910111213141 . . . Like π, e, and Liouville's number, Champernowne's number is transcendental.

Champernowne's number continues to fascinate me. It is a transcendental that we know is "normal" in base 10, which means that any finite pattern of numbers occurs with the frequency expected for a completely random sequence. In fact, David Champernowne showed that not only will the digits 0 through 9 occur exactly with a 10 percent frequency in the limit, but each possible block of two digits will occur with 1 percent frequency in the

limit, each block of three digits will occur with 0.1 percent frequency, and so on.

Some cryptographers have noted that Champernowne's number does not trigger some of the traditional statistical indicators of nonrandomness. In other words, simple computer programs, which attempt to find regularity in sequences, may not "see" the regularity in Champernowne's number. This deficit reinforces the notion that statisticians and cryptographers must be very cautious when declaring a sequence to be random or patternless. It is not known whether the digit sequences of π and e are normal.

2.14 Copeland-Erdös constant.
The number 0.23571113171923 . . . is the famous Copeland-Erdös constant, created by concatenating the prime numbers 2, 3, 5, 7, 11, . . . In 1945, Arthur Copeland and Paul Erdös proved that this number is normal in base 10, which, as we said for the Champernowne number, means that any finite pattern of numbers occurs with the frequency expected for a completely random sequence. See A. H. Copeland and P. Erdös, "Note on Normal Numbers," *Bulletin of the American Mathematics Society* 52 (1946): 857–60.

2.15 Thue constant.
The Thue constant is an example of an irrational, transcendental number that is *not* normal. For this binary number, the nth digit is 1 if n is not divisible by 3 and is the complement of the ($n/3$)th bit if n is divisible by 3. (In base 10, this number is 0.85909979685470310490357250 . . .) You can also generate this constant by mapping $0 \rightarrow 111$ and $1 \rightarrow 110$. To create the growing

string of characters, simply start with 0, do the substitutions, and watch the Thue constant emerge:

$$0$$
$$111$$
$$110110110$$
$$110110111110110111110110111$$

$$\cdots$$

2.16 Exclusionary squares. I learned about exclusionary squares from my colleague Andy Edwards. The other 6-digit example is $203,879^2 = 41,566,646,641$. Problems like this seem to be best solved using brute-force computer methods. If we do not require the digits to be distinct, Ilan Mayer has found several exclusionary cubes, such as $6,378^3 = 259,449,922,152$ and $7,658^3 = 449,103,134,312$. If we do not require that all the digits be unique, Jonathan Dushoff notes that we can find exclusionary squares of any length; for example, we can experiment with strings of 3s, such as $3,333,333^2 = 11,111,108,888,889$. German Gonzales tells me that he has found 168,569 exclusionary numbers from 1 to 1,000,000 of various orders. For example, here is an exclusionary number of the 83rd order: $2^{83} = 9, 671, 406, 556, 917, 033, 397, 649, 408$.

2.17 The grand search for isoprimes. Here is a list of other isoprimes (in base 10): 11; 111; 1,111,111,111,111,111,111; and 11,111, 111,111,111,111,111,111. In the world of factoring and primality testing, 11 is also called a repunit (repeated unit) prime. All repunit primes in base 10 can only be composed of 1s. The next such number has 317 digits:

11,111,111,111,111,111,111,111,
111,111,111,111,111,111,111,111,111,111,
111,111,111,111,111,111,111,111,111,111,
111,111,111,111,111,111,111,111,111,111,
111,111,111,111,111,111,111,111,111,111,
111,111,111,111,111,111,111,111,111,111,
111,111,111,111,111,111,111,111,111,111,
111,111,111,111,111,111,111,111,111,111,
111,111,111,111,111,111,111,111,111,111,
111,111,111,111,111,111,111,111,111,111,
111,111,111,111,111,111

The next such number has 1,031 digits. After that, the next two isoprimes that are believed to be prime, but are not proven such, contain 49,081 digits and 86,453 digits. Chris Caldwell has interesting Web sites on prime numbers for further exploring: primes.utm.edu/ and primes.utm.edu/glossary/page.php/ Repunit.html.

Do you think humanity will ever find larger isoprimes? Regarding the oscillating bit prime, in 1991, Harvey Dubner discovered a prime number with a total of 5,114 digits that is composed of only 1s and zeros. The precise number is $(10^{5114} - 10^{2612} + 9)/9$. Amazing. I do not know whether the 0s and 1s oscillate in any particular pattern in this large number.

2.18 Special augmented primes. Colleagues have determined that there is a general set of numbers of the form 90909 . . . 91 that sometimes consists of special augmented primes (SAPs). For example, one colleague discovered that 909090909090909091—when augmented by 1s to form 19090909090909090911—is a factor of $10^{19} + 1$. Other colleagues have searched for SAPs of higher order. For example, 137 corresponds to a SAP of order 2 because 21372/137 = 156.

2.19 Triangle of the Gods.

By computer search, my colleague Daniel Dockery found the following smallest prime number of this kind in Row 171:

```
12345678901234567890l234567890
l2345678901234567890l234567890
l2345678901234567890l234567890
l2345678901234567890l234567890
l2345678901234567890l234567890
   l23456789012345678901
```

The largest known prime number of this kind occurs in Row 567 and ends in the digit 7. When you perform such searches, note that you can immediately eliminate numbers ending in the even digits and the number 5.

We can ask many questions. What percentage of prime numbers do you expect as we scan more rows in the mysterious triangle? If you could add one digit to the beginning of each number in order to increase the number of primes, what would it be? If you could add one digit to the end of each number in order to increase the number of primes, what would it be?

2.20 Body weights.

This means that if you gained or lost weight, you would not change weight smoothly, but your weight would jump up or down by increments of 3.1415 . . . pounds. The largest biological effect of this strange quantization would be for the newborn, where a 3-pound difference would have the most profound and perhaps fatal effect. In other words, if this quantization became commonplace, many newborns would die. Could a premature infant weighing π pounds survive? (Of course, I'm not implying that there is something special about π in this question, because a 3-pound quantization would have similar effects.)

2.21 Jesus and negative numbers.

No. The concept of negative numbers started in the seventh century. At this time, we first see negative numbers used in bookkeeping in India. The earliest documented evidence of the European use of negative numbers occurs in the *Ars magna*, published by the Italian mathematician Girolamo Cardano in 1545. Al-Khwarizmi, who was born in Baghdad, discovered the rules for algebra around A.D. 800. Obviously, there is quite a bit of surprisingly simple mathematics that was not around in Jesus's time.

2.22 Jesus notation.

Some scholars, such as Daniel Dockery, have claimed that Jesus spoke Aramaic, and we expect that Jesus used the Aramaic/Hebrew number system, where alphabetic characters also served as their numbers. Because some of the apocryphal and the pseudepigraphic infancy gospels tell tales of Jesus having discussed the symbolism of the Greek and related alphabets, one might also argue that he could have written using the Greek number system, which likewise used its alphabet for numerical digits.

If one considers the text of the New Testament as definitive, reliable, or historical, all numbers that appear in passages with references to Jesus in the four gospels are written out in Greek (e.g., eis/mian [one], duo/duos [two], treis/trisin [three], tessares [four], hex [six], hepta [seven], okto [eight], heptakis [seven times], ennea [nine], deka [ten], eikosi pente [twenty-five], triakonta [thirty], hekaton [one hundred], hebdomekontakis heptai [seventy times seven], dischilioi [two thousand], pentakischilioi [five thousand], etc.). Most numbers in the text of the Bible tend to be written out, though there are a few exceptions, such as the infamous 666 of the Apocalypsis, written with the three Greek letters chi, xi, and the antiquated sigma. In the Greek numeral system, the letter chi has a value of 600, xi 60, and the sigma/digamma a value of 6, so that the three letters appearing together as a number have the combined value of 666.

2.23 Jesus and multiplication.

Very likely. In Matthew 18:22, we find, "*legei auto ho Iesous Ou lego soi eos heptakis all' eos hebdomekontakis epta*." Or, in Jerome's Vulgate, "*dicit illi Iesus non dico tibi usque septies sed usque septuagies septies*." Today we translated this as "Said Jesus: To you I say not 'til seven times,' but 'until seventy times seven.'" Because both seven and seventy can have symbolic meanings, the meaning may not be literal, but, nevertheless, it is an example of multiplication.

The Bible does not make it clear whether Jesus or his listeners would have been able to give the exact answer. Much earlier, in Leviticus 25:8, we find "Seven weeks of years shall you count—seven times seven years—so that the seven cycles amount to forty-nine years." Therefore, we know these people could do at

least 7×7. However, we must not lose sight of the possibility that the biblical translators introduced the terms.

In addition, conversion between monetary systems like Roman sesterces, Jewish shekels, and Persian darii probably required notions of multiplication and division. Jesus was probably aware of the concept of debts and interest charged on debts.

Jesus would not have used a symbol for zero, because neither the Hebrew, the Aramaic, nor the Greek number systems had a character representing the number 0, as it was not required by their nonpositional number systems.

2.24 The digits of π.

Certainly, if we assume that modern mathematical conjectures are correct. Pi contains an endless number of digits with what mathematicians conjecture to be a "normal" or "patternless" distribution. We can even search for some of the first few consecutive runs, using computer searches that are available on the Web. The string 123 is found at position 1924, counting from the first digit after the decimal point. The "3." is not counted. The string 1234 is found at position 13,807; 12345 is found at position 49,702; and so forth. You can do further searches of this kind at Dave Anderson's π Web site: www.angio.net/pi/piquery.

2.25 Living in the π matrix.

I believe so, although many people have debated me on this subject. Recall that the digits of π (in any base) not only go on forever but seem to behave statistically like a sequence of uniform random numbers. In short, if the digits of π are normally distributed, somewhere inside the endless digits of π is a very close representation for all of us—the atomic coordinates of all our atoms, our genetic code, all our thoughts, all our memories. Thus, all of us are alive and, it is hoped, happy, in π. Pi makes us live forever. We all lead virtual lives in π. We are immortal.

You can read a large group discussion of this topic, which I initiated at my Web site, sprott.physics.wisc.edu/pickover/pimatrix.html. At my Web site, I state my controversial opinion: "This means that romance is never dead. Somewhere you are running through fields of wheat, holding hands with someone you love, as the sun sets—all in the digits of pi."

2.26 Numbers and sign language. The person would sign a 1, followed by a C (made with curved fingers in the shape of a C) to signify the Roman numeral C, which stands for one hundred.

2.27 Adding numbers. My hero, the mathematical prodigy Karl Friedrich Gauss (1777–1855), the son of a bricklayer, discovered that he could sum the numbers from 1 to n using the formula $n(n + 1)/2$. Thus, if we want to sum 1 to 1,000, we simply compute $1,000 \times (1,001)/2 = 500,500$.

Little Gauss demonstrated his approach at age ten, when he quickly solved a problem that had been assigned by a teacher to keep the class busy. The teacher had asked the students to find the sum of the first 100 integers, and he was amazed that Gauss could add the terms so quickly. In fact, the teacher assumed that Gauss was wrong.

2.28 The mystery of 0.33333. The reason that we find $1 = 0.9999 \ldots$ is that it is true. There are numerous mathematical ways to show this that involve the sum of an infinite series, but my favorite way doesn't require too much math. Consider that any two distinct (different) real numbers must have another number in-between them. However, there is no number between 1 and $0.9999 \ldots$ Thus, 1 and $0.9999 \ldots$ are not different numbers.

2.29 The grand Internet undulating obstinate number search. My friend Daniel Dockery has computed the following obstinate number:

999999999999999999999999999999
999999999999999999999999999999
999999999999999999999999999999
999999999999999999999999999999
9999999999999999999999999999
9999999999999999999999999999
999999037

It leaves a composite residue for all 621 possible powers of 2 that can be subtracted from it.

The smallest difference between adjacent obstinates is 2; for example, 905 and 907 are both obstinate by this definition and have a difference of 2. Because obstinate numbers must be odd, 2 is the smallest difference.

Do any obstinates undulate? Certainly. Here are just a few: 6,161; 14,141; 39,393; 91,919; 1,313,131; 1,818,181; 7,070,707; 7,474,747; 7,676,767; 7,979,797; 59,595,959; 73,737,373; and so on.

How are obstinates distributed through the numbers as we scan ever larger numbers?

2.30 Mystery sequence. The missing number is 49. To create this sequence, I listed the numbers 1 through 13. Underneath this list, I listed the numbers 13 through 1. Then, I just multiplied the numbers in each column:

```
 1  2  3  4  5  6  7  8  9 10 11 12 13
13 12 11 10  9  8  7  6  5  4  3  2  1
13 24 33 40 45 48 49 . . .
```

You can also solve this another way, simply by adding 11, 9, 7, and so forth. These numbers represent the differences between consecutive terms.

2.31 Strange code. ...--, obviously. Here, we are using the Morse code, invented by Samuel Morse (1791–1872), in which letters and numbers are represented by dots and dashes: (0, -----), (1, .----), (2, ..---), (3, ...--), (4, ...-), (5,.....), (6, -....), (7, --...), (8, ---..), and (9, ----.).

2.32 Mystery sequence. 1. Starting with 1, I continue to add 8. However, if my number ever gets greater than 22, I then subtract 22, and continue. I would be interested in hearing from those of you who got a different answer, using another kind of reasoning. One of my colleagues arrived at an answer of 23 by examining the differences between consecutive terms, which follow the sequence +8, +8, −14, +8, +8, −14 . . . It's also easy to get 23 by viewing the sequence as three interleaved sequences with constant difference 2.

Of course, given a sequence of n arbitrary numbers, it is *always* possible to justify *any* other integer as the next number in the sequence by writing a polynomial equation of order $n + 1$. What I seek are very simple recipes. I am also interested to see which reader recipes are most common.

2.33 Mystery sequence. Write down all the numbers from 0 to 19. Start at 1. Jump 7. Repeat. When you get to 19, go back to the start of the list. Once you land on a number in your original list, it gets removed so that it is not used again as you traverse the numbers. You can imagine the list as numbers being around the circumference of a circle as you go round and round. Here is the sequence that is produced as a result: 1, 8, 15, 3, 11, 19, 9, 18, 10, 2, 14, 7, 5, 4, 6, 13, 0, 12, 16, 17.

2.34 Mystery sequence. The sequence lists the prime numbers (numbers divisible only by themselves and 1), starting at 2 and then lumping their digits into sets of 4:

$$2, \ 3, \ 5, \ 7, \ 11, \ 13, \ 17,$$
$$19, \ 23, \ 29, \ 31, \ 37, \ 41, \ 43$$

2.35 Time-travel integer. The most recent year is 864. Alas, Pete won't have too many amenities, but at least it is the most recent date with these characteristics. One colleague told me that this problem is too trivial to include in this book. If you agree, try this on a young student, and see how long it takes him or her to arrive at an answer.

2.36 Mystery sequence. The next number is 7,776. $1^0 = 1$; $2^1 = 2$; $3^2 = 9$; $4^3 = 64$; $5^4 = 625$; $6^5 = 7,776$.

2.37 Ostracism. Number 55. All the others are 1 less than a square number (4, 9, 16, 25, 36, 49, 64)—that is, a number produced by squaring an integer. For example, $24 = 5^2 − 1$.

2.38 Mystery sequence. The sequence consists of prime numbers, starting at 17, plus 1. Here are the original prime numbers: 17, 19, 23, 29, 31, 37, 41, . . .

2.39 Mystery sequence. Lizzy plays 4 notes. She is just multiplying the digits of each number to get the next.

2.40 Cellular communication. This is the Morse-Thue sequence, which has dozens of fascinating properties. There are many ways to generate the Morse-Thue sequence. One way is to visualize this as a sequence of 0s and 1s. Start with a zero and then repeatedly do the following replacements: $0 \rightarrow 01$ and $1 \rightarrow 10$. In other words, whenever you see a 0 in a row, you replace it with a 01 in the next row. Whenever you see a 1, you replace it with a 10. Starting with a single 0, we get the following successive "generations":

$$0$$
$$0\ 1$$
$$0\ 1\ 1\ 0$$
$$0\ 1\ 1\ 0\ 1\ 0\ 0\ 1$$
$$0\ 1\ 1\ 0\ 1\ 0\ 0\ 1\ 1\ 0\ 0\ 1\ 0\ 1\ 1\ 0$$

The next line should be

$$0\ 1\ 1\ 0\ 1\ 0\ 0\ 1\ 1\ 0\ 0\ 1\ 0\ 1\ 1\ 0$$
$$1\ 0\ 0\ 1\ 0\ 1\ 1\ 0\ 0\ 1\ 1\ 0\ 1\ 0\ 0\ 1$$

where the 0s are replaced by red and the 1s by blue.

I'm in love with the Morse-Thue sequence. Each generation can be formed by appending its complement. For example, 0 1 1 0 1 0 0 1 is just 0 1 1 0 placed next to its complement,

1 0 0 1. There can never be more than 2 adjacent terms that are identical. For example, we'll never see a 111, no matter how large we let the sequence grow. Hundreds of Web pages are devoted to this subject.

2.41 Mystery sequence. The next number is 92. These are pentagonal numbers. If balls are piled so that each layer is a pentagon, then the total number of balls in each successive pile follows this sequence (figure A2.1). The general formula for the nth number in the sequence is $(1/2) \times n \times (3n - 1)$. The first few are 1, 5, 12, 22, 35, 51, 70, and 92. Curiously, all numbers of such a type end in 0, 1, 2, 5, 6, or 7. This problem can also be solved simply by examining the differences between the numbers (4, 7, 10, 13, 16, 19 . . .). So, the next difference is 22, and 22 + 70 = 92.

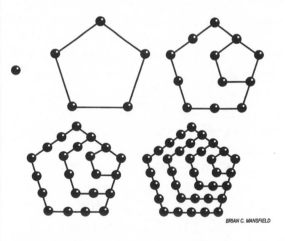

BRIAN C. MANSFIELD

Figure A2.1   Pentagonal numbers.

2.42 Mystery sequence. The solution is 89,793. The ith term of the sequence is the next i digits of the number π (π = 3.14159265358979323846 . . .).

2.43 Ostracism. This does not belong in the list: $15 \times 87 = 1{,}305$. In all other products, the digits on the left side of the equation also appear on the right side. Another answer might be 1,827, because this is the only product that does not have 5 as a factor.

2.44 The amazing 5. The four facts are: 5 is the hypotenuse of the smallest *Pythagorean triangle*, a right-angled triangle with integral sides; 5 is the smallest *automorphic* number; 5 is probably the only odd *untouchable number*; and, there are 5 Platonic solids: the tetrahedron, the cube, the octahedron, the dodecahedron, and the icosahedron. (All the faces of a Platonic solid must be congruent regular polygons.)

2.45 Mystery sequence. The value of the missing digit is 7. The solution relates to the number of segments on a standard calculator display that are required to represent the digits starting with 0:

2.46 Mystery sequence. The solution is 1828. The *i*th term of the sequence is the next *i* digits of the number e ($e = 2.7182818284 \ldots$). The number e, like π, is transcendental and consists of a never-ending string of digits.

2.47 Mystery sequence. The violinist is simply marking every prime number (numbers divisible only by themselves and 1) with a short note. So the second, the third, the fifth, the seventh, (and so on) are short:

2.48 Mystery sequence. This strange sequence lists the indices of the prime Fibonacci numbers. For example, the third, the fourth, and the fifth Fibonacci numbers ($F3$, $F4$, $F5$) are primes. (The Fibonacci sequence is 1, 1, 2, 3, 5, 8, 13 . . . , where each number is the sum of the previous two.)

2.49 The Lego sequence. Each element $a(n)$ is the number of stable towers that can be built from n Lego blocks.

2.50 Vampire numbers. I discuss vampire numbers in detail in my book *Wonders of Numbers*. Here are some other four-digit vampires:

$$21 \times 60 = 1{,}260 \quad 15 \times 93 = 1{,}395$$
$$30 \times 51 = 1{,}530 \quad 21 \times 87 = 1{,}827$$
$$80 \times 86 = 6{,}880$$

In fact, there are many larger vampire numbers. Here's a beauty for you:

$$1{,}234{,}554{,}321 \times 9{,}162{,}361{,}086 =$$
$$11{,}311{,}432{,}469{,}283{,}552{,}606$$

2.51 Jewel thief. We have 15 emeralds, 12 diamonds, and 9 rubies altogether. This means that the maximum number of gems that could be out of the bag without having any matched sets would be 2 of each set for each kind: 10

emeralds, 8 diamonds, and 6 rubies. Thus, withdrawing 25 gems would guarantee that there is at least one matched set (small, medium, and large) of one of the types of gems.

Notice that if you are creative and suggest that the thief can tell sizes by touch alone, then the answer is 12, because he can tell by touch which are the biggest!

Let us assume that he can't tell size by touch because he is wearing gloves. With respect to the second question, the thief has 12 large gems altogether and 36 gems total. The only way to guarantee that all the large gems have been randomly selected and pulled from the bag is to remove all 36 of the gems. Any gem left in the bag at random could be a large gem, so all must be removed.

Perhaps a more difficult question to ponder is, What is the probability that the first 3 gems he withdraws from the bag will make a complete set? We believe the answer to be 18/595, which is about equal to 0.03, very slim odds. The first gem drawn must be some size of some type. The probability of it being an emerald is 15/36, a diamond is 12/36, and a ruby is 9/36. If the first gem is an emerald, the probability of the second gem also being an emerald but of a different size is 10/35, and of the third being an emerald of the third size is 5/34. So, the probability of getting a set of emeralds is $15/36 \times 10/35 \times 5/34 = 25/1,428$. Similarly, the probability of getting a set of diamonds would be $12/36 \times 8/35 \times 4/34 = 16/1,785$, and the probability of getting a set of rubies would be $9/36 \times 6/35 \times 3/34 = 9/2,380$. Adding up these probabilities together gives us $25/1,428 + 16/1,785 + 9/2,380 = 18/595$.

2.52 Palinpoints of arithmetical functions. Here is one example for $n = 21$: Prime(21) = 73, Prime(12) = 37. Some people have used the Mathematica software package to seach for more examples. In 2003, Jens Kruse Andersen found the largest known palinpoint of $f(n)$ = Prime(n): $n = 8,114,118$ with *Prime*(n) = 143,787,341. (For more information, see Joseph Pe's article on palinpoints in his Number Recreations Page at www.geocities.com/windmill96/numrecreations.html. Also see Carlos Rivera, "The Palinpoints," www.primepuzzles.net/puzzles/puzz_194.htm.)

2.53 Dr. Brain's Mystery sequence. Number 28. These are triangular numbers. They can be formed by adding the series $1 + 2 + 3 + 4 + 5$ . . . Here's another way to visualize triangular numbers:

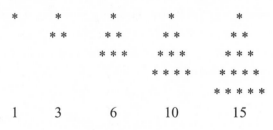

```
 *        *         *          *            *
         * *       * *        * *          * *
                  * * *      * * *        * * *
                            * * * *      * * * *
                                        * * * * *

 1        3         6         10           15
```

2.54 Poseidon's sequence. The number 77 is the smallest number with persistence 4. In particular, 77 creates the sequence 77, 49, 36, 18, 8. It turns out that 277777788888899 is the smallest number with persistence 11.

2.55 Mystery sequence. From day to day, each 🐧 is replaced with an ape-duck 🐒🐧 combination, and each ape 🐒 is replaced with a single duck 🐧 .

2.56 Constructions with I, 2, and 3. The following answers use very few numbers to represent the target number. I do not actually know whether these are answers with the fewest possible digits.

$$(3 + 2) \times 2^3 = 40$$

$$2^{(2 \times 3)} - 3 = 61$$

$$2^{(2^3)} + (2 \times 3) + 1 = 263$$

$$(2 + 2) \times ((3 + 2)^3) = 500$$

I would be happy to hear from readers regarding their own solutions.

2.57 Mystery sequence. The number is 15,625. The numbers go as $1^2, 2^3, 3^4, 4^5, 5^6, \ldots$

2.58 Blue liquid. Because –40 degrees Centigrade equals –40 degrees Fahrenheit.

2.59 Grasshopper sequences. Problems like these both delight mathematicians and drive them wild. We can even consider related sequences. For example, the grasshopper sequence $(x \rightarrow 2x + 2; x \rightarrow 5x + 5)$ yields a repeat after three generations. The grasshopper sequence $(x \rightarrow 2x + 2; x \rightarrow x + 1)$ yields a repeat after four generations. I have not yet found a solution to the $(x \rightarrow 2x + 2; x \rightarrow 6x + 6)$ problem or the related sequences: $(x \rightarrow 2x + 2; x \rightarrow 4x + 4)$ or $(x \rightarrow 2x + 2; x \rightarrow 7x + 7)$. I go into greater detail regarding these sequences in my books *Computers and the Imagination* and *Wonders of Numbers*, and similar kinds of problems are described in Richard Guy's "Don't Try to Solve These Problems!" in *The American Mathematical Monthly* 90, no. 1 (January 1983): 35.

As this book goes to press, my colleague Mark Ganson claims to have found a repeat in $(x \rightarrow 2x + 2; x \rightarrow 7x + 7)$ after 17 generations. This 17th generation has $2^{17} - 1 = 131,071$ elements, with just one repeat in it! The repeated value is 1,814,526.

2.60 The loneliness of the factorions. Two tiny factorions are 1 = 1! and 2 = 2! The largest known factorion is 40,585; it can be written as 40,585 = 4! + 0! + 5! + 8! + 5! (The factorion 40,585 was discovered in 1964 by R. Dougherty using a computer search.) Various proofs have been advanced indicating that 40,585 is the largest possible factorion, and that humans will never be able to find a greater factorion.

2.6I 69,696. This is the largest undulating square number known to humanity. A square number is an integer of the form $y = x^2$. For example, 25 is a square number. And 69,696 is a square number because $69,696 = 264^2$. Undulating numbers are of this form: abababab . . . For example, 171,717 and 28,282 are undulating numbers. Do you think humanity (or aliens) will ever discover an undulating square larger than 69,696?

2.62 The beauty of I53. The number 153 is a member of the class of numbers that are the sums of powers of their digits. In other words, these are n-digit numbers that are equal to the sum of the nth powers of their digits. For example, $153 = 1^3 + 5^3 + 3^3$. Variously called narcissistic numbers, "numbers in love with themselves," Armstrong numbers, or perfect digital variants, these kinds of numbers have fascinated number theorists for decades.

The English mathematician G. H. Hardy (1877–1947) noted that "There are just four numbers, after unity, which are the sums of the cubes of their digits. . . . These are odd facts, very suitable for puzzle columns and likely to amuse amateurs, but there is nothing in them which appeals to the mathematician."

The largest narcissistic number discovered to date is the incredible 39-digit number:

$$11513221901876399256509$$
$$5597973971522401$$

(Each digit is raised to the 39th power.) I believe it is impossible to beat this world record. What would Godfrey Hardy have thought of this multidigit monstrosity?

The number of narcissistic numbers has been proved finite. They can't have more than 58 digits in our standard base-10 number system. As one searches for larger and larger narcissistic numbers in other number systems, will they eventually run out in all number systems?

2.63 Ten silver boxes. Here is the only possible answer.

0	1	2	3	4	5	6	7	8	9	Row 1
6	2	1	0	0	0	1	0	0	0	Row 2

This problem is well known in the mathematical literature. A more advanced problem, which I designed, continues this process to generate a sequence in a recursive fashion. In this variation, each row becomes a "starting point" for the next. For example, start with the usual 0 through 9 digits in Row 1:

0	1	2	3	4	5	6	7	8	9	Row 1
6	2	1	0	0	0	1	0	0	0	Row 2
										Row 3

Now use Row 2 as the starting point, and your next task is to form Row 3, Row 4, and so on, using the same rules. For example, a digit in the first space of Row 3 would indicate how many 6s there are in Row 3's 10-digit number. The second entry in Row 3 tells how many 2s to expect in Row 3, and so forth. Can you find a Row 3 or a Row 4?

2.64 Erdös equation. The solutions are $25 = 4! + 1 = 5^2$, $121 = 5! + 1 = 11^2$, and $5041 = 7! + 1 = 71^2$. I do not know if there are any others. The mathematician Paul Erdös long ago conjectured that there are only three such numbers. Erdös offered a cash prize for a proof of this. Currently, it is not known whether the particular equation $m^2 - 1 = n!$ has only finitely many solutions. Various researchers have searched for solutions with incredibly large values for n and m, but so far none have been found.

2.65 Perrin sequence. The sequence produced by $u(n + 3) = u(n) + u(n + 1)$ is called the "Perrin sequence," after the French mathematician who wrote about this subject in *L'Intermédiaire des Mathématiciens* in 1899. Another French mathematician, Edouard Lucas, was the first to study this sequence in 1876. Both gentlemen were unable to find any counterexamples to the notion that this sequence always produced prime numbers.

Bill Adams and Dan Shanks, two American mathematicians, rediscovered the sequence in the 1980s and quickly found the first counterexample at $n = 271,441$, by computer.

You can learn more about the Perrin sequence from these papers:

William Adams and Daniel Shanks, "Strong Primality Tests That Are Not Sufficient," *Mathematics of Computation* 39, no. 159 (July 1982); G. C. Kurtz, Daniel Shanks, and Hugh C. Williams, "Fast Primality Tests for Numbers Less than 50×109," *Mathematics of Computation* 46, no. 174 (April 1986); Hugh C. Williams, *Edouard Lucas and Primality Testing* (Canadian Mathematical Society Series of Monographs and Advanced Texts, Vol. 22) (New York: Wiley, 1998).

2.66 Square numbers. The only such numbers are 2 and 9. For example, $9^2 = 81$ and $9 \times 2 = 18$.

2.67 $\sqrt{2}$ is irrational. Let us assume for the moment that $\sqrt{2}$ is rational—which is the opposite of what we expect to prove. This means we are saying that $\sqrt{2}$ can be expressed as the ratio a/b, where a and b are integers having no common factor.

Given that $\sqrt{2} = a/b$, we can square both sides of the equation to get $2 = a^2/b^2$, which means that $a^2 = 2b^2$. Because any integer multiplied by 2 is an even number, a^2 must be even. In fact, a itself must be even because if it were odd, a^2 would also be odd.

Given our assumptions, we now know that a is even, which means that a can be rewritten as $2r$, where ra is an integer having half the value of a. Making substitutions, we have $(2r)^2 = 2b^2$, or $4r^2 = 2b^2$, or $b^2 = 2r^2$. This also means that b is even, as we reasoned previously for a.

Wait just a second! If a and b are both even, they have a common factor of 2. But we previously defined a and b as having no common factor, so our original assumption that $\sqrt{2}$ can be expressed as the ratio a/b (where a and b are integers having no common factor) is false. Thus, we have shown that $\sqrt{2}$ is irrational.

Here's a simpler approach for showing the irrationality of $\sqrt{2}$. Consider the aforementioned step, $a^2 = 2b^2$. Any factor in the prime factorization of a^2 (and, also, of b^2) must have an even power because of the square, so the prime factor 2 appears an even number of times in a^2 (example: $8^2 = 16 = 2^4$), but an odd number of times in $2b^2$, implying that a^2 is not equal to $2b^2$, a contradiction.

2.68 Fibonacci snakes. To answer this question, write down the number of pairs in each generation. First write the number 1 for the single pair you bought from the pet shop. Next write the number 1 for the pair they produced after a year. The next year both pairs have young, so the next number is 2. Continuing this process, we have the sequence of numbers 1, 1, 2, 3, 5, 8, 13, 21, 34, 55, 89, 144, 233, 377, . . . This sequence of numbers is called the "Fibonacci sequence," after the wealthy Italian merchant Leonardo Fibonacci of Pisa. As we said in the definition of the Fibonacci sequence, these numbers play important roles in mathematics and nature. These numbers are such that after the first two, every number in the sequence equals the sum of the two previous numbers $F_n = F_{n-1} + F_{n-2}$.

2.69 Palindromes on parade. Following are the number of 1-digit palindromes, the number of

2-digit palindromes, and so on (the left-most digit is not allowed to be zero).

Digit	Form	W(P)
1 digit	*a*	9
2 digits	*aa*	9
3 digits	*aba*	9×10
4 digits	*abba*	9×10
5 digits	*abcba*	$9 \times 10 \times 10$
6 digits	*abccba*	$9 \times 10 \times 10$

Using this table, you can easily see that

$$W(10^n) = 2[10^{(n-1)/2} - 1] + 9 \times 10^{(n-1)/2}$$
$$\text{for } n \text{ odd}$$

$$W(10^n) = 2[10^{n/2} - 1] \text{ for } n \text{ even}$$

We can conjecture that

$$\lim_{P \to \infty} \frac{\log W(P)}{\log(P)} = \frac{1}{2}$$

2.70 Alpha and omega. When Dr. Brain's wife says, "Simply by looking at the product, I can't tell what alpha and omega are," what she really means is that the product does not automatically determine any set of two mystery factors such as alpha and omega. This means that omega and alpha cannot both be primes. Why? If the mystery numbers were two primes (5 and 11, for example), then Mrs. Brain would know them at once on being given the product (55). However, Mrs. Brain says that she can't tell what alpha and omega are.

We also know that none of the prime factors of either alpha or omega can be bigger

than 47. To see this, consider a case where one of them is 53, a prime number. The smallest number that this is a factor of is 53, the next smallest is 106, and so on. Since we have been told that both of our mystery numbers are between 2 and 99 inclusive, 53 must be one of the numbers. The other one must be the product, divided by 53. Once Mrs. Brain has figured out that 53 is a factor, she would be able to determine the other number. Since she couldn't, there can't be any prime factors bigger than 47. (I learned about this kind of problem from Professor Edsger W. Dijkstra of the Netherlands, olimu.com/Notes/ProductAndSum.htm.)

2.71 Does pi contain pi? No. Several of my colleagues have noted that if pi contained pi, at the point where the "contained pi" started, pi would become a repeating decimal. Then pi would not be irrational. Hence, pi cannot contain pi. Is there any way that a number can "contain itself" other than it having a pattern that repeats itself?

Here's another way to look at it. If a number contained itself starting after *n* digits, then the first *n* digits would be the same as the second *n* digits. Similarly, the second *n* digits of the *copy* would be the same as the first *n* digits of the copy, which means, since the second *n* digits of the copy are the third *n* digits of the original number, the third *n* digits of the original number would be the same as the first *n* digits. And so on, and so on, to infinity. The number would be a repeating decimal of period *n*.

In the original question, I said that pi "probably" includes every possible string of finite digits because this assumes that the distribution of pi's digits is normal. Currently, this is only a conjecture with some good evidence to back it up.

2.72 Word frequency.
Here are the occurrences, based on my searching with the Google search engine in the year I wrote this book. The number of occurrences is in units of millions. Note the spikes at ten, twelve, and twenty.

Number	Occurrence	Number	Occurrence
one	571	seventeen	8
two	360	twelve	16
three	183	thirteen	6
four	163	fourteen	4
five	115	fifteen	9
six	90	sixteen	5
seven	73	seventeen	4
eight	42	eighteen	4
nine	30	nineteen	3
ten	80	twenty	24

2.73 Bad luck.
The traditional explanation is that 13 people were at the Last Supper, which heralded the death of Jesus. The traitor Judas was said to be the 13th apostle. The "unluckiness of 13" started to spread through European culture in the Middle Ages. Today, many buildings do not have a 13th floor—the numbering skips from 12 to 14. See the next entry on paraskevidekatriaphobia.

2.74 The height of irrationality.
Let $A = \sqrt{2}^{\sqrt{2}}$. If A is rational, it is the desired example. On the other hand, if A is irrational, then consider $A^{\sqrt{2}}$. By one of the standard exponent laws, we have $(x^m)^n = x^{mn}$. Hence

$$A^{\sqrt{2}} = \left(\sqrt{2}^{\sqrt{2}} \right)^{\sqrt{2}} = \sqrt{2}^{\sqrt{2} \times \sqrt{2}} = \sqrt{2}^2 = 2$$

So, this is the desired example that an irrational number raised to an irrational power need not be irrational (James F. Hurley, *Litton's Problematical Recreations* [New York, N.Y.: Van Nostrand Reinhold, 1971]).

2.75 A common fear.
The term *paraskevidekatriaphobia* was coined by Dr. Donald Dossey and refers to extreme fear of Friday the 13th. Paraskevidekatriaphobia is related to triskaidekaphobia, the fear of the number 13. Interestingly, the number 13 was considered a lucky number in ancient Egypt and China and in the Mayan civilization. According to Robert Todd Carroll, the author of *The Skeptic's Dictionary*, 13 may have a bad reputation because Loki, the Norse god of evil, started a riot when he came to a banquet at Valhalla attended by 12 gods. The ancient Hebrews considered 13 unlucky because 13 was designated by the 13th Hebrew letter, which corresponded to the word *mem*, meaning "death." Friday may be considered unlucky because Jesus is thought to have been crucified on a Friday, the Roman day for executions.

Researchers writing in the *British Medical Journal* have compared the ratio of traffic volume to vehicular accidents on two different days, Friday the 6th and Friday the 13th, over a period of years. Surprisingly, they found that the number of hospital admissions due to accidents was significantly higher on Friday the 13th than on "normal" Fridays. For more information, see T. Scanlon, R. Luben, F. Scanlon, and N. Singleton, "Is Friday the 13th Bad for Your Health?" *British Medical*

Journal 307, no. 6919 (December 18–25, 1993): 1584–86.

According to Dr. Donald Dossey, as many as 21 million people suffer from paraskevidekatriaphobia in the United States alone.

2.76 Prime test. Yes, but only if the person is lucky and decides to check to see if the big number is divisible by 3, which is one of the easiest tests of divisibility. Because a prime number can't be divisible by 3, if we should find that this number is, it will not be a prime.

A number is divisible by 3 if its digits sum to a number that is divisible by 3. To see if our number is divisible by 3, first strike out all zeros and any other single digits that are divisible by 3, which quickly takes us from the original number

$$5,2\cancel{30,096,303,003,}1\cancel{96,309,630,96}7$$

to

$$5217$$

These digits sum to 15, which is divisible by 3. Thus 5,230,096,303,003,196,309,630,967 cannot be a prime number.

2.77 Never prime? No. Here is a counterexample: $1^2 + 2^2 + 5^2 + 7^2 = 79$. However, as my colleague Nick Hobson notes, if we have $a \times b = c \times d$ ($a, b, c, d > 0$), it can be shown that $a^2 + b^2 + c^2 + d^2$ is composite. In fact, the result remains true if we replace the exponent, 2, with any non-negative integer! Examples:

$24 = 3 \times 8 = 4 \times 6 \rightarrow$
  $3^2 + 4^2 + 6^2 + 8^2 = 5 \times 25$

$25 = 1 \times 25 = 5 \times 5 \rightarrow$
  $1^2 + 5^2 + 5^2 + 25^2 = 676 = 2 \times 338$

$26 = 1 \times 26 = 2 \times 13 \rightarrow$
  $1^2 + 2^2 + 13^2 + 26^2 = 850 = 2 \times 425$

$60 = 5 \times 12 = 6 \times 10 \rightarrow 5^5 + 6^5 + 10^5 + 12^5$
  $= 359,733 = 3 \times 119,911$

2.78 Integer hoax. Martin Gardner flustered readers in the April 1975 issue of *Scientific American* by saying (in jest) that $e^{\pi\sqrt{163}}$ is an integer, and various authors printed this "fact" in their later books. However, this number is only very "close" to being an integer. Its exact value is

$$262537412640768743.$$
$$9999999999992500\ldots$$

This is sometimes called a Ramanujan number, although Ramanujan (1887–1920) did not give this particular example. He did give other impressive examples of almost integers, such as $e^{\pi\sqrt{58}}$.

2.79 Cube square puzzle. N is 69. So $69^2 = 4,761$ and $69^3 = 328,509$. I believe there may be various observations that we can make to make this problem easier to solve. For example, it appears that N must be two digits, because N (2 digits), S (4 digits), and C (6 digits) are the only possible numbers to make this work. In turn, this means that N must be greater than 31 and less than 100.

2.80 Cube numbers. N is $216 = 6^3 = 3^3 + 4^3 + 5^3$.

2.81 0s and 1s. The missing equivalence is $1110 = 10$. I am simply replacing the 1s with their corresponding prime number and then adding each prime number. Here are the first few prime numbers: 2, 3, 5, 7, 11, 13, 17, 19, 23, 29, 31, 37, . . . So, for example, 01 = 3, 011 = 5 + 3 = 8, and 1,001 = 2 + 7 = 9.

2.82 Cube mystery. A number like 17 is one of only 6 other numbers that are equal to the sum of the digits of their cube. For example, $17^3 = 4,913$, and $4 + 1 + 9 + 3 = 17$.

2.83 Consecutive prime numbers. They are 2 and 3.

2.84 Alien ships. 3,456. To determine the number of ships in the sequence, start with 3, then perform the following operations: multiply by 2, multiply by 3, multiply by 4, repeat . . .

2.85 Trains with crystal balls. There are 138 crystal balls in the last car. Start with 42. After this, I simply added the digits together to get the first digit or digits in the next number. I also incremented the last digit of the number by 1 to obtain the last digits in the next number. So, for example, 4 + 2 = 6, which is the first digit of 63. The 3 is obtained simply by incrementing the 2 in 42. Next, we add 6 + 3 to get 9 and then append 4 as the right-most digit, and so forth.

2.86 Mr. Zanti's ants. The number in the last sequence is 27,306. Start with $n = 1$. The generating formula is $n \leftarrow 2(2n + 1)$. So, this means that 1 becomes 6, 6 becomes 26, and so forth.

2.87 Prime king. The deck now contains 32 cards, assuming a standard pack. Note that cards like the jacks and the aces don't show numbers on their faces. The only cards missing are the kings, the 2s, the 3s, the 5s, and the 7s.

2.88 Pi deck. Any shuffling of cards with numbers will produce a digit string found somewhere in the endless digits of pi. If we wish, we can analyze this further. Since all possible arrangements are found somewhere in pi, it comes down to determining how many unique arrangements (shufflings) are possible. By removing the aces, the kings, the queens, and the jacks, we are left with a deck of 36 cards. There are 9 cards (2 through 10) for each of the 4 suits (hearts, diamonds, spades, and clubs). The number of unique arrangements (ignoring suits) is $(4 \times 9)/24^9 =$ 140,810,154,080,474,667,338,550,000,000, each of which produces a string of 40 digits of pi, since the 10s each produce 2 digits: 1 and 0.

3. Algebra, Percentages, Weird Puzzles, and Marvelous Mathematical Manipulations

- - - - -

3.1 ½ puzzle. The number −1.

3.2 Shrunken heads puzzle. Gary had 10 shrunken heads. Joan had 14 shrunken heads.

3.3 Loyd's Leaning Tower of Pisa. In theory, the ball travels 218.777 . . . feet or 218 feet, 9 and ⅓ inches. Here's the reasoning that Loyd never gave. Let's generalize the problem to an initial drop of n feet. In Loyd's case, $n = 179$. During the initial drop, the ball falls n feet. Then it goes back up $n/10$ feet, then falls $n/10$ again, giving us $2(n/10)$ feet. To this distance, we must add the next up/down distance, which is $2(n/100)$ feet. Thus, for each successive rising and falling, we must add $2 \times n/(10^b)$, where b is the number of times the ball has fallen.

In a moment, I'll explain why the sum of all the $2n(1/10^b)$ distances converges to $2n(1/9)$, which is $39.777 \ldots$ feet when $n = 179$ feet. For now, you can see that to get the total distance traveled, we add to this the initial drop of 179 feet to arrive at 218.7777 $\ldots$ feet.

Notice that the ball goes though an infinite series of hops, each time adding to the distance traveled. Of course, this analysis omits any frictional effects. Also, at some point, the tiny bounces traveled become subatomic, and quantum effects would be considered. Can you imagine how much more difficult this would be if the ball is strange and its successive heights go as the Fibonacci sequence?

How do we know that $n/10^b$ converges to $1/9$? Start by observing that the sum looks like $S = 1/10 + (1/10)^2 + (1/10)^3 \ldots$ If $r = 1/10$, we get $S = r + r^2 + r^3 \ldots$ There exists a simple formula for determining to what value this kind of infinite sum converges, and we can derive this formula.

Consider this sum of the form $S = r + r^2 + r^3 \ldots$ where $r < 1$. Multiply both sides of the equation by r to yield $rS = r^2 + r^3 + r^4 \ldots$. The right-hand side of the equation is simply $S - r$, so we get $rS = S - r$. Solving for S allows us to determine that the sum of an infinite geometric series of this form is

$S = r/(1 - r)$ for $r < 1$. In our case, $r = 1/10$, so $S = 1/9$.

3.4 Chimpanzees and gorillas. The chimpanzee is 45 years old. The gorilla is 9 years old. Let the age of the chimpanzee be C and the age of the gorilla be G. We may solve this by solving two equations in two unknowns:

$$C = 5G$$
$$C - 3 = 7(G - 3)$$
$$5G - 3 = 7G - 21$$
$$-2G = -18, \; G = 9$$

3.5 RED + BLUE. Here is one possible solution. Can you find others?

$$RED + BLUE = BUOL$$
$$123 + 4{,}562 = 4{,}685$$

My colleagues believe that there are 260 unique solutions for integer values greater than zero.

3.6 Loyd's "teacher" puzzle. Loyd tells us there are infinite pairs of numbers that have the same sum and product. If one number is a, the other number can be found simply by dividing a by $a - 1$. Here is why. Combine the two formulas so that you have $ab = a + b$ and solve for b.

$$ab = a + b$$
$$ab - b = a$$
$$b(a - 1) = a$$
$$b = a/(a - 1)$$

As one example, if a is 10, then $b = 10/9$.

3.7 Worms and water. People are always amazed at the answer: the 1,000 pounds of worms now weigh a mere 200 pounds. We can solve this using algebra. The solid part, S, of the worms starts out as 1 percent of 1,000, or $S = 10$ pounds. After an hour, we know that the same solid ($S = 10$ pounds) of wriggling worms is now 5 percent of the worm weight W. This means that $S = 10$ pounds $= 0.05 \times W$ pounds, so W pounds $= 10 / 0.05$ pounds $= 200$ pounds. Your poor worms are probably quite thirsty now.

3.8 Loyd's mixed tea puzzle. To solve this problem, we can find two different integers (corresponding to the different edge lengths of the green and the black tea boxes) so that when their cubes are added, the result can be evenly divided by 22 and also have a cube root that is an integer. This problem can thus be expressed as $a^3 + b^3 = 22c^3$, where a, b, and c are integers. In other words, we must find a number, c, that is the sum of cubes of two numbers, which if divided by 22 gives another number that is the cube of a third number, which is the length of the one side of the 22 combined boxes of tea. By specifying terminating decimals, we are ensuring that a, b, and c are not irrational. This, in turn, allows us to restrict a, b, and c to integers, or, if they aren't integers, we can multiply the equation by the cube of the greatest common denominator. In addition, by dividing out any common factors that a, b, and c have, we can say there exists a "minimal" solution such that a, b, and c are relatively prime.

Loyd's answer is that a cube 17.299 inches on the side and a cube 25.469 inches on the side have a combined volume of 21,697.

794418608 cubic inches, which is exactly equal to the combined volume of 22 cubes, each 9.954 inches on the side. Therefore, the green and the black teas must have been mixed in the proportion of $17,299^3$ to $25,469^3$.

My friend Mark Ganson used a computer to solve this problem. He wrote to me,

> One possible solution is where the smaller green tea chest has a side length of 17,299 and the larger black tea chest has a side length of 25,469. The 22 chests each have a side length of 9,954. Thus, $17299^3 / 25469^3 \sim 0.313348$, the proportion of green tea to black. I found the solution through a brute force search. Initially, I created a list of the first 40,000 perfect cubes and another list of these same cubes $\times 22$. Then I looped through the first list of cubes, creating a third list each time. This third list was the sum of each cube and the nth cube in the original list. Each time through the loop I checked to see if the 2nd and 3rd list intersected each other, which indicated when a cube $\times 22$ equaled the sum of 2 other cubes.

How did someone of Loyd's time arrive at an answer without a computer?

3.9 Fragile fractions. The solution is $\frac{1}{2} = 1/2^2 + 1/3^2 + 1/4^2 + 1/5^2 + 1/7^2 + 1/12^2 + 1/15^2 + 1/20^2 + 1/28^2 + 1/35^2$. This is the *only* solution of which I am aware for $x < 100$. It's possible to extend the problem to other fractions— for example, $1/3 = (1/2)^2 + (1/4)^2 + (1/8)^2 + (1/24)^2 + (1/28)^2 + (1/30)^2 + (1/40)^2 + (1/56)^2 + (1/84)^2$. Can you find a set of fragile fractions for $1/5$?

My colleagues and I do not know whether solutions exist for all of the first 100 rationals of the form $1/n$ (such as 1/2, 1/3, 1/4, . . . 1/98, 1/99, 1/100) up to $n = 100$, given the domain of the first 100 rational squares of the form $1/n^2$ (such as $1/2^2$, $1/3^2$, $1/4^2$, . . . $1/98^2$, $1/99^2$, $1/100^2$) up to $n = 100$. Perhaps the only way to answer this question is through a brute force search—by systematically testing the sums of all subsets of those first 100 rational squares until we find solutions for all of the targeted values (1/1 through 1/100). This approach is challenging because so many subsets exist, in particular, $2^{100} - 1 = 1,267,650,600,228,229,$ $401,496,703,205,375$ subsets. Even with a computer that could perform 1 billion of these subset sums per second, it would take 4×10^{13} years to check them all. However, this number may be reduced substantially by using intelligent searches.

3.10 Square pizzas.

Abraham's task is impossible. Let V be the sum of all the numbers in the stacks of all corner vertices. Suppose that Abraham is working with N different squares. The sum of all the numbers at the vertices of all the squares is $N \times (1 + 2 + 3 + 10)$ or $16N$. Can the sums of each vertex stack sum to 67?

Let M be the common sum of each vertex stack. Then $4M = 16N,$ which leads to $M = 4N$. Note that $4N$ is always an even number. No matter how many squares Abraham has, the common sum in each stack would have to be an even number. Thus, there is no way that Abraham can get each vertex stack to sum to 67.

Another approach is to start with 67, which implies that the sum of all pizza numbers is $4 \times 67 = 268$. Since each square has a sum of 16 on its surface, 268/16 has to be an integer, which is not possible.

3.11 The problem of the rich jeweler.

The jeweler has a total of 8,615,889 precious objects! My colleague David Jones was the first to be able to solve this, and he has written a detailed analysis at his Web site, www. nightswimming.com/math/. To solve this problem, he used the Microsoft Excel spreadsheet program. Jones first notes that there are three types of objects: alloys, gems, and spices. Each type has four colors: beige, green, pink, and yellow. Each item will be referred to with two letters, the first by color and the second by type. For example, pink spices will be denoted by PS, beige alloys are denoted by BA, and so on.

The second paragraph of the problem talks exclusively about gems. We start by examining the gems, and we will deal with alloys and spices later. An elementary interpretation of the paragraph yields the following equations, labeled A, B, and C.

$$A: PG = (1/4 + 1/3)BG + GG$$

$$B: BG = (1/7 + 1/3)GG + YG/2$$

$$C: GG = (1/5 + 1/11)(PG + BG)$$

Algebraic manipulation allows us to rewrite these equations as

$$A: (7/12)BG + GG - PG = 0$$

$$B: -BG + (10/21)GG + (1/2)YG = 0$$

$$C: (16/55)BG - GG + (16/55)PG = 0$$

The alloys and spices equations are

$$J: PA = (1/2 + 1/5)(PA + PG + PS) + 2YA$$

$$K: BA = (1/3 + 1/2)GA$$

$$L: GA = (1/5)PA + (1/6)YA$$

$$M: YA = (1/8 + 1/3)(GA + GG + GS)$$

$P: PS = (1/2)(BA + BG + BS)$

$Q: GS + YS = BS + PS$

Using linear algebra, David finds a minimum solution:

Color	Type	Variable	Amount
Blue	Gems	BG	98,280
Green	Gems	GG	63,840
Pink	Gems	PG	121,170
Yellow	Gems	YG	135,760
Blue	Alloys	BA	871,035
Green	Alloys	GA	1,045,242
Pink	Alloys	PA	4,802,600
Yellow	Alloys	YA	508,332
Blue	Spices	BS	105
Green	Spices	GS	6
Pink	Spices	PS	484,710
Total			8,615,889

This is quite a lot of precious objects! In addition, we may wonder how much space would be required to store all of this wonderful material. David says that the total number of gems is 419,050; the total number of alloys is 7,227,209; and the total number of spices is 969,630. Sample gems and rocks often come in ½-inch cubes, so we will assume that a gem or an alloy sample is ⅛ in.^{3} in volume. Powder samples are often stored in test tubes ⅜ inch across and 2 inches tall, so we assume ⁹⁄₃₂ in.^{3} for spices. This totals up to a net volume of just under 711 cubic feet. Assuming that the walls are 8 feet high, as in most houses, you would need a room with 89 square feet of floor space just to store all of the objects. Keep in mind that this does not count whatever boxes or materials you use to store the objects or the space you would need for a human being to walk through to retrieve items. I would think that a room the size of a normal high school chemistry lab would be sufficient.

3.12 A huge number? At first glance, this looks like a huge number. After all,

$$a_{97} = 97^{96^{95^{\cdots^{3^{2^1}}}}}$$

This number is unimaginably bigger than a googol. However, one of the terms in the long expression $(a_{97} - a_1) \times (a_{97} - a_2) \times (a_{97} - a_3) \times \ldots \times (a_{97} - a_{98}) \times \ldots (a_{97} - a_{99})$ is $(a_{97} - a_{97})$, and thus the answer is zero. I learned about this sort of problem from William Poundstone's *How Would You Move Mount Fuji?* which is a collection of puzzles that the Microsoft Corporation uses when interviewing job candidates.

3.13 x^0? Examine the powers of any number, say 4:

$$4^3 = 64$$
$$4^2 = 16$$

To get from 4^3 to 4^2, we divide 4^3 by 4. Next consider

$$4^1 = 4$$

We divide by 4^2 by 4 to get 4^1. And finally we divide 4^1 by 4 to get $4^0 = 1$. This approach works for any non-zero number.

3.14 Winged robot. Yes. Danielle's father happens to be 52 years old, and her mother's father, her maternal grandfather, is 52 years old.

3.15 Fraction family. The odd one out is 74/372. For all other fractions, if we removed numbers that appear in both the numerator and the denominator, these fractions would equal a fourth: 1/4, 2/8, 3/12, 4/16, 8/32.

3.16 Tarantulas in bottles. Point to the middle bottle and say, "Does the bottle on the right contain a live tarantula?" If the answer is "yes," then the bottle on the right contains a live tarantula. If the answer is "no," the bottle on the left contains a live tarantula. Notice that if you are pointing to a live tarantula, then Tiffany is telling the truth, and if you are pointing to the dead tarantula, the answer doesn't matter anyway because you will pick one of the other two, which are both live tarantulas.

3.17 Large ape. The day after tomorrow.

3.18 Ψ substitution. Number 1. Here's why: $58 \times 13 = 754$; $27 \times 13 = 351$; $65 \times 13 = 845$; $17 \times 13 = 221$.

3.19 Angel number-guessing game. We need to search for numbers such that $(a + b) \times 1000 = a \times b$. One such answer is 1,250 and 5,000. There are twenty-five possible unique solutions for positive integers. The largest number of possible apples or grapes is 1,001,000.

One way to solve this is to solve for b: $(a + b) \times 1,000 = a \times b$, or $1,000a + 1000b = ab$, or $1,000(a/b) + 1,000 = a$, or $a/b = (a - 1,000)/1,000$, or $b = 1,000a/(a - 1,000)$. Since we are interested in positive solutions, this means that $a > 1,000$. We also know that because $a + b = b + a$ and $a \times b = b \times a$, half of the solutions are found simply by switching a and b.

If $a = b$, the original equation yields $(a + a) \times 1,000 = a \times a$ or $2,000a = a^2$ or $a = 2,000$ and $b = 2,000$. Here are all the (a, b) solutions: (1,001,000; 1,001), (501,000; 1,002), (251,000; 1,004), (201,000; 1,005), (126,000; 1,008), (101,000; 1,010), (63,500; 1,016), (51,000; 1,020), (41,000; 1,025), (32,250; 1,032), (26,000; 1,040), (21,000; 1,050), (16,625; 1,064), (13,500; 1,080), (11,000; 1,100), (9,000; 1,125), (7,250; 1,160), (6,000; 1,200), (5,000; 1,250), (4,125; 1,320), (3,500; 1,400), (3,000; 1,500), (2,600; 1,625), (2,250; 1,800), and (2,000; 2,000).

Another solution starts from $(a - 1,000)(b - 1,000) = 1,000,000$. Then $a - 1,000$ takes on all factors of $1,000,000 = 2^6 \times 5^6$, which can easily be written out.

3.20 Starship journey. Kirk requires 30 hours. Eck requires 23 hours. Let Kirk's ship travel at speed u, and Eck's travel at speed v, in arbitrary units. Let the time of their meeting be t hours. We know that distance is time × speed. Thus, we have the following distances covered:

Kirk, before meeting: tu	Kirk, after meeting: $17u$
Eck, before meeting: tv	Eck, after meeting: $10v$

Note that the distance that Kirk travels before meeting Eck is the same as the distance that Eck travels after meeting Kirk. Thus, we then have $tu = 10v$ and $17u = tv$. We can manipulate the formulas to solve for t. For example, $t = 10v/u$ and $u = tv/17$. Thus $t = 170v/tv$, which means $t^2 = 170$, and t is approximately equal to 13. Thus Kirk's ship requires $17 + 13$ hours = 30 hours, and Eck requires $10 + 13 = 23$ hours.

3.21 Missing numbers. Here is one solution. Are there others?

13	8	9	10
5	6	33	11
4	19	7	19
3	2	10	1

We believe that there are 32 different solutions for this problem. If we randomly scrambled the digits 1 through 11 to fill the white cells, the probability of producing a correct answer is 32/11! = 32/39,916,800, which simplifies to 1/1,247,400—a roughly one in a million shot.

3.22 Journey. Let the distance between the city and New York be D miles. We know that time equals distance divided by velocity. This means that it takes $D/21$ hours to travel to New York and $D/3$ hours to return. The total time for the trip is $D/21 + D/3 = 8D/21$ hours to travel $2D$ miles, or $4/21$ hours to travel 1 mile. Therefore, the average speed is $21/4$ or 5.25 miles an hour.

3.23 Four intelligent gorillas. The successful gorilla received 2,468 votes. Let the winner receive w votes. Then the other three candidates received $w - 888$, $w - 88$, and $w - 8$ votes. So the total number of votes is $4w - (888 + 88 + 8)$, which we know equals 8,888. Hence, $w = (8,888 + 888 + 88 + 8)/4 = 2,468$.

3.24 George Nobl: A true story. He'd have to mix 10 pounds of peanuts. Let x be the pounds of peanuts Noah would have to mix with all the cashews. The cost of the peanuts is $2.50 \times x$. Thus, the total weight, w, of the final mix is $20 + x$. The equation to solve is $3.55 \times 20 + 2.50 \times x = 3.20 \times (20 + x)$, or $71 + 2.5x = 64 + 3.2x$, or $7x = 0.7x$, or $x = 10$ pounds of peanuts (Reference: NPR radio, "Math Teacher's Mission: To Make Equations Fun Sidewalk Instructor Fights Epidemic of 'Math-aphobic' Americans," www.npr.org/programs/morning/features/2002/apr/math/).

3.25 Human hands. He owns 4 hands and uses 3 shelves. Let H represent the number of hands and S the number of shelves. We know that: $H - 1 = S$ (all the hands except one had a shelf), and $H/2 = S - 1$ (half the number of hands filled all the shelves but 1). Solving for H and S, we get 4 hands and 3 shelves.

3.26 Ages. Teja: 5; Danielle: 11; Nick: 21; Pete: 31; Mark: 41.

3.27 Three arrays. Here is one solution. Can you find others? (Obviously, the numbers inside a particular array might be shuffled around to produce a valid solution.) How fast could you do this? There are 6 missing numbers, giving 6! = 720 permutations. I tried all 720 and found 8 that worked. All 8 are just variations on the given solution. Each 2×2 grid can be arranged in 2 different ways, giving $2 \times 2 \times 2 = 8$ ways to do this.

1	2
10	11

8	9
12	5

6	4
7	3

3.28 Perfect cubes? No, they cannot be perfect cubes. First, consider that if integers A and B are perfect cubes, then $A \times B$ must be a perfect cube. Why? Let $A = x \times x \times x$, and $B = y \times y \times y$. Then we have $A \times B = x \times x \times x \times y \times y \times y$, or, rearranging, $(x \times y) \times (x \times y) \times (x \times y)$, which is a perfect cube.

So, we can multiply our starting numbers $n + 3$ and $n^2 + 3$ to get $(n + 3)(n^2 + 3) = n^3 + 3n^2 + 3n + 9 = (n + 1)^3 + 8$. This means $(n + 1)^3 + 8$ must be a perfect cube, and we can see that $(n + 1)^3$ is a perfect cube.

These two perfect cubes must differ by 8. However, you can see that perfect cubes grow very rapidly: 1, 8, 27, 64, 125, ... Thus, there are no perfect cubes that differ by 8 for positive n. Therefore, $n + 3$ and $n^2 + 3$ cannot both be perfect cubes. (This puzzle was inspired by Nick Hobson, www.qbyte.org/puzzles/.)

3.29 Four pets. In 84 days; 84 is the lowest common multiple of 6, 4, 3, and 7. The lowest common multiple (LCM) of a group of numbers is the smallest number that is a multiple of each number in the group. One way of finding the LCM is to list the multiples of each number, and then pick the smallest number that is common to all numbers in the group. You can learn about other ways of finding the LCM on the World Wide Web.

3.30 Fractional swap. $1{,}534/4{,}602 = 1/3$. How long did you take to solve this? What method is the most efficient to use when solving problems of this kind? Do you think there is more than one answer? Recently, I found four different answers: 1354/4062, 1534/4602, 0454/1362, and 0544/1632. However, the last two solutions involve swapping 3 numbers, not 2. I tried all 8! permutations, which required only 0.93 seconds of actual processing time.

3.31 Rational roots. When I posed this problem to colleagues at the end of 2003, I was not certain that anyone could solve it. However, Graham Cleverley found a solution to this for $n = 113569/14400$. This meets the criteria of the problem because

$$n = 113569/14400 = 337^2/120^2$$

$$n - 7 = 12769/14400 = 113^2/120^2$$

$$n + 7 = 214369/14400 = 463^2/120^2$$

Other solutions are possible, but we think this is the smallest solution. Cleverley notes that if you can find integer solutions for the variables p, q, r, and s such that $(p^2 - 7q^2)/2pq = (r^2 + 7s^2)/2rs$, then n is the left-hand (or right-hand) side of the equation squared. The previous solution occurs with $p = 20$, $q = 3$, $r = 15$, and $s = 4$.

Cleverley used a Microsoft Excel spreadsheet and varied r and s until he found values for which $r^4 + 42r^2s^2 + 49s^4$ is square, the requisite condition for the previous equation having integers p, q, r, s.

It is intriguing that today mathematicians of all degrees of sophistication can solve problems like this using spreadsheets. It took Cleverley about an hour of programming and experimenting to find a solution.

3.32 Donald Trumpet. William got $200,000. ($200,000 + $800,000 = $1,000,000.)

3.33 Rubies and emeralds. The combined weight is 20 pounds. Let R = the weight of one ruby and E = the weight of one emerald. We know that $3R + 2E = 32$ and $4R + 3E = 44$, which means that the weight of a single ruby is 8 pounds, and the weight of a single emerald is 4 pounds. So, $2R + E = 20$ pounds.

Perhaps a more elegant solution is to note that removing one ruby and one emerald (4R + 3E → 3R + 2E) reduces the weight by 12 pounds (44 to 32), so doing the same again (3R + 2E → 2R + E) reduces by the same amount, to 20 pounds. Using this logic, there is no need to calculate the individual weight of each stone.

3.34 Crow and eagle.
The crow gets 14 and ²⁄₇ worms (approximately 14.2857 worms). The eagle gets 85 and ⁵⁄₇ worms (85.7143 worms).

3.35 Number grid.

8	+	2	+	2	=	6
×		+		×		
7	×	4	+	6	=	34
+		+		−		
3	×	5	+	9	=	24
=		=		=		
59		11		3		

3.36 Four digits.
Here is one solution: $(23 + 2) \times 5 = 125$. Are there other solutions? Recent studies suggest that 8 possible solutions exist if we require that A, B, C, and D are unique. These solutions are {A,B,C,D} {8, 3, 2, 1}, {7, 9, 2, 1}, {2, 9, 4, 1}, {2, 3, 5, 1}, {6, 0, 4, 2}, {6, 7, 5, 3}, {9, 0, 5, 4}, and {8, 9, 5, 4}.

3.37 Wizard's card.
Number 23. In puzzles of this kind, if the two numbers have a common divisor, then no numbers that are not a multiple of that divisor can be obtained. Otherwise, if one integer number is x and another is y, then the largest impossible score is $xy - x - y$. So, in our case, we have $9 \times 4 - 9 - 4 = 23$.

3.38 Insert symbols.
This was the first solution I came up with: $87 + 65 - 43 - 21 = 88$. In fact, there are at least 9 solutions, including $87 - 6 + 5 - 4 + 3 + 2 + 1$, $87 + 6 - 5 + 4 - 3 - 2 + 1$, $8 + 7 + 65 + 4 + 3 + 2 - 1$, $87 + 6 - 5 - 4 + 3 + 2 - 1$, $87 - 6 + 5 + 4 - 3 + 2 - 1$, $87 + 6 + 5 - 4 - 3 - 2 - 1$, $8 - 7 + 65 + 4 - 3 + 21$, $8 - 7 + 65 + 43 - 21$, and $87 + 65 - 43 - 21$.

At least 45 different solutions exist, if we were to allow the additional use of × and ÷. If we wish to study a related problem, 8 7 6 5 4 3 2 1 = 1, we find only one solution using + and − symbols: $8 - 76 + 5 + 43 + 21 = 1$.

3.39 Missing numbers.
Here is one solution. Are there others?

2	3	5	6	16
4	2	4	7	17
2	0	5	8	15
2	1	5	9	17
10	6	19	30	18

3.40 New Guinea sea turtle. I am 10 years old, and my turtle is 40 years old. Here's one way to solve this. Let x = my age. Let $4x$ = the turtle's age. Then,

$$4x + 20 = 2(x + 20)$$

$$4x + 20 = 2x + 40$$

$$2x = 20$$

$$x = 10$$

3.41 Find A and B. The answer is $42^2 = 1,764$.

3.42 Prosthetic ulnas. She weighs 160 pounds. The equation is $x = 120 + x/4$.

3.43 Ladybugs. The relevant equation is $L + 10 = 5L - 2$. The answer is 3. You don't have to denote the number of grasshoppers by G.

3.44 Nebula aliens. The radius is 18 feet. The area of a sphere is $4\pi r^2$ and the volume is $(4/3)\,\pi r^3$. According to our problem, this means that both $4r^2$ and $(4/3)r^3$ must be between 1,000 and 9,999. From the area condition, we get $50 > r > 15$, and from the volume condition we get $20 > r > 9$. Next, note that $(4/3)r^3$ can be an integer only if it is divisible by 3. This means $r = 18$ feet. Is it possible to determine the sphere's radius if the surface area and the volume are both five-digit integers times π?

3.45 Martian females. In the crater, 3 females and 3 males are married. Let's refer to the females as women and males as men, even if they don't quite look like us. The number of married men and married women must be equal. Thus, we have $3/7\ W = \frac{1}{2}M$. Multiply both sides of the equation by 14 to yield $6W = 7M$, and thus the least number is $W = 7$, $M = 6$. This gives 13 Martians total, of which 6/13 of the crater's inhabitants are married (i.e., 3 wives of the 7 women and 3 husbands of the 6 men). There are exactly 3 married couples. For this problem, we assume that marriages are between male and female Martians. If I do not ask for the least number, there are an infinite number of solutions to $6W = 7M$.

3.46 Dimension X. A third of 12 would be $5\frac{1}{3}$. Assume that there is a strange factor that causes $\frac{6}{2}$ to be equal to 4 instead of 3. This factor is $1.3333\ldots$ or 1 and 1/3. Therefore, $(12/3) \times 1.3333\ldots$ is $5\frac{1}{3}$. Of course, this is one of those frustrating puzzles when more than one answer is possible. For example, we might consider the idea that 1 is added to each division operation so that $^{12}\!/_3 = 5$. Platonists may disapprove of this puzzle and say that the eternal idea "6" is 6 in any dimension whatsoever.

If my wording bothered you, you can recast the question as the following when presenting it to friends. My calculator has been acting up lately. For instance, every time I divide 6 by 2, it gives the answer 4. What do you think it would give if I divided 12 by 3? That stops all of the philosophical arguments about what I might mean by "is" in "half of 6 is 4" or indeed "half," because there is no reason that in this nearby dimension, "half" should mean what you get if you divide by 2.

3.47 Wine and vinegar. Monica carries $10 \times 3 \times (\frac{1}{4}) = 7.5$ cups. William carries $5 \times 4 \times (\frac{3}{4}) = 15$ cups. Thus, William carries $15 - 7.5 = 7.5$ cups more than Monica does.

3.48 Unholy experiment. There are 40 humans and 30 rabbits. Let h be the number of humans, and let r be the number of rabbits. Thus, we have two equations with two unknowns.

$$r + h = 70$$

$$4r + 2h = 200$$

We can multiply the first equation by 2 and subtract it from the second:

$$2r + 2h = 140$$

$$4r + 2h = 200$$

Thus, $2r = 60$; $r = 30$; $h = 40$. Are there other answers to this problem?

3.49 Space race. The Americans and the Russians traveled different routes between the planets, and one route was shorter than the other. (You didn't need much algebra for this one!)

3.50 Odin sequence. Indeed, every starting two-digit number in the Odin sequence returns to itself eventually. Number 13 requires 20 steps to return; that is, the 21st number is the same as the first. All other two-digit starting numbers require 20 steps to return, except for multiples of 5, which behave differently and require fewer steps.

3.51 Tanzanian zoo. Question 1: Bill has 18 giraffes and 4 ostriches. Each animal has 2 hind legs, so 22 heads means 44 hind legs total. The remaining 36 legs must be the front legs of 18 giraffes. Therefore, 18 heads belong to the giraffes, and the remaining 4 heads belong to the ostriches. Question 2:

Altogether, Bill has 3 birds in his "vast" collection! Two of his birds are not geese, but 1 is. Two birds are not ostriches, but 1 is. Two birds are not swans, but 1 is. This adds up to 3 birds, 1 of each type.

3.52 Fluid pool. It would take 12 minutes. The creatures will rule Earth. This is a classic problem that can be analyzed in a simple manner. If only Hose A is open, the portion of the pool filled by Hose A is $T/30$, where T is time. (Notice how after $T = 30$ minutes, the pool would be filled.) The portion of the pool filled by Hose B is $T/20$. When both hoses are open and the pool is completely filled, the equation to solve is

$$T/30 + T/20 = 1$$

Multiply by 60 to get

$$2T + 3T = 60 \text{ or } T = 12$$

This means that the creatures will fill the pool in 12 minutes.

We can also think of many more complicated variations. For example, Hose A would fill the pool in 30 minutes by itself. Hose B would fill the pool in 20 minutes by itself. The creatures fill the pool in 3 minutes using Hose A, Hose B, and Hose C simultaneously. How long would it take to fill the pool using Hose C by itself?

Another question to ponder is what the approximate color of the fluid would be.

3.53 Madagascar death snail. The snail can't go fast enough to average 60 feet per hour for both times around, no matter how hard it tries. Here's why.

$$60 \text{ feet/hour} = 1 \text{ foot/minute}$$

To average 60 feet per hour, the snail would have to travel around the 1-foot track *twice in 2 minutes*. First, let us compute how much time the snail has taken to travel one lap. At 30 feet/hour (or ½ foot a minute), the snail has taken 2 minutes to complete the first lap, because the track is 1 foot long.

As we said, if we want the snail to travel at an average speed of 60 feet/hour (1 foot/minute), this requires that the two laps be completed in 2 minutes. If the snail needs more than 2 minutes to travel the two laps, it is traveling at a speed slower than 60 feet an hour. However, the snail has already used up 2 minutes to do one complete lap. This means it has used up all of its time. It can't possibly complete two laps in 2 minutes. This means there is no way that it can achieve an average speed of 60 feet/hour.

3.54 Pulsating brain.

Yes. Let us assume that there are x astronomical objects in the jar. Then, the number b of black holes is more than $(70/100)x$ and less than $(75/100)x$. We can rewrite this as

$$\frac{70x}{100} < b < \frac{75x}{100}$$

We can rewrite this as the following by multiplying by 100:

$$70x < 100b < 75x$$

or

$$14x < 20b < 15x$$

We want to find the smallest integer x such that $20b$ is in-between $14x$ or $15x$, or, equivalently, the smallest possible number x such that the interval from $14x$ to $15x$ includes a multiple of 20. How might we do this? First, divide by $20b$ and recast into the form

$$(14x)/(20b) < 1 < (15x)/(20b)$$

Multiply the left inequality by 20/14: $x/b < 20/14$.

Multiply the right inequality by 20/15: $20/15 < x/b$.

Combine the two: $20/15 < x/b < 20/14$.

$$20/15 < x/b < 20/14$$

or

$$4/3 < x/b < 10/7$$

Now, for positive fractions r/s in which $r/s < u/v$, we also have $r/s < (r+u)/(s+v) < u/v$, and so, $x = 14$, $b = 10$, or $x = 7$, $b = 5$ is one solution.

3.55 Grotesque vessel for total human harvest.

The sphere to hold humanity is about 1.00 kilometer (0.6 miles) in diameter. Let us estimate the answer in such a way that Queequeg does not choose too small a vessel and end with a mess pouring out the top. First, assume that an average human weighs about 75 kg (165 pounds). Next, assume that liquefied homogenized human specific gravity is about 0.9 (all the air removed, but the higher density of bones not quite offsetting the lower density of fat). Assume that water weighs 1 kg/liter, that the human population is 6.37×10^9, that there are 10^{12} liters in a cubic kilometer, and that the diameter of a sphere is $2 \times (3 \times \text{volume}/4 \times \pi)^{1/3}$. Given these assumptions, the sphere to hold humanity is about 1.00 kilometer (0.6 miles) in diameter. Excellent storage efficiency!

3.56 *Kama Sutra* **puzzle.** Number 28. When the man tells the woman how good she is with the method of inversion, he is referring to the idea of working a problem backward, so that when, for example, a divide operation is given, we actually multiply, and so forth.

To solve the *Kama Sutra* problem, we start with the answer 2 and work backward. When the problem says divide by 10, we multiply by 10. When we are told to add 8, we subtract 8. When told to find the square root, we take the square, and so forth. So we have $[(2 \times 10) - 8]^2 + 52 = 196$. Instead of "multiplying this number by itself," we take the square root, which is 14. We proceed as follows:

$$\frac{(14)(3/2)(7)(4/7)}{3} = 28$$

3.57 Alien robot insect. In 9 months. The insect appears to be traveling upward by 4 feet each month. In 8 months, it will have traveled 32 feet from the base of the wall. However, on the ninth month, it will travel upward 8 more feet and reach the top of the wall before slipping down.

3.58 Goa party. You see 9 tricycles and 4 bicycles. As confirmation, note that $9 \times 3 + 4 \times 2 = 35$ wheels. We can solve this in the following manner. Let $y =$ the number of bicycles and x the number of tricycles. Then we solve for x and y in the following:

$x + y = 13$ describes number of vehicles

$2y + 3x = 35$ describes number of wheels

$x =$ number of trikes

$y =$ number of bikes

We know that $3x + 2y = 35$ and $x + y = 13$. Thus, we can multiply $x + y = 13$ by 2 to yield $2x + 2y = 26$. We subtract this from $2y + 3x = 35$, yielding $x = 9$. This implies 9 tricycles and 4 bicycles.

3.59 Gelatinous octopoid. He was an active octopoid for 4.41 centuries. The problem may be solved by using the following equation: $a/3 + a/5 + a/7 + 10 = a$, where a denotes Dr. Eck's age in centuries. The equation may be solved as follows. Multiply each term by 105 to get

$$35a + 21a + 15a + 1{,}050 = 105a$$

$$71a + 1{,}050 = 105a$$

$$1{,}050 = 34a$$

$$a = 1{,}050/34 = 30.88 \text{ centuries}$$

Therefore, Dr. Eck's life spent as an active octopoid is $= 30.88/7 = 4.41$ centuries.

3.60 Cloned Jefferson in Rome. The magnificent floor has a column at each corner. Each side has 13 columns, excluding the corner columns. Thus, the total number of columns used is $4 + (4 \times 13) = 56$. Notice that the 0.3 dimensions were not needed to solve this. Here's a diagram showing a smaller example in which Jefferson counts 4 columns on a side:

3.61 Hamburgers in space. In elementary calculus class, we learn that the "derivative" of a distance function can be used to give us the speed of a moving object. Given $f(t) = 3t^2 + 4t^3$, we can find the derivative $f'(t) = 6t + 12t^2$. When $t = 100$, the derivative is 120,600. This means that the hamburger is traveling at 120,600 miles per second. Of course, this is outrageously fast, considering that the speed of light is 186,000 miles per second in a vacuum! In real life, Britney's hamburger (or anything else she tried to accelerate to this speed) would have been damaged long before it reached this speed.

While on the topic of derivative, note that the mathematics professor Hugo Rossi once said, "In the fall of 1972 President Nixon announced that the rate of increase of inflation was decreasing. This was the first time a sitting president used the third derivative to advance his case for reelection" (Hugo Rossi, "Mathematics Is an Edifice, Not a Toolbox," *Notices of the AMS* 43, no. 10 (October 1996): 1108).

3.62 Spores of death and madness. First, let's start with our exponential growth equation: $N(t) = N_0 e^{kt}$. The current time is time $t = 0$, and N_0 is 10. So we have $N(t) = 10e^{kt}$.

I'd like to first determine the value for k. We know that tomorrow at this time, which corresponds to 24 hours, the population will be 2 million bacteria:

$$N(24) = 2,000,000 = 10e^{k(24)}$$

or

$$200,000 = e^{k(24)}$$

Let's solve for k.

$$\ln(200,000) = 24k$$

$$k = \ln(200,000)/24 \approx 0.508$$

Now that we have the value for k, we can substitute this value in our original equation.

$$N(t) = 10e^{0.508t}$$

This equation tells us the number of bacteria in the milk at any time. When will the milk have reached its maximum of 6 billion bacteria?

As we think about this, the bacteria have already started their rapid multiplication. Notice that people around us are beginning to gag from the stench, which resembles that produced by a rotting water buffalo corpse, or so I'm told. I have never actually encountered a rotting water buffalo corpse.

We can quickly solve this problem by plugging 6 billion into our equation.

$$6,000,000,000 = 10e^{0.508t}$$

$$600,000,000 = e^{0.508t}$$

$$\ln(600,000,000) = 0.508t$$

$$t = \ln(600,000,000)/0.508 \text{ or } 39.7 \text{ hours}$$

In 39.7 hours, there will be 6 billion bacteria in the milk. I don't want to be around here then!

Note that in real life, bacteria won't continue to grow exponentially, because they need to eat and excrete. Growth tends to slow when creatures are trying to subsist on their own excreta.

3.63 An ancient problem of Mahavira. The girl originally had 148,608 pearls on her necklace. That's a big necklace! Let's reflect on the problem: $\frac{1}{6}$ fell on the bed; $\frac{1}{3}$ scattered toward her. This means that the remaining pearls that are neither on the bed nor near her are $\frac{1}{2}$ of all the pearls. The remaining pearls

are halved six times, so $((1/2)^7)x = 1{,}161$, where x is the total number of pearls. Thus x is 148,608. The Indian woman must have had a huge necklace!

My colleague Jaymz provides another view of the solution. We assume that the pearls on the bed, the pearls that scattered toward the lady, and those mysterious remaining scattered pearls that factored with a sum of diminishing fractions are all "scattered" pearls. Assume that the 1,161 unscattered pearls included all but the previous. Let T be the total number of pearls, and let S be the total number of scattered pearls. Thus, we have

1. Number scattered toward lady $= T/3$

2. Number scattered on the bed $= T/6$

3. Number scattered up to this point $= (T/3 + T/6)$

4. Factor of "what remained" up to this point $= T - (T/3 + T/6)$

5. Mysterious sum of diminishing fractions: $(1/2 + 1/2^2 + 1/2^3 + 1/2^4 + 1/2^5 + 1/2^6)$

6. Total of scattered pearls: $S = (T/3 + T/6) + (T - T/3 - T/6)((1/2 + 1/2^2 + 1/2^3 + 1/2^4 + 1/2^5 + 1/2^6)$, which reduces to $S = T(1/3 + 1/6) + T(1 - 1/3 - 1/6)[(32 + 16 + 8 + 4 + 2 + 1)/64]$, or $S = T(1/2) + T(1/2)(63/64)$, or $S = T(64 + 63)/128$, or $S = T(127/128)$, which provides us with a first equation. As for a second equation, we have $T - S = 1161$.

Combining the two final equations, we obtain $T(1 - 127/128) = 1{,}161$, or $T(1/128) = 1{,}161$, or $T = 1{,}161 \times 128 = 148{,}608$ original pearls. Quite a necklace, and worth quarreling over!

3.64 Alien slug. Dr. Oz is 20.44 years old and his wife is 17.66. Let Dr. Oz's age be x and his wife's age be y. Then

$$x - 6 = 10 + y - (x/2 + 3)$$

$$y - 12 = 3 + x - (2y/3 + 6)$$

The $2y/3$ term comes from the fact that when the wife was a third of her present age, this occurred precisely $(2/3)y$ years ago. These equations reduce to

$$(1) \ \ 3x - 2y = 26$$

$$(2) \ \ 5y - 3x = 27$$

Adding these two equations gives

$$3y = 53$$

$$y = 17.66666$$

Placing $y = 17.66666$ into (2) gives us $3x = 61.33333$ or $x = 20.4444$.

3.65 Lucite pyramids. Alpha has 40 beetles. Omega has 30 beetles. Let alpha have x beetles and omega have y beetles. The number taken by omega is $y/3$, which leaves alpha with $x - y/3$. Alpha retrieves the same number, which is $(x - y/3)/3$, so we have

$$(1) \ \ y = 70 - x$$

$$(2) \ \ y/3 = (x - y/3)/3$$

From (2), we have $4y = 3x$ (3). From (3) and (1), we get $x = 40$ and $y = 30$.

3.66 *Bakhshali* manuscript: A true story. The solution is 2 men, 5 women, and 13 children. We can let the number of men, women, and children be m, w, and c, respectively. Two formulas describe our situation.

$$m + w + c = 20 \text{ and}$$
$$3m + (3/2)w + (1/2)c = 20.$$

We may multiply the first equation by 3 and subtract the second equation to yield

$$3 \times [m + w + c = 20]$$
$$-[3m + (3/2)w + (1/2)c = 20]$$
$$= (3/2)w + (5/2)c = 40$$

So $3w = (80 - 5c)$. The quantity $80 - 5c$ must be a multiple of 3 and of 5. Our choices are 15, 30, 45, 60, and 75. Therefore, $c = 1, 4, 7, 10,$ or 13. We can attempt to list all possible values of m, w, and c. The following is a table of possible values of c, and the corresponding values of m and w:

c	w	m
1	25	–6
4	20	–4
7	15	–2
10	10	0
13	5	2

The only valid combination is $c = 13$, $w = 5$, and $m = 2$.

3.67 Luminescent being. N is 81. ($\sqrt{81} = 9 = 8 + 1$) If you experiment further, you'll see that no larger values of N could exist because the sum of the digits, M, becomes much too small to sum to N.

3.68 Mathematical romance. The answer is 128. I do not know whether 128 is the only number with these characteristics. My colleague Joseph Pe used Mathematica to search for additional solutions. He discovered that 128 is the largest such number less than $2^{45,000}$. As the number of digits increases, it becomes increasingly unlikely that a larger such number will be found, because the probability that a random string of digits contains only 1, 2, 4, and 8 very rapidly approaches 0 as the length of the string increases.

3.69 Harmonic series. Although the divergence of the harmonic series is not immediately apparent, we can demonstrate this fact fairly easily. Given the harmonic series

$$H = \sum_{n=1}^{\infty} \frac{1}{n} = 1 + \frac{1}{2} + \frac{1}{3} + \frac{1}{4} \ldots$$

we can begin with the ½ term and group terms—1, 2, 4, . . . terms at a time:

$$H = \tfrac{1}{1} + (\tfrac{1}{2}) + (\tfrac{1}{3} + \tfrac{1}{4}) +$$
$$(\tfrac{1}{5} + \tfrac{1}{6} + \tfrac{1}{7} + \tfrac{1}{8}) + \ldots$$

Next, replace each group in parentheses with the smallest fraction in the group, to get

$$H > \tfrac{1}{1} + (\tfrac{1}{2}) + (\tfrac{1}{4} + \tfrac{1}{4}) +$$
$$(\tfrac{1}{8} + \tfrac{1}{8} + \tfrac{1}{8} + \tfrac{1}{8}) + \ldots$$

Adding the terms in parentheses, we get

$$H > \tfrac{1}{1} + \tfrac{1}{2} + \tfrac{1}{2} + \tfrac{1}{2} + \ldots$$

Since we have an infinite number of ½s to be added, the result will be infinite. Since this smaller series diverges, the harmonic series (whose terms are larger than the new terms) must also diverge. Hence H is unbounded.

As I told you, the series diverges extremely slowly: after the first thousand of the terms, the partial sum is 7.485; after a million terms, it is 14.357; for the first billion terms, it is approximately 21; and for the first trillion terms, it is approximately 28. My colleague

Pete Barnes notes that to reach a partial sum of 100, we would have to add up 10^{43} terms of the harmonic series. A computer adding up one million terms per second would take about 10^{37} seconds to complete the job. Since the age of the universe is only about 10^{17} seconds, it would take quite a few universe lifetimes. To get to 1,000, we must add together over 10^{434} terms.

3.70 Harmonic series on a diet. Given

$$H = \sum_{n=1}^{\infty} \frac{1}{n} = 1 + \frac{1}{2} + \frac{1}{3} + \frac{1}{4} \ldots$$

We can remove all terms with 9s in the denominator and group as follows:

$J = (1/1 + \ldots + 1/8) + (1/10 + \ldots + 1/18 + 1/20 + \ldots + 1/28 + \ldots + 1/80 + \ldots + 1/88) + (1/100 + \ldots + 1/888) + \ldots$

In each set of parentheses, replace each term with the greatest term in that set of parentheses. This increases the sum of each set of brackets, and so J will be less than the sum of new sets of brackets. We then show that this greater sum is finite. Hence, by comparison, J is finite.

$J < (1 + \ldots + 1) + (1/10 + \ldots + 1/10) + (1/100 + \ldots + 1/100) + \ldots$

$J < 9(1) + 9^2(1/10) + 9^3(1/10^2) + \ldots$

$J < 9[1 + 9/10 + (9/10)^2 + \ldots]$

$J < 9/(1 - 9/10)$

$J < 90$

Hence the harmonic series with all the 9s removed converges! This proof applies equally for any missing digit.

You might think that the harmonic series with only odd terms converges: $1/1 + 1/3 + 1/5 + 1/7 + \ldots$. After all, it grows much slower than the standard harmonic series. Alas, the series diverges.

3.71 Shopping mall puzzle. The smallest value for N is $132 = 12 + 21 + 13 + 31 + 23 + 32$.

3.72 Castles and strings. Light the 10-minute string at both ends. The flames will die in 5 minutes. When this happens, light both ends of the 1-minute string. This will burn in 30 seconds. Thus, the total time elapsed is 5 minutes and 30 seconds.

3.73 Target practice. Here is one solution: $17 + 13 + 19 + 51$. Try this on friends. How long did it take them to find an answer? Other solutions include $\{22, 62, 9, 7\}$, $\{22, 20, 7, 51\}$, $\{15, 61, 17, 7\}$, $\{15, 23, 55, 7\}$, $\{15, 13, 63, 9\}$, $\{15, 13, 17, 55\}$, $\{61, 23, 9, 7\}$, $\{61, 13, 17, 9\}$, $\{61, 13, 19, 7\}$, $\{23, 13, 9, 55\}$, $\{23, 17, 9, 51\}$, $\{23, 19, 7, 51\}$, $\{13, 63, 17, 7\}$, $\{13, 17, 19, 51\}$, and $\{17, 9, 19, 55\}$.

3.74 Hobson gambit. The answer is $101 + 2/3$. Perhaps we can make an interesting point with the Hobson gambit. Problems that would have taken mathematical geniuses like Euler, Gauss, and Newton a lifetime to solve are now solvable in seconds by mathematical punks on every street corner.

For this problem, we have

$$a + b + c = 1,$$

$$a^2 + b^2 + c^2 = 15,$$

$$a^3 + b^3 + c^3 = 3$$

My colleague Mark Ganson used the following Mathematica software steps to obtain a solution:

```
Clear[a, b, c]

eqn = {

a + b + c == 1,

a^2 + b^2 + c^2 == 15,

a^3 + b^3 + c^3 == 3

}

NSolve[eqn, {a, b, c}, 200][[1]]
```

In the previous, the "200" indicates that the equation is to be solved using 200 digits of accuracy. We find

a = 0.89253182605396733441976900973253
10692439549024646489097681739675495575
28926708463520584967657846331256161955
25981285162044277081375689535498553288
96282582210326719059230379147684158
4901206735270950 . . .

b = −2.61062439447004233790802535634971
85491720273047056273796943382412811747
98088160245862497217001612441751133865
02431268444878763864970434429258632333
27374240019988458151117523454606140
9369936836225713 . . .

c = 2.71809256841607500348825634661718
74799280724022409784699261642737315995
08821075610656647540423149129189514312
42618416824435993051213539074273099443
645484179167212675588137319777645560
357869483516215 . . .

Once Mark had values for a, b, and c, he obtained

$$a^4 + b^4 + c^4 = 101.666666666 \ldots$$

If anyone today can obtain answers to problems like these in seconds, a century from now will ten-year-olds press buttons to solve challenges like Fermat's Last Theorem in seconds? Even if solutions are easier to find, each new computer-assisted discovery will bring fresh, tougher questions. What could legendary mathematicians like Euler, Newton, and Pythagoras have accomplished with the tools we have today?

3.75 Magic light board. Here is one solution. Can you find others? How many others exist?

r	y	o	g
b	i	v	t

3.76 Aqueduct. This is based on a classic puzzle. When the soldier is in the hut, sneak onto the aqueduct and walk toward Greece. After 9 minutes and 59 seconds, turn back and start walking toward Italy. When the guard sees you, he will assume you are coming from Greece and will order you to go back.

4. Geometry, Games, and Beyond

- - - - -

4.1 Martian bodies and Venn diagrams. There are 90 Martians with none of these body characteristics. One way to solve a problem of this type is to use Venn diagrams with three circles (figure A4.1). Circle A contains Martians with pointed ears ($a_1 = 600$). Circle B holds Martians with fangs ($a_2 = 300$). Circle C

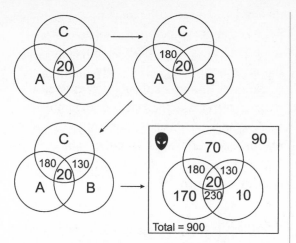

Figure A4.1 Martian bodies and Venn diagrams.

holds Martians with horns ($a_3 = 400$). A Venn diagram is usually drawn as a rectangle (which holds all the elements in a problem) surrounding circles that depict subsets of elements. As we mentioned previously, these diagrams are named after the English logician John Venn (1834–1923), who in 1880 popularized their use.

The practical use of Venn diagrams will be best understood by filling them with our Martians. In the top left figure in figure A4.1, we note that 20 individuals have all three body characteristics (a_1, a_2, a_3). Proceeding to the right in figure A4.1, we continue our analysis by noting that 200 creatures have a_1 and a_3. Of these creatures, 20 are already displayed in the Venn diagram, and 180 remain. Note that 150 creatures have a_2 and a_3, and 20 of these are already in the diagram, so 130 remain. Also, 250 creatures have a_1 and a_2. Of these, 20 are displayed in the figure and 230 remain. Now we must fill in the larger parts of each circle. For example, to find the remaining part of A, we know that $a_1 = 600$, so $600 - 180 -$

20 – 230 = 170. For B, we have 390 – 230 – 130 – 20 = 10. For C, we have 400 – 180 – 20 – 130 = 70.

Initially, I told you that we were examining 900 Martians. So to compute the number of Martians not contained in the 3 circles, we have 900 – 70 – 180 – 20 – 130 – 170 – 230 – 10 = 90. Thus, 90 Martians have neither fangs, nor horns, nor pointed ears.

Now do you realize how useful Venn diagrams can be?

4.2 Star of David. This is possible. As stated in a previous ●, you can traverse a graph by going through every segment just once only if the graph has less than 3 vertices of odd "valence." For this graph, *every* vertex has an even valence, so it is possible. If you haven't solved this, keep trying!

4.3 Mondrian puzzle. The mystery square has an area of 9.384 in whatever units of area are appropriate to the measurements. First, we can ignore the two side rectangles of Areas 4 and 3. They don't contribute to the solution. If we label each of the rectangle's horizontal dimensions (a, b, c, d) and vertical dimensions (e, f, g, h), we find that $a \times h = 10$, $c \times h = 13$, $b \times g = 8$, $b \times f = 7$, $a \times e = 9$, and $d \times e = 12$. The special unknown Mondrian square has area $\underline{cf}$.

Here's what we know:

$$a \times b \times c \times d \times e \times f \times g \times h = \\ ae \times bf \times ch \times dg = 9 \times 7 \times 13 \times 11$$

$$a \times b \times c \times d \times e \times f \times g \times h = \\ ah \times bg \times cf \times de = 10 \times 8 \times \underline{cf} \times 12$$

Thus, $9 \times 7 \times 13 \times 11 = 10 \times 8 \times \underline{cf} \times 12$, which means $cf = 9.384$.

4.4 Compact formula. Drum roll, please. . . . One answer is $e^{i\pi} + 2\phi = \sqrt{5}$. Isn't that a beauty?

4.5 Remarkable formulas involving I. Both of these formulas equal the golden ratio, 1.61803 . . . The second formula is called a *continued fraction* and has numerous uses for mathematicians and physicists. A general expression uses the letter b to denote the number in each denominator:

$$b_0 + \cfrac{1}{b_1 + \cfrac{1}{b_2 + \cfrac{1}{b_3 + \cfrac{1}{\dots}}}}$$

Manfred Schroeder remarked, "Continued fractions are one of the most delightful and useful subjects of arithmetic, yet they have been continually neglected by our educational factions." These typographical nightmares can be more compactly written as $[b_0, b_1, b_2, \dots]$. And the golden ratio can be represented as $[1,1,1,1, \dots]$. We can also write this more compactly as $[1,\underline{1}]$, where the bar indicates a repetition of the number 1.

Even though transcendental numbers do not have repeating numbers in their continued fractions, their continued fractions often display patterns. It is mind-boggling that the irrational number e ($e = 2.718281828 \dots$), unlike π, can be represented as a continued fraction with unusual regularity $[2, 1, 2, 1, 1, 4, 1, 1, 6, 1, 1, 8, 1, \dots]$; however, this initially converges very slowly because of the many 1s. In fact, the golden ratio, which contains infinitely many 1s, is the most slowly converging of all continued fractions. Schroeder notes, "It is therefore said, somewhat irrationally, that

the golden section is the most irrational number." The approximation of the continued fraction to the golden ratio is also worse than for any other number. Therefore, chaos researchers often pick the golden ratio as a parameter to make the behavior of simulations as aperiodic as possible. For more information, see Manfred Schroeder, *Number Theory in Science and Communication* (Berlin: Springer, 1986).

4.6 Goly. A golygon is a special kind of path on a plane that has an equally spaced grid of points. To create a golygon, start at one point, and take a first step one unit to the north or the south. The second step is two units to the east or the west. The third is three units to the north or the south. Continue until your path closes on itself and reaches the starting point. No crossing or backtracking is allowed.

Figure A4.2 is a typical golygon. The number of golygons of length $8n$ for the first few n are 4, 112, 8432, 909288, . . . , which approaches

$$\frac{3 \cdot 2^{8n-4}}{\pi n^2 (4n+1)}$$

for large values of n. Note the appearance of the ever-present pi in a formula that appears unrelated to a circle.

Golygons are more than just a geometric curiosity. Golygons, formally known as serial isogons of 90 degrees, have inspired countless puzzles and problems for research. For further reading, see L. Sallows, M. Gardner, R. K. Guy, and D. Knuth, "Serial Isogons of 90 Degrees," *Mathematics Magazine* 64 (1991): 315–24.

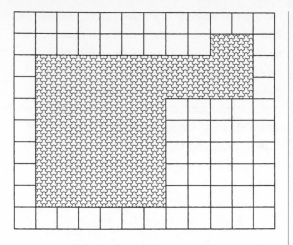

Figure A4.2　A typical golygon.

4.7 Special number. This number, discussed in chapter 2, is called the *golden ratio* and equals 1.61803 . . . . It appears in the most surprising places in nature, art, and mathematics. A golden rectangle has a ratio of the length of its sides equal to 1:φ. Some people have reported that the golden rectangle is among the most visually pleasing of all rectangles, being neither too squat nor too thin:

Golden Rectangle

Because $\phi = (1 + \sqrt{5})/2$ (and $\phi^2 = \phi + 1$) , it has some amazing properties: For example,

$$\phi - 1 = 1/\phi \qquad \phi\Phi = -1$$

$$\phi + \Phi = 1 \qquad \phi^n + \phi^{n+1} = \phi^{n+2}$$

where $\Phi = (1 - \sqrt{5})/2$. Both φ and Φ are the roots of $x^2 - x - 1 = 0$. The occurrence of the golden ratio in various areas of geometry becomes evident in the next question.

4.8 Mystery pattern. The Lute of Pythagoras (figure A4.3) is among mathematics' most beautiful recursive shapes and has many remarkable properties. You may wish to spend some time constructing this interesting shape and thinking about its features.

Begin your construction by drawing an isosceles triangle, such as you see marked *ABC* in figure A4.3. Next, construct the ladder steps marked *ED*, *FG*, and so on. There is a trick you must follow to locate the ladder steps: the length *AC* must equal *AE*, the length *DE* must equal *DG*, and so on. Thus we have *AC = DC = AE*, *DE = DG = EF*, and so on. Next create the pentagonal, starlike figures by

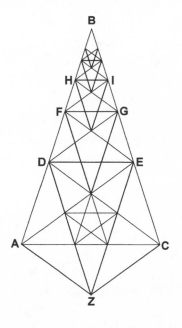

Figure A4.3　The Lute of Pythagoras.

joining the appropriate ladder points. A new point, Z, is located so that $CE = AD = AZ = ZC$. You can also recursively create star shapes for as far as your eye can see. Note that the initial isosceles triangle for the lute is a "golden triangle," that is, one where $BC/AC = AC/EC = \phi$, the golden ratio. Also, care must be taken during construction to make successive ladder length ratios equal, such as $CE/EG = EG/GI$. Given this construction, mathematicians have noted that the ratio of successive sides of the ladder $CE/EG = EG/GI \ldots$ is the golden ratio.

4.9 Cut the crescent to make a cross. Figure A4.4 shows Loyd's solution.

Figure A4.4   Sam Loyd's "Cross and Crescent" puzzle.

4.10 Torus versus marmoset. The torus home. I surveyed several dozen individuals. A majority preferred living in a torus. Most found the shape more practical and aesthetically pleasing than a marmoset. From a purely geometrical standpoint, the marmoset would probably have more surface area than a comparably sized torus and therefore would be more difficult to heat. By "surface area," I mean the area exposed to outside air. Body features such as the ears, the head, the neck, and the limbs increase the surface area of the marmoset relative to a smooth torus.

On the other hand, some respondents reasoned that a marmoset could be more novel than a torus and thus could be used to make money as a tourist attraction. Also, perhaps the marmoset shape would keep the owner more in touch with nature.

My colleague Marie S. writes,

I would choose a torus. I could have a "covered" patio in the center where I could exercise my dogs (and wouldn't need a fenced yard). If there were doorways in and out of the center, I would have quick and easy access to the different areas. Even without that, I could have easy access to the main living areas, access to kennel areas that could be separated from the main living areas, and access to a garage within the torus.

The shape would be reasonably efficient to heat and cool. A marmoset, on the other hand, would be more expensive to heat and cool because of the greater surface exposure. If the marmoset were in a standing position, it would have more than one story—and the height would also increase heating and cooling costs. If lying, it would be more spread out and have more difficult access to areas in the arms, legs, and head, as well as costing more to heat and cool because of the limited air circulation and loss through the extremities.

Actually, a torus with a covered interior atrium (with drainage) and both interior and exterior access doors might be really workable. The floor would be part way up the sides of the torus, and there would be good storage space like a basement underneath.

Depending on size, there could be a balcony with "bedrooms" as a second story, although I'd prefer everything on one level.

However, Nick H. writes, "I'd prefer the marmoset; the torus might roll away!"

4.11 Green cheese moon puzzle.

Sam Loyd's answer is shown in figure A4.5. He says, "By taking the best possible advantage of the crescent form of the moon, 21 pieces of green cheese can be cut for the hungry." Martin Gardner notes that for a crescent, the number of pieces increases as the number of cuts n increases: $(n^2 + 3n)/2 + 1$. Try the same experiment using a circle. How many pieces can you cut a circular region into, using 4 straight cuts? 5? How about a circle with a hole in the middle?

The crescent answer gives an equation for the maximum number of pieces that can be produced with n cuts of a 2-D crescent region. There are similar formulas for a 3-D doughnut and a sphere cut with n plane cuts. For a doughnut (or a torus), the largest number of pieces that can be produced with n cuts is $(n^3 + 3n^2 + 8n)/6$. Thus, a doughnut can be sliced into 13 pieces by three simultaneous plane cuts. For a sphere, the equation is $(n^3 + 5n) / 6 + 1$.

4.12 Circle crossing.

According to "Johnson's theorem," named after the American mathematician Roger Johnson (1890–1954), if three circles with identical radii pass through a common point, then their other three intersections lie on another circle with the same radius (see figure A4.6).

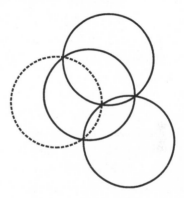

Figure A4.6   Circle crossing.

4.13 Sam Loyd's fifteen puzzle.

The "14-15" puzzle, or just the "fifteen puzzle," illustrated Loyd's interest in practical jokes (figure A4.7). Many people claimed to have solved the puzzle, but none could duplicate their achievement at collection time. Quite simply, Loyd knew this because the puzzle was impossible to solve. Mathematicians are

Figure A4.5   Sam Loyd's green cheese moon puzzle.

fascinated by the fact that for *some* initial arrangements, this rearrangement is possible, but for others, it is not.

Figure A4.7   Sam Loyd's fifteen puzzle.

This puzzle became an instant success, much like the Rubik's Cube 100 years later. The only way to get from the standard starting position to the finishing position is to physically lift the 15 and 14 tiles out of the frame and then swap them, an illegal move. Can you imagine how many hours people must have spent attempting to solve this properly?

It's fun to think about what initial arrangements can lead to solutions. Here is the strategy. If the sliding square containing the number i appears "before" (reading the squares in the box from left to right and top to bottom) n numbers that are less than i, then call it an inversion of order n, and denote it n_i. Then define

$$\gamma \equiv \sum_{i=1}^{15} n_i = \sum_{i=2}^{15} n_i$$

where the sum need run only from 2 to 15, rather than 1 to 15, since there are no numbers less than 1 (so n_1 must equal 0). If γ is

even, the set of positions is possible to create; otherwise, it is not. For example, in the following arrangement

2	1	3	4
5	6	7	8
9	10	11	12
13	14	15	

$n_2 = 1$ (2 precedes 1) and all other $n_i = 0$, so $\gamma = 1$ and the puzzle cannot be solved. I fondly recall as a boy giving a friend the previous arrangement to solve, never telling the friend that it was impossible.

The famous Russian puzzlist Yakov Perelman (1882–1942) quoted the German mathematician W. Arens regarding Loyd's fifteen puzzle:

> In the late 1870s, the Fifteen Puzzle bobbed up in the United States; it spread quickly, and owing to the uncountable number of devoted players it had conquered, it became a plague. The same was observed on this side of the ocean, in Europe. In offices and shops bosses were horrified by their employees being completely absorbed by the game during office and class hours. In Paris, the puzzle proliferated speedily from the capital all over the provinces. A French author of the day wrote, "There was hardly one country cottage where this spider hadn't made its nest lying in wait for a victim to flounder in its web (Yakov Perelman, *Fun with Maths and Physics* [Moscow: Mir Publishers, 1988]).

To learn more about this puzzle, see Eric Weisstein, *CRC Concise Encyclopedia of Mathematics* (New York: CRC Press, 1998).

4.14 Find the bugs! Here is one possible answer showing the locations of the bugs. Fourteen other solutions exist. Can you find any of the other solutions?

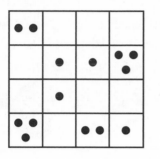

Most of my colleagues used computer programs to solve this—including the "solver feature" in Microsoft Excel. Others solved similar puzzles by writing a program that puts random integers into each cell until a viable solution is found that meets all of the row and the column sum requirements.

4.15 Tablet of Ezekiel. The diagonal line cuts the most tiles in the 16×19 array. If you were to draw large versions of these arrays and carefully draw diagonals, you would soon realize that a diagonal enters a new tile at the beginning and each time it crosses a horizontal or a vertical line. However, in situations where the diagonal enters exactly at the corner of a tile, the diagonal crosses two lines but enters only one tile. In other words, the diagonals of such tiles are on the main diagonal.

The number of tiles that a diagonal crosses is therefore the length A of one side of a face plus the length B of the other minus the greatest common divisor (GCD) of the sides' lengths: $A + B - \text{GCD}(A,B)$. The greatest common divisor of two integers is the largest number that divides both integers. For example, a 16×24 face would have $16 + 24 - 8 = 32$ crossed tiles, since 8 is the greatest common divisor of 16 and 24.

Of the three tablets given, we have

A	B	GCD	Number of Squares Cut by Diagonal
16	19	1	34
16	18	2	32
16	24	8	32

4.16 Strange dimension. You can design a computer program to solve this problem by representing the three religions as red, green, and amber squares in a 3×3 checkerboard. The program uses three squares of each color. Have the computer randomly pick combinations and display them as fast as it can, until a solution is found. The rapidly changing random checkerboard is fascinating to watch, and there are quite a lot of different possible arrangements. In fact, for a 3×3 checkerboard there are 1,680 distinct patterns. If it took your computer 1 second to compute and display each 3×3 random pattern, how long would it take, on average, to solve the problem and display a winning solution? (There is more than one winning solution.)

I believe there are twelve solutions for the problem that asks us to find an arrangement with no more than one type of church in any row or any column. Here are two of these twelve solutions:

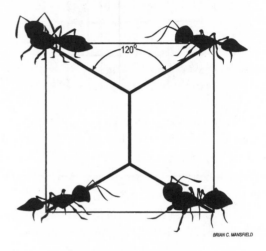

entries of N different elements, none of them occurring twice or more within any row or any column of the matrix. The integer N is called the *order* of the Latin square. Recently, the subject of Latin squares has attracted the serious attention of mathematicians, due to their relevance to the study of combinatorics and error-correcting codes.

4.17 Ant mathematics. Here is one solution with intersections at 120 degrees (figure A4.8).

BRIAN C. MANSFIELD

Figure A4.8 Ant puzzle.

It can be proven that for any valid solution, (1) one diagonal must contain the same religion in each of its cells, and (2) the other diagonal must contain unique religions in each of its cells.

For the second problem, here is one way to arrange the religions so that there are only two of the same religion in each row and each column:

Two in Each Row and Each Column

Try these problems on a few friends. Many people have difficulty visualizing the solution. The "strange dimension" problem can be thought of as a special problem in the remarkably rich mathematical area concerned with *Latin squares*. Latin squares were first systematically developed by the Swiss mathematician Leonhard Euler in 1779. He defined a Latin square as a square matrix with N^2

I believe this solution to be optimal, in the sense that it uses the least amount of tubing. Aside from using analytical approaches, my colleagues have found it even easier to use Mathematica to actually perform computer simulations with tubes and ants to suggest this answer. How would your answer change if we used a pentagon-shaped ant farm with five different ant colonies? Another variation is to place three ant colonies at three corners of a square, leaving one of the corners unoccupied. Which tube configuration would use the least amount of tubing?

4.18 Nursery school geometry. Here is one solution. How many other solutions can you find?

☹	☺					1
☹	☹					0
☺	☹		☹			1
☹	☹	☹	☺			1
☹	☺	☹		☹	☺	2
	☹		☺	☹	☹	1
1	2	0	2	0	1	1

4.19 Frodo's magic squares. Here is one solution:

16	3	2	13
5	10	11	8
9	6	7	12
4	15	14	1

10	18	1	14	22
11	24	7	20	3
17	5	13	21	9
23	6	19	2	15
4	12	25	8	16

4.20 Lost in hyperspace. Mathematical theoreticians tell us that the answer is one—infinite likelihood of return for a one-dimensional random walk. If you were placed at the origin of a 2-space universe (a plane) and then executed an infinite random walk by taking a random step north, south, east, or west, the probability that the random walk would eventually take you back to the origin is also one—infinite likelihood.

Our three-dimensional world is special: three-dimensional space is the first Euclidean space in which it is possible for us to get hopelessly lost. While executing an infinite random walk in a 3-space universe, you will eventually come back to the origin with a 0.34 or 34 percent probability. In higher dimensions, the chances of returning are even slimmer, about $1/(2n)$ for large dimensions n. The $1/(2n)$ probability is the same as the probability that you would return to your starting point on your second step. If you do not make it home in early attempts, you are probably lost in space forever.

For those of you who enjoy more mathematical rigor or staring at impressive-looking formulas, the precise probability for returning to your start point in a walk on a 3-D lattice is

$$1 - \frac{1}{u} = 0.3405373296\ldots$$

where

$$u = \frac{3}{(2\pi)^3} \int_{-\pi}^{\pi}\int_{-\pi}^{\pi}\int_{-\pi}^{\pi} \frac{dxdydz}{3 - \cos x - \cos y - \cos z}$$

$$= 1.5163860592\ldots$$

Some of you may enjoy writing computer programs that simulate your walks in confined hypervolumes and making comparisons of the probability of return. By "confined," I mean that the "walls" of the space are reflecting so that when you touch them, you are reflected back. Other kinds of confinement are possible. You can read more about higher dimensional walks in Daniel Asimov, "There's No Space Like Home," *The Sciences* 35, no. 5 (September–October 1995): 20–25.

4.21 Heterosquares. Here is one solution.

9	8	7
2	1	6
3	4	5

These kinds of squares are called heterosquares and were first considered with enthusiasm in the early 1950s. Royal V. Heath, an American magician and a puzzle enthusiast, was the first to prove that a 2×2 heterosquare formed with the numbers 1, 2, 3, and 4 is impossible.

Although this puzzle may seem difficult at first, and you might not be able to create a heterosquare at all, I believe there to be 24,960 possible solutions out of $9! = 362,880$ possible orderings. This means that the probability of a random ordering being a heterosquare is $24960/9! = 13/189$ or roughly 6.9 percent.

4.22 Charged array. Here is one solution. I believe there are a total of four solutions. Can you find the others?

0	–	0	–
0	–	–	–
+	+	+	+
+	–	0	0

4.23 Bouncing off the Continuum. For clarity, assume that the spaceship is located at point S, symbolized by the black dot next to the spaceship. Further assume that the star is located at the black dot next to the drawing of the star. Suppose that the aliens aim for some point, P, on the edge of the Continuum (figure A4.9). Reflect the aliens' original position in the edge of the Continuum. Then the distance SPA equals S'PA, and the latter will be a minimum when S'PA is a straight line. It follows that P is the point such that SP and PA make the same angle with the edge of the Continuum. (Note that Heron of Alexandria [A.D. 75] used a similar argument to conclude that when light is reflected from a mirror, the angles of incidence and reflection are equal.)

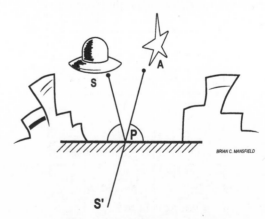

BRIAN C. MANSFIELD

Figure A4.9   Bouncing off the Continuum.

4.24. Alien heads. Here is one solution. Can you find others?

4.25 Relationship. Switch the 48 and the 28 cell colors. White cells are divisible by 3 and gray cells are not.

4.26 Sticky faces. One possibility is that Danielle (D) has 56 points, Cliff (C) has 74 points, and Pete (P) has 77 points. We don't know anything about the ordering of the faces in each row, but here is one possible solution that shows the number of each person's faces in each row. Is this the only answer?

10 points	C D D D
13 points	C DD P P
17 points	CCC PP P

We believe there are three possible solutions for how many of each player's faces fell in each row (ignoring the orderings), including the one given here. Here are a few clues to help you in your quest to find the other two solutions. In every possible solution, Danielle has at least two 3s, Cliff has at least one 10 and at least two 17s, and Pete has at least two 17s and at least one 13. It is not possible for all three players to be represented in all three rows. If all three players have at least one 17 each, then Cliff has all the 10s. If all three players have at least one 13 each, Danielle has no 17s. Can you find the other two solutions?

4.27 Bioterrorist puzzle. One solution is for Romeo to start at Square 1 and Juliet to wait at Square 2. Romeo makes the following moves:

$$1\text{-}6\text{-}11\text{-}16\text{-}17\text{-}12\text{-}13\text{-}8\text{-}7\text{-}2$$

One way to solve this problem is to write a computer program to search for possible paths. Another solution is simply to experiment yourself, using trial and error. I would be fascinated to hear from readers who have developed methods for finding solutions more efficiently than using trial and error.

My colleagues found solutions by drawing diagrams and placing coins on an arbitrary set of squares that added up to 93. Once the coins were placed, they explored potential solutions by moving coins in pairs (one +1, one −1) so that the total remained constant.

It would also be interesting to find the shortest solution for this puzzle, assuming that Romeo gets tired if he walks too far! One might modify the original puzzle to force this by giving Romeo a limited amount of fuel with which to make his hops. We believe that four solutions exist that require Romeo to take just five steps.

4.28 Magic square. Here is one solution. Are there others?

0	6▶	◀12	23	19
13▶	◀24	15	1▶	◀7
16	2	8	14	20
9▶	◀10	21	17▶	◀3
22	18▶	◀4	5	11

4.29 Rope capture. This is based on a classic problem. Tie the lower ends of the rope together near the floor. Climb one rope to the ceiling. Cut the other rope about a foot from the ceiling and let the remainder fall. Tie a loop in the 1-foot-long rope. Cut the remaining rope and thread it through the loop. Finally, hold both ends of the long rope and lower yourself to the ground.

Note that because Grand Central Terminal in New York City, the largest station in the world, covers 48 acres on two levels with 67 tracks, it should be relatively easy for you to hide from your captors once you are released.

4.30 Talisman square. Here is a solution.

1	5	3	7
9	11	13	15
2	6	4	8
10	12	14	16

These squares, known as *Talisman squares*, were invented by Sidney Kravitz, a mathematician from Dover, New Jersey. Here I show an example of a talisman square in which the difference between any number and its neighbor is greater than 1. The number 1, for example, has 3 neighbors (5, 9, and 11); 11 has eight neighbors (1, 5, 3, 13, 4, 5, 2, and 9).

Talisman squares have only been studied since the late 1970s, and no rules for constructing them are known.

4.31 Circle madness. In circle arrangements constructed in the way I described, the lines joining opposite points of contact cross at a single point (see figure A4.10). This is sometimes referred to as the "seven circles theorem" (C. J. A. Evelyn, G. B. Money-Coutts, and J. A. Tyrrell, "The Seven Circles Theorem," §3.1 in *The Seven Circles Theorem and Other New Theorems* [London: Stacey

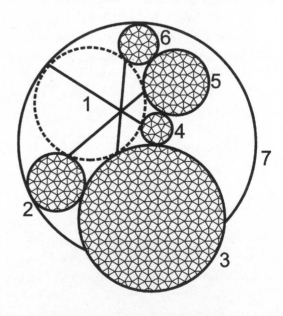

Figure A4.10   Circle madness answer.

International, 1974] 31–42; Stanley Ogilvy, *Tomorrow's Math: Unsolved Problems for the Amateur*, 2nd edition [New York: Oxford University Press, 1972]; and David Wells, *The Penguin Dictionary of Curious and Interesting Geometry* ([Middlesex, England: Penguin Books, 1991]).

4.32 Platonic solids. The ancient Greeks recognized and proved that there are only five Platonic solids: the tetrahedron, the cube, the octahedron, the dodecahedron, and the icosahedron (see figure A4.11).

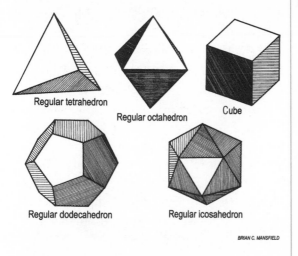

Regular tetrahedron

Regular octahedron

Cube

Regular dodecahedron

Regular icosahedron

BRIAN C. MANSFIELD

Figure A4.II   The five Platonic solids.

The following table shows each Platonic solid's name, the number of faces (F), the face shape, the number of faces at each vertex (NF), the number of vertices (V), the number of edges (E), and the "dual"—that is, the Platonic solid that can be inscribed inside by connecting the midpoints of the faces.

Platonic Solids

Platonic Solid	F	Shape of Faces	NF	V	E	Dual
Tetrahedron	4	Equilateral Triangle	3	4	6	Tetrahedron
Cube	6	Square	3	8	12	Octahedron
Octahedron	8	Equilateral Triangle	4	6	12	Cube
Dodecahedron	12	Regular Pentagon	3	20	30	Icosahedron
Icosahedron	20	Equilateral Triangle	5	12	30	Dodecahedron

The Platonic solids were known to the ancient Greeks and were described by Plato in his *Timaeus*, c. 350 B.C. Pythagoras of Samos, the famous mathematician and mystic, lived in the time of Buddha and Confucius, around 550 B.C., and probably knew about three of the five regular polyhedra. Historians now seem to agree that the cube, the tetrahedron, and the dodecahedron were known to the Greeks of this period.

4.33 Mystery pattern. Fractal tic-tac-toe by Patrick Grim and Paul St. Denis. As we discussed in the puzzle "So many Xs and Os" in chapter 1, the first player in tic-tac-toe, labeled X, has a choice of one of nine squares in which to place his or her mark. The opposing player O then has a choice of one of the remaining eight squares. On X's next turn, again he or she has a choice of seven squares, and so forth. There are thus a total of 9! possible series of moves (9 factorial:

$9 \times 8 \times 7 \times \ldots \times 1$), giving us 9! possible tic-tac-toe games. Some of these are wins for X, some for O, and some draws (wins for neither player).

In figure 4.19, the mathematicians Grim and St. Denis offer an analytic presentation of all possible tic-tac-toe games. Each cell in the tic-tac-toe is divided into smaller boards to show various possible choices. For more information, see Paul St. Denis and Patrick Grim, "Fractal Images of Formal Systems," *The Journal of Philosophical Logic* 26 (1997): 181–222. This piece also appears in Patrick Grim, Gary Mar, and Paul St. Denis, *The Philosophical Computer* (Cambridge, Mass.: MIT Press, 1998). See also Paul St. Denis and Patrick Grim, "Fractal Images of Formal Systems," www.sunysb.edu/philosophy/fractal/2Tic.html.

4.34 Omega sphere.
You should choose to toss the darts at the sphere. For geometrical reasons, your three impact points will always be on the "same half of the sphere (hemisphere)." Any two points on the sphere determine a great circle dividing the sphere into two hemispheres. The third point will be in one hemisphere or the other, unless you are so unlucky as to have it land exactly on the great circle. But that's worth the gamble. And, in any case, three points on a great circle are all considered to be located in the same (closed) hemisphere.

4.35 Cutting the plane.
Here is one possible solution. You can draw a line, so that all of the symbols that fall in the gray squares are on one side. Can you find other solutions?

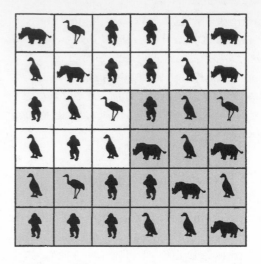

4.36 Twinkle, twinkle, little stars.
With seven star systems, at least one will always survive. There will be two star systems closer together than any other star pair, and these must have fired their missiles at each other. Continue this line of reasoning, and you will see why at least one star system must survive. What are the maximum and the minimum number of survivors for related problems, using different numbers of stars? How are your answers affected for two-dimensional, three-dimensional, and four-dimensional arrangements of stars?

4.37 Mystery triangles.
Figure A4.13 is one solution with nine triangles. Are there others? Can you do better?

Try this problem with larger arrays of crystals. Can you make any generalizations as to how the maximum number of triangles depends on the grid's dimensions? If you have access to computer graphics, extend your results to three dimensions, using triangles or tetrahedrons.

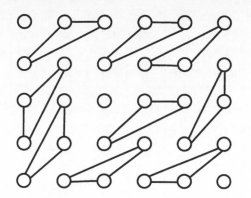

Figure A4.13   Mystery triangles.

4.38 Space station jam. The missing color is green. Consider the various rectangles and circles in this figure. Let's call these "bounding shapes." The color relates to the number of bounding shapes in which a plate is enclosed. (Plates are not just circles and rectangles but can have various hard-to-describe forms.) Yellow regions are enclosed by only one bounding shape. Green regions are enclosed by two bounding shapes. Red regions are enclosed by three. Blue regions are enclosed by four.

4.39 Grid of Gebeleizis. Essentially, it is certain that all lines through your initially selected dot will never meet another dot! (This assumes that you choose the direction of all the lines randomly. If, for example, the direction of a line were chosen precisely parallel to one of the cube's edges, then the probability of hitting another dot would be one.)

Georg Cantor (1845–1918) was interested in this kind of problem. For example, on a quite related topic, he showed that there is an infinity of rational numbers (terminating or recurring decimals like 0.666666 . . . , 0.5, and 0.272727 . . .) and an infinity of irrational numbers (non-terminating, nonrepeating decimals like pi and $\sqrt{2}$). As we have said, he conjectured that the infinity of irrationals is greater than the infinity of rationals. This gives a hint as to why we can place infinitely many lines between the precisely placed grid of points.

For a line to intersect another point, the slope of this line must be rational, because the differences in the x, y, and z points are all integers. If we choose random lines, we must note that there are many more lines that have irrational slopes and do not intersect points in the grid than those that do because of the infinite number of points in space between the points in the grid. Because there are infinitely more ways of not hitting a point than of hitting one, the odds of hitting a point are essentially zero.

4.40 In an ancient tomb. Figure A4.12 shows one solution. Many of my genius friends have told me that this puzzle was impossible to solve. However, if my friends looked at the puzzle a day later, they usually were able to solve it on the second day.

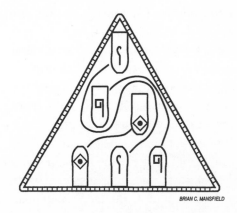

BRIAN C. MANSFIELD

Figure A4.12   In an ancient tomb.

4.41 Poor Pythagoras. The correct answer is c, but most people say it is a. Experiment with two quarters. I've heard rumors that a few years ago, the people who designed the SAT had the wrong answer to a similar problem.

4.42 Alien colonies. A bacterium's color is determined by the number and the size of the bacteria it touches. Denote the "base" color of a bacterium by "0." This number is incremented by 1 for each touching bacterium that is larger than the base bacterium, and decremented each time a smaller bacterium is attached. The base color is not incremented or decremented when a bacterium touches another bacterium of the same size. For example, a bacterium with 2 smaller touching bacteria is decremented by 2 and therefore has a "color" of "–2."

These structures are very beautiful when drawn in color, where colors may be assigned as follows: –3 and smaller numbers are yellow; –2 orange; –1 red; 0 black; 1 blue; 2 green; 3 and larger numbers are violet. I would be interested in hearing from readers who have designed larger colorful colonies using these rules.

4.43 Ant planet. Nadroj's quick way to tell whether he is inside or outside the Jordan curves is to count the number of times an imaginary line drawn from his body to the outside world crosses a wire. If the straight line crosses the curve an *even* number of times, the ant is outside the maze; if an *odd* number of times, the ant is inside.

Back in the real world, the French mathematician Marie Ennemond Camille Jordan (1838–1922) offered a proof of the same rules for determining the inside and the outside of these kinds of curves. (The proof was corrected in 1905 by Oswald Veblen.) Jordan had originally trained as an engineer.

4.44 Aesculapian mazes. Figure A4.14 shows one solution. How many other solutions can you find?

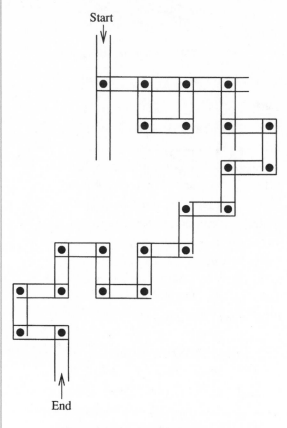

Figure A4.14 Aesculapian mazes.

4.45 Contact from Aldebaran. The chances are 100 percent that there was a spot along the spiral that Britney passed at exactly the same time on both days. To understand this, visualize Britney's initial trip and return trip taking place at the same time. In other words, she starts from the center at the same time her "twin" starts from the spiral outlet. At some point along the journey, they must meet as they pass each other. That will be the place and the time.

4.46 Detonation! Here is one solution (see figure A4.15). Can you find others?

THE SOLUTION

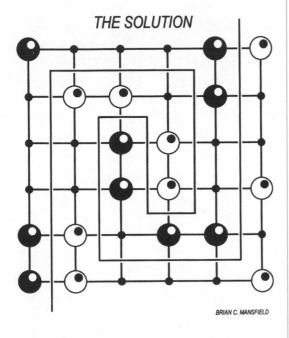

BRIAN C. MANSFIELD

Figure A4.15   Detonation!

4.47 Gear turns. The same speed—zero! The collection of gears cannot turn at all. The number of gears in the ring of gears in the middle is an odd number. An even number of gears is required for a ring of gears to spin because every touching gear must rotate clockwise, counterclockwise, clockwise, counterclockwise, and so on.

4.48 Infinitely exploding circles. By continually surrounding the shapes with circles, you might have guessed that the radii should grow larger and larger, becoming infinite as we continue the process. With a simple computer program, you can in fact show that the radius of a circle is always larger than its predecessor. (After all, the predecessor shapes are *enclosed* by the most recently added circles.) However, the assembly of nested polygons and circles will never grow as large as the universe, never grow as large as Earth, or never grow as large as a basketball.

Although the circles initially grow very quickly in size, the rate of growth gradually slows down, and the radii of the resulting circles approach a limiting value given by the infinite product: $R = 1/ [(\cos \pi/3) \times (\cos \pi/4) \times (\cos \pi/5) . . .]$ The limiting radius is: 8.7000366252081945 . . . .

4.49 Hexagonal challenge. Figure A4.19 shows two solutions. Do others exist?

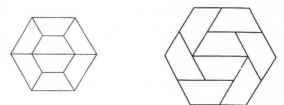

Figure A4.19   Hexagonal challenge.

4.50 Magic circles. Figure A4.16 shows one solution. These magic circles are about a century old and are from W. S. Andrews, *Magic Squares and Cubes* (New York: Dover, 1960), the second edition of which was originally printed in 1917.

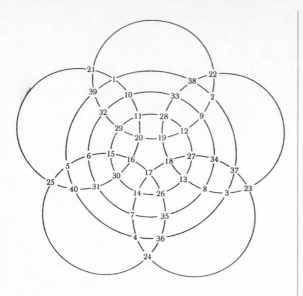

Figure A4.16   Magic circles.

4.51 Robotic worm. There seems to be more than one answer; for example,

CI	BI	C4	D4
DI	AI	B4	A4
A2	D2	A3	D3
B2	C2	B3	C3

AI	BI	C4	D4
DI	CI	B4	A4
A2	D2	A3	D3
B2	C2	B3	C3

One way to solve the problem is to start by inserting values that are forced. For example, we know that a "C" is above the "B" given in the original problem.

4.52 $y = x\sin(1/x)$. Because $\sin x$ is everywhere continuous, and $1/x$ is continuous for $x \neq 0$, it follows that the composite function $\sin(1/x)$ is continuous for $x \neq 0$. Therefore $\psi(x)$ is a simple mathematical function that is everywhere continuous, even though it has infinitely many oscillations in the neighborhood of $x = 0$, where the size of the oscillations becomes infinitely small. However, unlike the fractal Koch curve, $\psi(x)$ is differentiable (i.e., smooth) for $x \neq 0$, even though the frequency near $x = 0$ approaches infinity and the spacing between the maxima approaches zero. Don't you love mathematics? It really lets the soul soar.

4.53 Magic sphere. Figure A4.17 is one solution. Notice that pairs of numbers on each

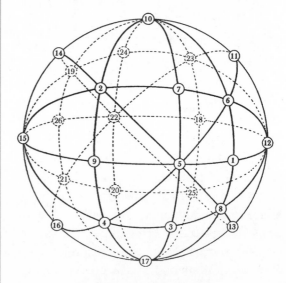

Figure A4.17   Magic sphere.

side of an imaginary diameter line sum to 27. As in the previous question, these magic spheres are about a century old and are from W. S. Andrews, *Magic Squares and Cubes*.

4.54 Continuity. A curve can be continuous even if it has points at which the derivative (or the slope) is not defined. In fact, some of these kinds of curves can be quite fascinating.

Curves can be continuous even if they have points at which they are not "differentiable." (A function is differentiable at a point x if its derivative exists at this point.) For example, $y = |x|$ is continuous everywhere—you can draw it without lifting your pen—even though the derivative is not defined at $x = 0$ where a sharp corner exists. Figure A4.20 is a plot of $y = |x|$, where the $|x|$ denotes the absolute value of x.

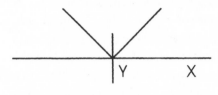

Figure A4.20   A graph of $y = |x|$.

Think of a seesaw and a long plank of wood atop the pivot point of a seesaw. Further visualize the seesaw teetering back and forth because there is no single tangent to a corner on which it balances. Similarly, a function is not differentiable at a point in which the derivative has no particular value.

I often think about curves as being differentiable everywhere only if they are smooth and have no sharp edges or jumps. On the other hand, my favorite curve that is nondifferentiable everywhere (because it is so pointy!) is the Koch curve (figure A4.21).

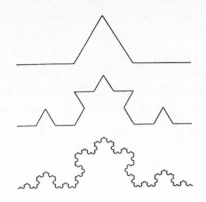

Figure A4.2I   How to create a Koch curve.

The edge of a Koch curve looks a little like a snowflake's edge. You can construct a Koch curve edge by continually replacing the center third of every straight line segment with a V-shaped wedge. After numerous generations, the length of the curve becomes so great that you could not carefully trace the path in your entire lifetime.

In 1904, the Swedish mathematician Helge von Koch proposed this curve, and it is considered to be a fractal curve because it displays similar structures as you magnify the crinkly edges. You can't draw tangent lines at any points, so the curve is nondifferentiable at every point.

Generally speaking, a fractal is an object or a pattern that exhibits similar structures at different-sized scales. Think of the leaves of a fern, the branchings of blood vessels, or the edges of a coastline. Sometimes, it's a little strange to imagine a fractal shape like the Koch curve with *infinitely* many bumps between bigger bumps, in a *finite* region in space, and realize that it is still a continuous curve.

4.55 Limits. The answer is ¼. Your friends may tell you that when x is 2, the denominator is zero, and thus the limit of this function must be undefined. But this is not so. Don't give up so easily. The limit can exist for some fractions with 0 denominators. First, try to simplify the fraction:

$$\lim_{x \to 2} \frac{x-2}{x^2-4} = \lim_{x \to 2} \frac{x-2}{(x+2)(x-2)} =$$

$$\lim_{x \to 2} \frac{1}{x+2} = 1/4$$

It is true that the *function* $(x-2)/(x^2-4)$ is not defined at $x = 2$. It has a "hole" at this point. At other points, the curve just looks like $1/(x+2)$. Remember, when you evaluate a limit, you are only concerned about values of a function *near* a particular point and not *at* the actual point.

4.56 Space station scordatura. In figure A4.18, the space station should be built at the intersection of lines AB and CD.

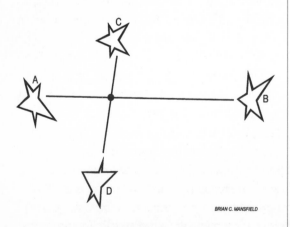

Figure A4.18 Space station scordatura.

4.57 The Procyon maneuver. The plantation owner ties the rope around the base of the ankh tower on the edge of the funnel and then carries the rope on a walk around the funnel's edge. As he completes half of his walk around the circular aperture, the rope begins to wrap around the central ankh on the cylindrical platform, and when he reaches his starting point, he ties the other end of the rope to the ankh on the edge of the funnel. Having created a rope bridge from ankh to ankh, he can pull himself across. The chicken skull serves no purpose.

4.58 Leaning tower of books. The stack of books will not fall if the following rule is met: the center of mass of all the books above any particular book must lie on a vertical axis that cuts through that particular book. This must be true for each book in the stack. For example, the top book in a stack can be made to clear the table if there are 5 books in the stack (identical book sizes are assumed). For an overhang of 10 books, you need 272,400,600 books. Although there is no limit as to how far out one can "travel" with the book stack, a great many books are required to do so. I discuss this classic physics problem further in my book *The Mathematics of Oz*.

4.59 Mystery swirly curve. This is the famous "Ikeda attractor." A deep reservoir for striking images is the *dynamical system*. Dynamical systems are models comprising the rules that describe the way some quantity undergoes a change through time. For example, the motion of planets about the sun can be modeled as a dynamical system in which the planets move according to Newton's laws. Figure 4.41 shows the behavior of mathematical expressions called *differential equations*. Think of a differential equation as a machine that takes in values for all the variables at an initial time and then generates the new values at some later time. Just as one can track the path of a jet by the smoke path it leaves behind, computer graphics provides a way to follow paths of particles whose motion is determined by simple differential equations. The practical side of dynamical systems is that they can sometimes be used to describe the behavior of real-world things, such as planetary motion, fluid flow, the diffusion of drugs, the behavior of inter-industry relationships, and the vibration of airplane wings. Often the resulting graphic patterns resemble smoke, swirls, candle flames, and windy mists.

The *Ikeda attractor* shown in figure 4.41 is an example of a strange attractor. As background, *predictable attractors* represent the behavior to which a system settles down or is "attracted" (for example, a point or a looping closed cycle). An example of a *fixed point attractor* is a mass at the end of a spring, with friction. It eventually arrives at an equilibrium point and stops moving. A *limit cycle* is exemplified by a metronome. The metronome will tick-tock back and forth, its motion always periodic and regular. A "strange attractor" has an irregular, unpredictable behavior. Its behavior can still be graphed, but the graph is much more complicated. With "tame" attractors, initially close points stay together as they approach the attractor. With strange attractors, initially adjacent points eventually follow widely divergent trajectories. Like leaves in a turbulent stream, it is impossible to predict where the leaves will end up, given their initial positions.

Here is an outline of computer code that will allow you to produce the Ikeda pattern, a representation of a dynamical system. Simply plot the position of variables j and k through the iteration. The variables *scale*, *xoff*, and *yoff* simply position and scale the image to fit on the graphics screen.

```
c1 = 0.4, c2 = 0.9, c3 = 6.0, rho = 1.0;
for (i = 0, x = 0.1, y = 0.1; i <= 3000; i++)
[
  temp = c1 - c3 / (1.0 + x * x + y * y);
  sin_temp = sin(temp);
  cos_temp = cos(temp);
  xt = rho + c2 * (x * cos_temp - y *
sin_temp);
  y = c2 * (x * sin_temp + y * cos_temp) ;
  x = xt;
  j = x * scale + xoff;
  k = y * scale + yoff;
  DrawDotAt (j,k)
]
```

The Ikeda attractor has been described in greater detail in this article: K. Ikeda, "Multiple-Valued Stationary State and Its Instability of the Transmitted Light by a Ring Cavity System," *Optical Communications* 30, no. 2 (1979): 257–61.

4.60 Kissing circles. For years, I've been fascinated by several simple recipes for graphically interesting structures that are based on the osculatory packing of circles. The problem of covering a finite area with a given set of circles has been popular in mathematics journals. As background, the densest packing of non-overlapping uniform circles is the hexagonal lattice packing, where the ratio of covered area to the total area (packing fraction) is $\phi - \pi/\sqrt{12} \approx 0.9069$. The limiting packing fraction for the nested hexagonal packing of circles, with *k different* circle sizes, is $\phi_k = 1 - (1 - .09069)^k$. This applies to cases where each of the uncovered areas, or *interstices*, is also hexagonally packed by smaller circles. For larger values of *k*, ϕ_k approaches unity.

Other researchers have usually defined a distribution of circles as *osculatory* if any available area is always covered by the largest possible circle. If the original area to be covered is a tricuspid area (figure A4.22), then the first circle to be placed must be tangent to the three original larger circles. This kind of packing is also often referred to as *Apollonian packing*. In contrast to past work, my criterion for osculatory packing is relaxed here: each successively placed circle on the plane need only be tangent to at least *one* previous circle ("tangent-1 packing"). To generate figure 4.43, I randomly place a circle center within the available interstice. The circle then grows until it becomes tangent to its closest neighbor. The process is repeated several thousand times.

One easy way to simulate this on a computer is to determine the distances d_i from the newly selected circle center to all other circles *i* on the plane. Let $\delta_i = d_i - r_i$, where r_i is the radius of circle *i*. Min $\{\delta_i\}$ is then the radius of the new circle. Note that if there exists a negative δ_i, then the selected center is within a circle on the plane. In this case, the circle center is discarded, and a new attempt to place a circle is made. Repeated magnification of the figures reveals the self-similarity of the figures; the figures are fractal and look the same at various-sized scales.

4.61 Heron's problem. Let *u* and *v* be the side lengths of one rectangle and *x* and *y* be the sides of the second rectangle. We know that

$$u + v = 3(x + y) \text{ (perimeter relationship)}$$

$$xy = 3uv \text{ (area relationship)}$$

How would you go about finding an integer solution for *x*, *y*, *u*, and *v*? According to David Wells, the author of *The Penguin Book of Curious and Interesting Puzzles* (Puzzle 21), Heron's ancient solution was a 53 × 54 rectangle and a 318 × 3 rectangle.

Today we know that there are many solutions to this problem, and Internet debates and contests regarding the puzzle are currently maintained by Ken Duisenberg (ken.duisenberg.com/potw/). In particular, see ken.duisenberg.com/potw/archive/arch96/960820.html. Computer searches are particularly effective in yielding numerous solutions. Such searches can be restricted by noting that $u + v$ and xy are multiples of 3. Large rectangle solutions include 15,012 × 17,578,886,877 and 5,859,498,852 × 135,111.

4.62 The game of elegant ellipses. Figure A4.23 shows one solution. How did you go about solving this? How much more difficult is this if we add more ellipses and more dividing lines? How would a computer solve this problem?

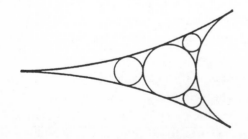

Figure A4.22   Tricuspid interstice.

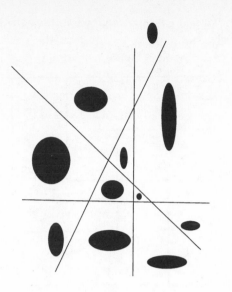

Figure A4.23 The game of elegant ellipses.

4.63 Dudeney's circles. Here is one solution that leaves behind 3 squares when perpendicular lines are drawn between the circles.

○ ○ ○ ...
○ ○ ○ ○
.... ○ ○

4.64 Eschergrams. In several of my books, I've discussed the use of computer graphics to represent noisy data. One method I've employed uses a pattern from the Dutch artist M. C. Escher. To produce figure 4.41, simply draw the generating tile (shown in figure A4.24) with a random orientation, and place it within the corner of a large square lattice. Successive adjacent tiles are added to the lattice for a particular row until it is filled, and a new row is started.

A seamless plane-filling pattern is created, no matter what tile orientation is used. I used just two orientations of the tile to create this

figure. You can see that there are some diamond-shaped objects in the pattern. As correlations within the data become greater, the number of diamonds decreases. A completely random tiling contains the maximum number of diamonds; in this case, "the diamond fraction" is approximately 5 percent (the number of diamonds in the pattern divided by the number of tiles).

Try using this tile to represent genetic sequence data symbolized by the four different letters G, C, A, and T. Genetic sequences can be represented by a collection of tiles with four different orientations. For example, what does the resulting tile pattern (an "Eschergram") tell you about the patterns, and the degree of randomness, within the first 1,000 bases of the AIDS virus? For more information on Escher tiles, see D. Schattschneider, *Visions of Symmetry* (Freeman: New York, 1990). For information on the use of tiles to represent noisy data, see C. Pickover, "Picturing Randomness with Truchet Tiles," *Journal of Recreational Math* 21, no. 4 (1989):

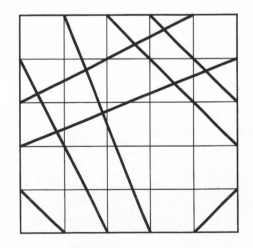

Figure A4.24 Eschergram.

256–59. Also, C. Pickover, "Mathematics and Beauty: Several Short Classroom Experiments," *American Math Society Notices* 38, no. 3 (March 1991): 191–95.

4.65 Dudeney's house. Figure A4.26 is one solution.

Figure A4.26   Dudeney's house.

4.66 Dudeney's 12 counters. Any two overlapping triangles could be used to create a solution (figure A4.25).

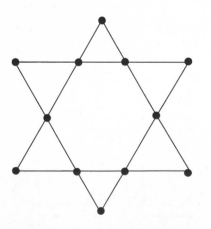

Figure A4.25   Dudeney's 12 counters.

4.67 Dudeney's cuts. Figure A4.27 is one solution. Are there others?

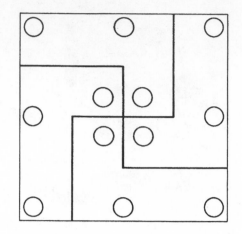

Figure A4.27   Dudeney's cuts.

4.68 $e^{\frac{1}{e}}$. Here is the question: For what value of x is $\sqrt[x]{x}$ a maximum? Note that

$$\sqrt[e]{e} \geq \sqrt[x]{x}$$

Isn't that great? The Swiss mathematician Jacob Steiner (1746–1827) was one of the first people to work on the $e^{\frac{1}{e}}$ problem. Interestingly, Euler proved that

$$y = x^{x^{x^{x^{x \cdots}}}}$$

has a limit if x is between $e^{\frac{1}{e}} = 0.065988\ldots$ and $\sqrt[e]{e} = 1.444667861\ldots$

4.69 Dudeney's horseshoe. The hard part of this problem is doing it in your head. It turns out that you just have to make two straight lines that intersect each other (figure A4.28).

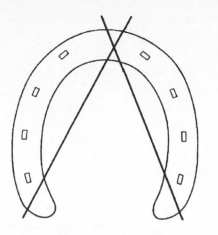

Figure A4.28 Dudeney's horseshoe.

4.70 Tunnel through a cube: A true story. $\frac{3\sqrt{2}}{4} = 1.060660\ldots$ In other words, a cube with a side length of $1.060660\ldots$ inches can pass through a cube with a side of 1 inch.

In the late seventeenth century, Prince Rupert, whose other titles were Count Palatine of the Rhine and the Duke of Bavaria, won a wager that a hole could be made in one of two equal cubes large enough for the other cube to slide through. I believe that the 1.060660 solution was not known until Pieter Niuewland solved it more than a century *after* Prince Rupert asked the question. For more information, see D. J. E. Schrek, "Prince Rupert's Problem and Its Extension by Pieter Nieuwland," *Scripta Mathematica* 16 (1950): 73–80 and 16 (1950): 261–67.

4.71 The mathematics of love. Bill was 565.7 feet north of Mark. Because Bill runs 3 times as fast as Mark and because Mark and Bill arrive at Shannon simultaneously, the distance from Bill to Shannon must be 600 feet (speed ×

time = distance). We can now use the Pythagorean theorem to compute the distance from Mark to Bill.

$$A^2 + B^2 = C^2$$

$$A^2 + 200^2 = 600^2$$

$$A^2 + 40,000 = 360,000$$

Now we need to solve for A. $A^2 = 320,000$

$$A = 565.7 \text{ feet.}$$

4.72 Antimagic square. Here is one solution.

6	8	9	7
3	12	5	11
10	1	14	13
16	15	4	2

Antimagic Square

4.73 Annihilation magic squares. Here is one solution.

8	-7	-6	5
-4	3	2	-1
1	-2	-3	4
-5	6	7	-8

4.74 Cramming humanity into small spots. To solve this problem, we can estimate the population at 6.4 billion, which corresponds to 2,066 square miles of land. Thus, everyone should be able to fit into any state's land area in the United States, except for Rhode Island or Delaware. They would all fit onto the island of Puerto Rico.

If we fashioned a cube and placed each person on one of the inside or the outside faces of the cube, the cube would be only 13 miles wide. They could all fit into Alaska, leaving 99.7 percent of the land unoccupied.

More precisely, let us assume that the number of people on earth (at the time of this writing) is 6,366,780,666. The area allocated per person is 9 square feet. The total area required = 6,366,780,666 × 9 = 57,301,025, 994 square feet. Total area in square miles is $(57301025994)/(5280^2) = 2,055$ square miles. This many people could fit in Rhode Island (1,545 square miles) if they sucked in their guts and occupied 2.6-by-2.6-ft. plots.

My colleague Nick Hobson notes that if we make use of the third dimension, 6.4 billion people, each allocated a $6' \times 6' \times 6'$ cube, would fit into a cube with a side that's just over 2.1 miles. This would comfortably fit on Manhattan Island, or it could be "dismantled" to fit in the Grand Canyon.

4.75 Chocolate cake computation. The circular cake is a better buy. The area A_1 of the circular cake is $(d/2)^2 \times \pi$. The area A_2 of the square is $(d-1)^2 = d^2 - 2 \times d + 1$. $A_1 = 56.74$. $A_2 = 56.25$.

However, the circular cake is not always a better buy than the square. Notice that at about $d = 8.78942582838$, the areas are equal: $A_1 = 60.67515473587$ and $A_2 = 60.67515473587$. If d grows larger than 8.789425, the square becomes a better buy.

The general solution for the special value of d is

$$d = \frac{4 + 2\sqrt{\pi}}{4 - \pi}$$

We obtain this by solving for $(d/2)^2 \times \pi = (d-1)^2$ and finding the points where both sides of the equation are the same.

$$\pi(d/2)^2 = (d-1)^2$$
$$(\pi d^2)/4 = d^2 - 2d + 1$$
$$\pi d^2 = 4d^2 - 8d + 4$$
$$d^2(4 - \pi) - 8d + 4 = 0$$

By the quadratic formula, we get

$$d = [8 \pm \sqrt{(64 - 16(4 - \pi))}]/2(4 - \pi)$$
$$= [8 \pm 4\sqrt{(4 - (4 - \pi))}]/2(4 - \pi)$$
$$= [4 \pm 2\sqrt{(\pi)}]/(4 - \pi)$$
$$= [4 \pm 2 \times 1.772453851]/0.858407$$

which yields

$$d \approx 8.789426 \text{ or } 0.530159$$

In theory, d values between these two bounds yield a circular cake with a bigger area, and outside of them, the square area is bigger. However, we exclude the solution $d = 0.530159$ because it yields a negative cake area. The quantity $d - 1$ must be greater than 0 to be physically meaningful.

4.76 Bears in hyperspace. Yes, the polar bear could be crammed into an 11-dimensional sphere of radius 6 inches. To help you understand this answer, consider the act of stuffing rigid circular regions of a plane into a sphere.

If the circular disks are really two dimensional, they have no thickness or volume. Therefore, in theory, you could fit an infinite number of these circles into a sphere—provided that the sphere's radius is slightly bigger than the circles' radius. If the sphere's radius were smaller, even one circle could not fit within the enclosed volume, since it would poke out of the volume. Therefore, the volume of a polar bear *could* reside comfortably in an 11-dimensional sphere with a 6-inch radius. In fact, an infinite number of polar bear volumes could fit in an 11-dimensional sphere. However, you could not physically stuff a polar bear into the sphere because the bear has a minimum length that will not permit it to fit. Indeed, a polar bear volume equivalent could be contained within the sphere, but to do so would require the bear to be first put through a meat grinder that produces pieces no larger than the diameter of the sphere. It would help if the polar bear could be folded or crumpled in higher dimensions like a piece of paper.

4.77 Hypersphere packing.
For circles, we know that the answer is 6 (figure A4.29). For spheres, the largest number is 12, but this fact was not proved until 1874. In other words, the largest number of unit spheres that can touch another unit sphere is 12. For 4-D hyperspheres, the number is 24, and 196,560 24-dimensional spheres can kiss a central 24-dimensional sphere of the same radius. The number of equivalent hyperspheres in n dimensions that can touch an equivalent hypersphere without any intersections is sometimes called the Newton number or ligancy. Newton correctly believed that the kissing number in three dimensions was 12.

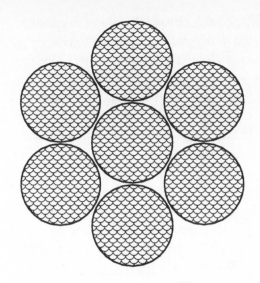

Figure A4.29 In two dimensions, a circle can make contact with six other circles of the same size. What happens in higher dimensions?

4.78 Lattice growth.
It turns out that these lattice numbers grow very quickly, but you might be surprised to realize just how quickly. The formula describing this growth is fairly simple for an $n \times n$ lattice: $L(n) = n^2 (n + 1)^2/4$. The sequence goes 1, 9, 36, 100, 225, 441 . . .

Let's jump up a dimension. Can you find a formula for the number of boxes for 3-D lattices? Can you find a formula for the number of hyperboxes in hyperlattices of any dimension?

Did you know that a small $n = 2$, 7-dimensional lattice ($2 \times 2 \times 2 \times 2 \times 2 \times 2 \times 2$) has over 1,000 hyperboxes? If we think of each box as a container, an $n = 9$ hyperlattice in the 50th dimension can hold each electron, each proton, and each neutron in the universe (each particle in its own cage). Does this all boggle your mind?

4.79 In the garden of the knight. Here is the solution (Ed Pegg Jr., "Math Puzzles," www.mathpuzzle.com/29Jun2003.html).

4.80 In the garden of the knight. Here is the solution from Jean-Charles Meyrignac, who has done extensive computer analyses of

these kinds of problems (Ed Pegg Jr., "Math Puzzles," www.mathpuzzle.com/29Jun2003.html).

4.81 In the garden of the knight. The challenge was to place 48 knights on a 10-by-10 chess board so that each knight attacks two, and only two, other knights. This is one of twenty possible solutions from Jean-Charles Meyrignac (Ed Pegg Jr., "Math Puzzles," www.mathpuzzle.com/29Jun2003.html).

4.82 Polyhedral universe. A dodecahedron. According to *Science* magazine (Charles Seife, "Polyhedral Model Gives the Universe an Unexpected Twist," 302 [October 10, 2003]: 209), a team of scientists from France and the United States studied the measurements from the Wilkinson Microwave Anisotropy Probe (WMAP) satellite to reach a surprising conclusion: the universe might be finite and twelve-sided. Although the hypothesis is being challenged, its proponents say that it matches the data. Also, opposite faces of the dodecahedron correspond in unusual ways to each other. In fact, these faces are actually the same face so that a spaceship flying out of one side of the universe winds up flying back into the other side.

To make a finite dodecahedral space, one glues together opposite faces of a slightly curved dodecahedron—a soccer ball–like shape with twelve pentagonal sides. Of course, such gluing is difficult to imagine in our ordinary 3-D space. See also Erica Klarreich, "The Shape of

Space," *Science News* 164, no. 19 (November 8, 2003): 296–97.

In the spirit of full disclosure, I should note that scientists' theories about the shape of the universe change almost every month. In April 2004, Frank Steiner at the University of Ulm in Germany suggested that the universe is shaped like a medieval horn— a very long funnel. In Steiner's model of the universe, technically known as a Picard topology, the universe is infinitely long in the direction of the funnel's spout, but so narrow that the universe has finite volume.

4.83 Tank charge. Here is one path. Can you find others?

S	I	2	3	4
10	9	8	6	5
12	II	7	17	18
13	14	15	16	19
E	23	22	21	20

4.84 Molecular madness. Figure A4.30 shows one solution. How many solutions exist? When solving this puzzle, my colleagues often start by searching for those atoms near the periphery of the molecule that sometimes have fewer bonding options than interior atoms do. I notice that some friends seem to solve this by pure intuition, in ways I don't fully understand. Others attempt to rigorously analyze the possibilities and get "stuck" when they encounter so many initial options.

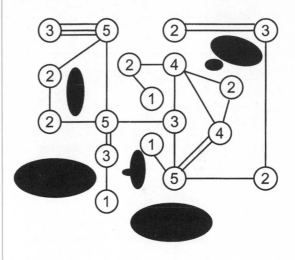

Figure A4.30   Alien molecule confounds scientists.

4.85 Clown with balls. The side length is 11.464 . . . My colleague Mark Ganson discussed this puzzle of his at my discussion group, and I like the puzzle for its simplicity and because it can be solved in so many different ways. Here is one way. Consider a ball at the bottom left-hand corner. Let the center of the ball be O, the tangent point at the bottom T, and the left triangle corner C. OC bisects the angle at C, so angle OCT is 30 degrees; OTC is 90 degrees; COT is 60 degrees; OT is 1, and TC is tan(60) or 1.732 . . . Therefore, the length of one side of the triangle is $8 + 2 \times TC = 11.464$ . . .

4.86 Stained glass squares of Constantine. There are 16 squares, counting the black frame around the window. I find that most people who casually study the window find only about 12 to 14 squares. How can our perceptual systems be so different? How many squares did you find? Let's see how well your friends can do on this puzzle. If you are a teacher, how many squares can your students find? How does the number of perceived squares depend on a person's age, gender, occupation, emotional state, IQ, caffeine level, and culture?

4.87 Chocolate puzzle. Here is the solution:

		4					
	6					6	
					8		2
		6					4
	6				4	4	
					3		
6			5				

Most of my brainy friends at work solved this in about 5 minutes. One way to approach the problem is to simply eliminate possibilities. For example, you can examine the 6 in the bottom-left corner, which seems to have two possible rectangles that can be associated with it. However, we can eliminate the vertical 1×6 rectangle by noting that this would not permit a legal rectangle at the upper left of the grid. The solution is unique.

4.88 Jerusalem crystal. Follow the black circles to see one possible path (figure A4.31). How many other solutions exist?

Figure A4.31 Jerusalem crystal.

5. Probability: Take Your Chances

5.1 Gambling. You can easily turn Donald's unfair coin into a fair one. Get ready to toss the coin twice. Before doing so, you can select heads followed by tails (HT), and Donald can select tails followed by heads (TH). If both tosses give the same result, discard the results, and toss the coin twice again. The first time the coin lands either TH or HT, you can determine who wins the bet!

If you don't see why this is fair, let's pretend, for example, that the biased coin lands tails 80 percent of the time. Even so, sometimes we will have TH or HT. There is no reason that either of these will be favored over the other, and, in fact, they should each occur with equal frequency. Thus, by monitoring "HT" versus "TH," we can create an unbiased selection process. If each toss is independent, then the probabilities for each pair are

$$p(\text{TT}) = 80\% \times 80\% = 64\%$$

$$p(\text{TH}) = 80\% \times 20\% = 16\%$$

$$p(\text{HT}) = 20\% \times 80\% = 16\%$$

$$p(\text{HH}) = 20\% \times 20\% = 4\%$$

So, 32 percent of the time you have a result. Otherwise, you toss the coin two more times.

5.2 Monkeys typing *Hamlet.* Typewriters have about 44 keys. *Hamlet* has about 32,301 words or 176,522 characters, including spaces. If a typewriter with 44 keys is used, the odds of getting one character correct (the first one, being "H" for *Hamlet* in the title) is 1 out of 44. Getting two characters correct would clearly be less likely, and, in fact, the probability would be $(1/44) \times (1/44)$. Getting all 176,522 characters correct would be 1 out of $44^{176,522}$. A computer keyboard has about 80 keys, giving us the lower odds of 1 out of $80^{176,522}$.

5.3 Many monkeys typing *Hamlet.* Each monkey still individually has the same odds, 1 out of $44^{176,522}$, which we discussed in the answer to the previous problem. But each day the probability that one of the monkeys will succeed doubles, so that on day two the odds become 2 out of $44^{176,522}$, and on day three the odds become 4 out of $44^{176,522}$, and so on, day by day. On some future day, d, the number of monkeys 2^d will exceed $44^{176,522}$, and on that day, the odds are very good (i.e., close to 1 out of 1) that one of those poor monkeys will succeed. Of course, at this point the visible universe would be flooded with monkeys, not leaving much room for humanity to have fun. (Naturally, I've made some assumptions in this interpretation. Can you name a few?)

5.4 The Book of Everything That Can Be Known.
When a book is flipped open, it displays one page on the left and another on the right. So, if the statement is anywhere on the two displayed pages, we've found the statement. *The Book of Everything That Can Be Known* is 10,000 pages long, so it has 5,000 different openings. The chance of flipping the book open to the fact "the universe is finite in spatial extent" is therefore about 1 in 5,000. On the average, you would need 5,000 random flips to find this fact. There's no guarantee, however, that you will find it in thousands of random flips. As William Poundstone notes for a similar problem in his book *How Would You Move Mount Fuji?* no matter how many times you flip the pages, you can never be 100 percent sure of flipping so that a particular 2 pages are displayed.

In fact, the chances of flipping to the *wrong*

2 pages on any given try is 4,999/5,000 because there are exactly 10,000 pages and exactly 5,000 openings to flip to. The chances that the first n flips will all fail to open to "the universe is finite in spatial extent" is then $(4,999/5,000)^n$. The chance that you will flip to the *correct* opening in n flips or less is $1 - (4,999/5,000)^n$. Thus, the odds of flipping to "the universe is finite in spatial extent" is about 50 percent on the 3,470th flip. Half of the time, you'll hit "the universe is finite in spatial extent" in 3,470 flips or less.

If we wish to be more precise about our assertion that a 10,000-page book has 5,000 different openings, note that if you open a book at the front, it has only one page on the right. At the back, the book has one page on the left. So, for example, we would find that a 4-page book has 3 ways of opening it. A 10,000-page book has 5,001 ways. So, more precisely, the aforementioned probability is 1/5,001, not 1/5,000, and on average, you would have to randomly flip the book open 5,001 times.

5.5 Bookkeeper arrangements. The solution is $10!/(2!2!3!) = 3,628,800/24 = 151,200$. If you didn't get this, reread the previous entry.

5.6 Subway odds. Turn over any three slips of paper. Next continue to turn papers until you reach a number higher than the highest number on the first three slips of paper. This is the number you stop on and show your captor. If you are about to turn over all the papers, stop at (and choose) the last one turned. This system gives you maximum odds of escaping, which are slightly better than one in three.

5.7 The ancient paths of Dr. Livingstone. There are 6 ways to get from A to B and 4 ways to get from B to C. Using the multiplication principle in "⟐ Combinatorics defined," we find $6 \times 4 = 24$ possible paths from A to C.

5.8 Organ arrangements. We have 5 organs. As we learned, the organs have $5! = 5 \times 4 \times 3 \times 2 \times 1 = 120$ permutations. The number of permutations of any 3 organs from the 5 is $_5P_3 = 5!/(5 - 3)! = 120/2 = 60$ permutations. If you didn't get this answer, reread the previous entries.

5.9 Mombasa order. Because the wife has chosen the last city, we are left with a permutation of 6 cities, which is $6! = 720$ different ways they can order their visits.

5.10 Diamond arrangements. We find $(n - 1)!$ arrangements, given n different objects arranged in a circle. Thus, the movie star has an amazing $15! = 1,307,674,368,000$ different possible arrangements of the diamonds on the bracelet. If diamonds were cheap, it would be very easy to give a different arrangement of the 16 diamonds to everyone on Earth and still have many more arrangements left over.

5.11 Marble maze. One way to solve this maze puzzle is to realize that at each marked intersection (or decision point), the marble has a 50-50 choice (or 1/2 chance) of going right or left. There are three paths by which the marble can get to the winning exit (1-2-6-8, 1-2-4-6-8, and 1-3-5-4-6-8). All we need to do is multiply the 1/2 probabilities at each decision point along each path to find the likelihood of traveling a particular path. So for the three paths we have $(1/2 \times 1/2 \times 1/2 \times 1/2) + (1/2 \times 1/2 \times 1/2 \times 1/2 \times 1/2) + (1/2 \times 1/2 \times 1/2 \times 1/2 \times 1/2 \times 1/2)$. Thus, we find $1/16 + 1/32 + 1/64$ which is about 0.11—only about an 11 percent chance of winning!

Here's a new one for you to ponder from my colleague Derek Ross. Consider the following super-simple maze:

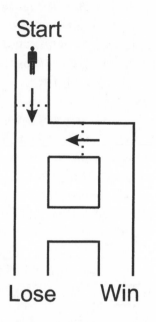

Figure A5.1   Super-simple maze.

You can go any direction: north, east, south, west. There are two one-way doors, indicated by dotted lines. The arrow indicates the direction of the door. Dr. Brain has been getting you drunk on wine. Next, he drops you at the start gate of the tunnel. Since it is completely dark, and you are mentally disoriented, you are capable of making only random turns whenever you encounter a fork in the tunnel. The problem is to find the probability that you will eventually crawl out of the exit marked "Win."

5.12 Sushi combinatorics. By the multiplication principle we discussed in " Combinatorics defined," we can make $3 \times 2 \times 5 = 30$ different kinds of sushi.

5.13 Turnip permutations. The word *turnip* has 6 letters. Thus, the letters have $6! = 6 \times 5 \times 4 \times 3 \times 2 \times 1 = 720$ permutations.

5.14 The Africa gambit. My friend Nick Hobson notes that the answer depends on the map projection that is used. The African mainland has an area of about 11.6 million square miles, out of about 197 million square miles for all of Earth. So, throwing the dart at an equal-area projection map, such as the "Eckert IV," will give a probability of about 11.6/197 = 5.9 percent, or 1/17.

The Eckert IV projection, presented by Max Eckert in 1906, is a "pseudocylindrical equal-area" projection. The central meridian and all parallels are straight lines; other meridians are equally spaced elliptical arcs. On the other hand, the more familiar Mercator projection, originally produced in the sixteenth century for navigation purposes, stretches higher latitudes east-west, and therefore, for example, exaggerates the size of North America and Europe. In such maps, the probability of hitting Africa would be much less than 5.9 percent.

5.15 Sushi gambit. Place 1 piece of octopus sushi on one dish and the other 9 pieces of sushi on the other dish. We can determine the chances that Dr. Sushi will eat an octopus sushi as follows. He will choose the dish containing 1 octopus sushi with 50 percent probability. We must add to this the probability that he will eat octopus sushi from the other dish.

The other dish has 5 tuna and 4 octopus sushi, so the chances of selecting an octopus from this dish are 4/9. Thus, the probability that he will eat octopus is $0.5 + 0.5 \times 0.44 = 0.722$ or about 72 percent.

5.16 Alien gambit.

Write all the possible sequences of button presses in order to compute the probabilities of your turning green. For discussion, name the three buttons "explode (E)," "disable (D)," and "happy (H)." Here are the six different possible scenarios, with $B1$, $B2$, and $B3$ indicating the effects of pressing Buttons 1, 2, and 3:

	B1	B2	B3	Outcome
1	E	D	H	You turn green.
2	E	H	D	You turn green.
3	D	E	H	You stay flesh colored.
4	D	H	E	You stay flesh colored.
5	H	E	D	You turn green.
6	H	D	E	You stay flesh colored.

I calculate six different possible orders in which you could press the three buttons. Of these, there are three scenarios in which you turn green and three in which you do not. This would appear to suggest that you have a 50-50 chance of turning green. If you want to be more realistic, the chances are probably *greater* that you will turn green, because in Line 6 you might be so happy that you don't care whether you turn green, and that means you may not press any button again to disable the timer that opens the capsule. That leaves 3 sure greens, 2 sure not greens, and 1 uncertain. Thus, you are somewhat more likely to turn green.

5.17 Cards of Minerva.

The probability that Zeus obtains more blacks than Minerva is 1/2. Either Zeus throws more blacks than Minerva, or Zeus throws more whites than Minerva. By symmetry, these two mutually exclusive possibilities occur with equal probability. Therefore, the probability that Zeus obtains more blacks than Minerva is 1 out of 2 (1/2). Are you amazed that this probability seems to be independent of the number of cards, n? You can learn more about problems of this kind from Richard Epstein, *The Theory of Gambling and Statistical Logic* (New York: Academic Press, 1995).

An example may clarify. Suppose $n = 1$. Minerva throws "no blacks" with probability 1/2 and one black with probability 1/2. Zeus throws "no blacks" with probability 1/4, one black with probability 1/2, and two blacks with probability 1/4. We can illustrate this in a table.

	Minerva Throws	
Zeus Throws	**0 blacks** ($p = 1/2$)	**1 black** ($p = 1/2$)
0 blacks ($p = 1/4$)	1/8	1/8
1 black ($p = 1/2$)	1/4	1/4
2 blacks ($p = 1/4$)	1/8	1/8

Examining the table, we can see that Zeus throws more blacks (bold cells) half the time, the same number 3/8 of the time, and fewer 1/8 of the time.

5.18 Card shuffles.

Only eight consecutive perfect shuffles are required to restore a deck of cards to its original order. Let us consider the second card from the top of the deck. The shuffles move it to the following sequence of positions: 2, 3, 5, 9, 17, 22, 14, 27, and back to 2. This works for any card in the deck. For further reading, see Robert Ehrlich, *What If You Could Unscramble an Egg?* (New Brunswick, N.J.: Rutgers University Press, 1996).

5.19 Piano probability. First, let us assume that the keys have equal area so that, for example, you wouldn't be more likely to press a white note than a black note. Then, the four most equally likely notes are A, A#, B, and C, since on a standard 88-key piano, there are 8 of each of those notes, and 7 of all the others (12 notes in an octave; 8 of them occur 7 times, the other four 8 times: $8 \times 7 + 4 \times 8 = 56 + 32 = 88$). For example, there is an 8/88 or a 1/11 chance of striking an A key but only a 7/88 chance of striking a D key. Of all the black keys, you are most likely to strike the A# key.

5.20 Martian shuffle. There are six possible orders of the cards: *MPD*, *MDP*, *PMD*, *PDM*, *DMP*, and *DPM*. Thus, one in six shufflings will return your original order. With ten cards in the deck, there would be 3,628,800 possible orderings. If you shuffled this deck as fast as you could, without stopping for eating and sleeping, it would still take you over a month before you got your original deck back, based on chance alone.

5.21 Air molecules. Consider one air molecule in the room. For simplicity, let's pretend that a molecule can jump from its current position to any position in the room. By chance alone, it has a 10 percent or 1/10 probability of being in a predefined corner room volume

that is 1/10 the volume of the room. If you consider two molecules moving in the room, the chance of both being in the corner is much smaller, only $(1/10) \times (1/10)$ or 1 percent. Ten molecules have a

$$\frac{1}{10{,}000{,}000{,}000}$$

chance of being in the corner. One hundred molecules have a

$$\frac{1}{\text{googol}}$$

chance of being in the corner, where a googol is 1 with 100 zeros. It has been estimated that we would have to wait 10^{80} times the age of the universe for all 100 molecules to migrate to the corner of the room by chance alone. In *What If You Could Unscramble an Egg?* Robert Ehrlich notes that there are 10^{27} air molecules in a typical room. The probability of all of them going to the corner at one moment due to random motion is

$$\frac{1}{10^{10^{27}}}$$

The number is so tiny that it is about equal to the odds of the Statue of Liberty jumping into the air. (Robert Ehrlich, *What If You Could Unscramble an Egg?* [New Brunswick, N.J.: Rutgers University Press, 1996].) Of course, this is just an approximation, because, in real gases, molecules diffuse by randomly floating through space, and future positions depend on current positions.

5.22 Chess moves. There are 400 possible board configurations. Each side has 8 pawns and 2 knights, each of which has 2 possible moves, giving 20 potential moves for Player 1, against the same for Player 2. Thus, there are 20 times 20, or 400 possible configurations.

5.23 Hannibal's organs.
The probability of removing a brain is now quite high, much higher than 50 percent. The chance is two out of three (2/3) or 66.6 percent. Denote the brain added to the jar as "Brain 2." After you remove the brain from the jar, there are three equally likely states: Brain 1 in jar, Brain 2 outside of jar; Brain 2 in jar, Brain 1 outside of jar; kidney in jar, Brain 2 outside of jar. In two out of three cases, a brain remains in the jar.

Let me clarify by adding my assumptions. When Hannibal reaches into a jar containing a brain and a kidney, assume that it is equally as likely that he will withdraw a brain as it is that he will withdraw a kidney. Also, it is equally likely that the jar initially contains a brain or a kidney. (This exercise in "conditional probability"—that is, the probability of pulling a second brain, "conditioned by" or "given" that the first pull was a brain—always leads to much debate from readers. Write to me if you need more information.)

5.24 Alien pyramid.
Yes. Of the remaining 9 pyramids, there is a 5/9 = 55.55 percent chance of selecting an even one and a 4/9 = 44.44 percent chance of selecting an odd one, so this pyramid is more likely to be even than odd.

5.25 Rabbit math.
The probability is 1 out of 3 (1/3). There are three possible pairings of the brown and polka-dot rabbit, given that at least one of the rabbits is male:

Brown	male	male	female
Polka-dot	male	female	male

Thus, the probability is 1 out of 3 (1/3) that both rabbits are male.

5.26 Card deck cuts.
This is a beautiful and fun trick question. Try it on friends. Think about it. This approach does not really shuffle the deck at all! The deck will not be randomized.

My mathematician friends tell me that technically speaking, one cut will create some degree of randomization, but it's not the kind of randomization most players would want. No further randomization will take place regardless of how many additional cuts are made, and in some special cases the minor initial randomization will be lost if the shifting of the cards in the pack reaches its original position.

5.27 Russian American gambit.
Both humans are liars! If the woman was lying, then there might be two Russians. If the man was lying, there might be two Americans. However, we are told that one of the humans is a Russian and one is an American, and the truthful robot tells us that at least one of them is lying. Therefore, both are lying.

5.28 Car sequence.
The three cars nearest you can be in 3! = 6 different arrangements. For example, if we represent the dates as numbers 1 through 3, we have these 6 arrangements:

123, 132, 213, 231, 312, 321

Of these arrangements, only one arrangement will be in descending order. Thus, the chances are 1 in 6.

5.29 Quantum kings and the death of reality.

Monica's second hand—the one with the king of spades—is more likely to have another king. My colleague Mark Ganson reasons in the following manner. Consider the similar case of a deck of 4 cards—1, 2, 3, 4—dealt to a hand that contains just 2 cards. The deck of 4 cards can be dealt into hands of 2 cards each 6 different ways:

$$\{1, 2\}, \{1, 3\}, \{1, 4\}, \{2, 3\}, \{2, 4\}, \{3, 4\}$$

Now, let's refer to Cards 1 and 2 as "small" cards and Cards 3 and 4 as "large" cards. "Do you have a small card?" is equivalent to, "Do you have a king?" and, "Do you have a 1?" is equivalent to, "Do you have the king of spades?"

Notice that these hands all contain a small card:

$$\{1, 2\}, \{1, 3\}, \{1, 4\}, \{2, 3\},\{2, 4\}$$

These hands contain a 1:

$$\{1, 2\}, \{1, 3\}, \{1, 4\}$$

This is the only hand that contains both small cards:

$$\{1, 2\}$$

The probability of having $\{1, 2\}$ is 1 out of 3 (1/3) for hands containing the 1 card. The probability of having $\{1, 2\}$ is 1 out of 5 (1/5) for hands containing a small card. Thus, if you have the king of spades, you are more likely to have at least one other king than if you simply have a king.

Here's another way to consider this problem. Note that the probability of having the king of spades is higher for hands that contain more than one king than it is for hands that have only one king. If you've got only one king, it is probably not the king of spades. But if you have three kings, it is more probable than not that one of these kings will be the king of spades.

5.30 AIDS tests.

No. The chances are 50-50 that you don't have AIDS—no better than the flip of a coin! Suppose your company has 10,000 people. We assumed that 98 percent (9,800 people) do not have AIDS, and 2 percent (200 people) actually have AIDS.

Of the 9,800 uninfected people, 98 percent of them (9,604 people) will test negative, and 196 will test positive. Of the 200 infected people, 98 percent of them (196) will test positive, and 4 will test negative. Thus, we have:

196 tests are false positives

196 are true positives

The ratio of false positives to true positives is 1:1.

To better see why the chances are so poor of you having AIDS, despite the positive AIDS test, let's consider a more extreme "theoretical" example of AIDS testing. Assume we know that exactly one person out of Earth's six billion has AIDS, and we have a test that is 99 percent accurate. The test says you have AIDS. Would you think it more likely that you're that one person out of six billion who has AIDS, or is it more likely that you're among the 1 percent who got false positives?

5.31 Woman in a black dress.

The probability is $5/2^3 = 5/8$. The total number of possible sequences from n card tosses is $2n$. Let $f(n)$ be the number of sequences of reds and blacks, of length n, in which two consecutive reds do not appear. The probability that no two consecutive reds occur in n card tosses is $f(n)/2^n$.

Let's explore the meaning of $f(n)$. For one card, $f(1) = 2$ because both choices (a red face R or a black face B) do not show two consecutive reds. For two cards, we have $f(2) = 3$ because three of the four possibilities do not contain RR: RB, BR, and BB. With three cards, we have RRR, BBB, RBR, RRB, BRR, RBB, BBR, and BRB. Here $f(n) = 5$. If we experiment a few more times, we find that $f(n)$ follows a sequence that many math nerds will recognize: 2, 3, 5, 8, 13, and so on, which is the Fibonacci sequence, where each term is the sum of the previous two. Thus, it is possible to compute the probabilities of larger sets of cards by dividing appropriate values of the Fibonacci sequence by 2^n.

5.32 Random chords. The probability is 1 out of 3 (1/3). This answer applies only when you use the definition of "random chord" that I gave in the question. Other definitions will give different answers. If you're a computer programmer, you can actually simulate this by randomly selecting chords and determining how many of them are shorter than the radius of the circle. I did this for 100,000 trials and found the ratio to be 0.33. Therefore, we can say that 1/3 of all the possible chords of a circle are less than the radius. If you are a gambling person, you should not bet that a randomly selected chord is smaller than the circle's radius!

We can solve this by using some simple

geometric diagrams. Consider a circle centered at C. Choose a point on the circle, Q. Construct chords PQ and QR such that they are equal in length to the radius CQ. Just to make it obvious, construct line segments CP, CQ, and CR, forming two equilateral triangles, CPQ and CQR. From geometry, we know that these two adjacent triangles fill 120 degrees of the circle's sweep, leaving 240 degrees of the circle remaining. Any random point on the circle, X, will enjoy a 1/3 probability of falling within the 120 degrees and a 2/3 probability of falling without. Chord QX then has a 1/3 probability of being shorter than the radius. (The process of randomly selecting chords on a circle to solve this problem leads to some famous paradoxes. As I said, the term *random chord* is ambiguous unless the exact procedure for producing it is defined.)

5.33 Bart's dilemma. Here is one clever solution. Bart should say, "You will shave my head."

5.34 Multilegged creatures. First, note that Mr. Hundred cannot have 100 legs, so he must have either 80 or 90 legs. Because the robot with 90 legs replies to Mr. Hundred's remark, Mr. Hundred cannot have 90 legs. Therefore, Mr. Hundred has 80 legs. Now consider Mr. Ninety. He cannot have 90 legs because this would match his name. Mr. Ninety has 100 legs.

5.35 Three fishermen. One fish is sufficient. Look at one fish from the bucket labeled "flounder and mackerel." Say that it's a flounder. Because each bucket is labeled incorrectly, the bucket cannot be the "flounder and mackerel" bucket; therefore, it must be the flounder bucket. The bucket labeled "mackerel" must contain mixed fish, and the bucket labeled "flounder" must actually be the mackerel bucket.

5.36 Mummy madness. The probability is 4/7. There are seven possible pairings of the cat, the hyena, and the mouse. Of these, 4 of the cases present a cat with a missing tail. In the following table, *M* signifies that the mummy is missing a tail, and *T* signifies that it has a tail.

cat	M	M	T	T	M	T	M
hyena	M	T	M	T	T	M	M
mouse	T	M	M	M	T	T	M

5.37 Heart attack. The chances are zero! If four of the hearts are returned to their correct bodies, then five of the hearts must also have been returned to their correct bodies.

5.38 Sushi play. The chances are 1 out of 3 (1/3). Nick has these numbered sushi pieces: 1, 2, 2, 3, 4, 5, 6, 7, 8, 9, 10. Because you know that the mystery sushi is not odd, we have one of these in the box: 2, 2, 4, 6, 8, 10. Thus, 2 of the 6 sushi pieces have the number 2 on them. This means the chances are 1 out of 3 (1/3) that the sushi with a 2 is in the box.

5.39 Floating boat game. Don's, Melissa's, and Carl's odds of "winning" are 4 to 2 to 1. Here's why. Don has a 1 out of 2 (1/2) chance of making his boat float on his turn. For Melissa to win on her turn, Don must sink and Melissa must win, which requires us to multiply (1/2)(1/2) = 1/4. For Carl to win, Don must lose, Melissa must lose, and Carl must win, which has a chance of (1/2)(1/2)(1/2) or 1/8.

5.40 Clown's dreams. Believe it or not, both aquaria are equally contaminated. The juice contains exactly as much vinegar as the vinegar contains juice. Perhaps the best way to visualize this is to put 6 purple balls in an aquarium (to represent grape juice) and 6 white balls in another aquarium (to represent vinegar). Let's assume that your glass cup holds 3 balls. You take 3 white balls and add them to the aquarium containing 6 purple balls. The grape aquarium now contains 6 purple balls and 3 white balls. Next "stir" the balls in the grape aquarium. If you dip your cup into the contaminated aquarium, on average your cup will contain 2 purple balls and 1 white ball. Add these to the vinegar aquarium. Each aquarium will now have 4 of one ball and 2 of the other.

Another explanation: If you moved 1 ounce of grape juice to the vinegar aquarium, you must have moved 1 ounce of vinegar to the grape juice aquarium, because the total amount of liquid in each has not changed.

5.41 Lottery. Because there is no connection between one lottery number and the next one, you won't do any better if you play different numbers instead of always playing the same one.

5.42 Fossil lock. Put the fossil into the sphere, secure it with one of your locks, and send the sphere to Homer. Homer should then attach one of his own locks and return the sphere. When you receive it again, remove your lock and send it back. Now, Homer can unlock his own lock and retrieve the fossil.

5.43 Nontransitive dice. Examine figure A5.2. Proceeding clockwise around the circle, the die with 6s and 2s beats the die with 5s and 1s, which in turn beats the die with 4s and 0s, which beats the die with all 3s, which beats the die with 6s and 2s. In particular, each die, in order, is twice as likely to beat its partner. For example, comparing the die with all 3s and the die with 6s and 2s, we find that Bill loses when Monica throws a 6 (which occurs 2/6 of the time), and he wins when she throws a 2 (which occurs 4/6 of the time), so he is twice as likely to beat her than she is to beat him. Whichever die Monica selects, Bill chooses the die at the tail of the arrow.

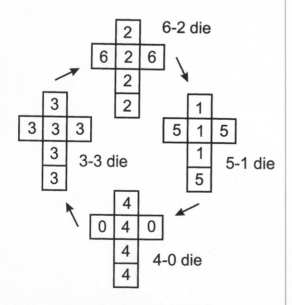

Figure A5.2 Nontransitive dice.

Companies like Grand Illusions manufacture these and other nontransitive dice.

The toy collector Tim Rowett uses a set of three nontransitive dice, each die having two different numbers:

Die 1: 1, 4, 4, 4, 4, 4

Die 2: 3, 3, 3, 3, 3, 6

Die 3: 2, 2, 2, 5, 5, 5.

Here, Die 1 beats Die 2 in 25 out of 36 throws; Die 2 beats Die 3 in 21 out of 36 throws, and Die 3 beats Die 1 in 21 out of 36 throws.

You can read more about nontransitive dice at many Web sites, including Ivars Peterson, "Tricky Dice Revisited," *Science News Online* 161, no. 16 (week of April 20, 2002); www.sciencenews.org/20020420/mathtrek.asp; and Grand Illusions, "Magic Dice, or Nontransitive Dice," www.grand-illusions.com/magicdice.htm.

5.44 Tic-tac-toe. Many people whom I surveyed next chose Cell 5 for X; however, other moves would be better. For example, if you were to play 4 or 6, this would give O three losing moves out of a possible six. If O plays to 5, 6, or 9, O loses (assuming that X plays correctly). If O plays to 1, 3, or 7, then it could go either way, but if both play correctly, it's a draw.

5.45 Farmer McDonald's moose. He must retrieve 201 gloves. If old McDonald was having bad luck, he could conceivably take out 100 red left-handed gloves and then 100 blue left-handed gloves. After these 200 were removed from the sewer, the next one would have to be either the right-hand or the left-hand match.

5.46 Black versus red. The probability that the other side is black is 2 out of 3 (2/3), good odds. To help solve this, draw all of the possible scenarios where Brad sees a black side of the bill. For example, if you label the three black sides $B1$, $B2$, and $B3$, then in 2 of the 3 cases in which Brad sees a black side, the other side will also be black.

5.47 Brunhilde's moose. The chances are 4 out of 25 (4/25). The chance that she will hit an O is 2/5. The chance that she will do this twice in a row is $(2/5) \times (2/5) = 4/25$ or 16 percent.

5.48 Board game of the gods. It would take 4.21875 coin tosses. We may create a table that shows each possible winning situation and the probability of getting it. For example, one path to the solution is to hop 2, hop 2, hop 2, which has a probability of 1/8 because during each hop, we have a 1/2 probability of hopping by 2. This represents the first entry in the following table. Similarly, we can compute the other probabilities for each path.

Path	Probability	Path	Probability
222	1/8	11112	1/32
1122	1/16	11121	1/32
1212	1/16	11122	1/32
1221	1/16	11211	1/32
1222	1/16	11212	1/32
2112	1/16	12111	1/32
2121	1/16	12112	1/32
2122	1/16	21111	1/32
2211	1/16	21112	1/32
2212	1/16	111111	1/64
		111112	1/64

Notice that there is one length-3 solution, with a probability of 1/8. There are nine length-4 solutions, with odds of 1/16. There are nine length-5 solutions, with a probability of 1/32. There are two length-6 solutions, with a probability of 1/64. The formula to combine these various probabilities is

$$(1 \times 3 \times 1/8) + (9 \times 4 \times 1/16) + (9 \times 5 \times 1/32) + (2 \times 6 \times 1/64) = 135/32 = 4.21875$$

My friend Jon Anderson has performed computer simulations of this game and found that they confirmed the solution that was arrived at by analytical means. At my request, he also performed simulations for games in which the fifth chamber has a chute that moves the player back to Chamber 1 whenever he lands on Chamber 5. In this more complicated game, the average number of tries needed to win rises from 4.21875 to 9.6, based on both simulations and analytical considerations.

5.49 Chimps and gibbons. The odds are 1 out of 3 (1/3). We have three possibilities. The primate is Chimp Gibbon, and he called out his first name. The primate is Chimp Chimp, and he called out his first name. The primate is Chimp Chimp, and he called out his last name. All of these possibilities are equally likely. In one out of the three possibilities is the primate Chimp Gibbon.

Sometimes my readers argue with me over problems of this kind that involve conditional probabilities. Colleagues tell me that this may be an interesting problem to which we can apply Bayes's theorem: the probability of a hypothesis given a relevant event is the probability of the event given the hypothesis times the initial probability of the hypothesis divided by the initial probability of the event. This is usually written as $P(H|E) = (P(H) \times P(E|H))/P(E)$. In our case,

$P(H)$ = 1/3 (probability that the chimp's last name is Gibbon)

$P(E)$ = 3/6 = 1/2 (probability that one of the names is Chimp)

$P(E|H)$ = 1/2 (probability that the chimp will say "Chimp," given his last name is Gibbon)

$P(H|E)$ is therefore $(1/3) \times (1/2)/(1/2) = 1/3$.

5.50 Cars and monkeys. Yes, we can make a determination. The woman driving Barbara's car and with Claire's monkey can't be Barbara or Claire; thus, she is Andrea. If Barbara is driving Andrea's car, Claire is driving her own. But she is not, so Barbara is driving Claire's car, and Claire is driving Andrea's.

5.51 Guessing numbers. Let us suppose that Bill called out "3,500,002." This means you must guess the year before his number and the year after his number—namely, 3,500,001 and 3,500,003. It's virtually certain that you will win.

5.52 Martian identities. There are 520 Martians, at most. The first position of the card could contain 1 of 26 letters. The second position could contain 10 possible digits. $26 \times 10 = 260$. We must double this number for all the cases in which the number appears first and the letter appears last.

5.53 Robotic ants. Since Punisher always points to a tunnel that terminates in a warrior ant, the odds favor you if you switch to another tunnel. Many of my colleagues did not believe my answer, so they wrote a computer simulation that proved me correct. Alternatively, it may be helpful for you to imagine that there are 1,000 tunnels, and Punisher points to the 998 tunnels that terminate in a warrior ant. Now would you switch your choice of tunnels? (What assumptions are being made in this puzzle?)

Let's look at this from another perspective. Assume that Tunnel 1 is the tunnel to freedom. Here are the possibilities before our captor points to a sticky tunnel:

- You choose Tunnel 1. If you don't switch, you win.
- You choose Tunnel 2. If you don't switch, you lose.
- You choose Tunnel 3. If you don't switch, you lose.
- You choose Tunnel 4. If you don't switch, you lose.
- You choose Tunnel 5. If you don't switch, you lose.

  Odds of winning: 1/5 = .2

Now let's examine the possibilities after your captor points to a sticky tunnel:

- You choose Tunnel 1. If you switch, you lose.
- You choose Tunnel 2. If you switch, you have 1/3 chance of winning.
- You choose Tunnel 3. If you switch, you have 1/3 chance of winning.
- You choose Tunnel 4. If you switch, you have 1/3 chance of winning.
- You choose Tunnel 5. If you switch, you have 1/3 chance of winning.

Odds of winning: $4/5 \times 1/3 = 4/15 = 0.2666$
. . .

This puzzle is very similar to the famous Monty Hall problem that is discussed in great detail in the literature. These kinds of problems are most relevant when you're considering the best strategy to use if the "game" is played repeatedly.

5.54 Venusian insects. One hundred percent of them eat insects. The problem states that all Venusians eat four of the five foods, so how could a Venusian avoid the insects? Here's another way to look at it. Select a Venusian at random. Either he eats beetles or he does not. If he eats beetles, he is eating an insect. If he doesn't eat beetles, then he can choose from the other four groups, which includes ants. Therefore, he is still eating an insect.

5.55 Domain names. Consider that there are 38 possible characters, if you count the various symbols on the keyboard in addition to the 26 letters. You just take the number of possible characters and raise it to the number of characters in the domain name, so there are 38 to the third, or 54,872 three-character possibilities; and 38 to the fourth, or 2,085,136 four-character possibilities; and $38^5 = 79,235,168$ five-character possibilities. Of course, many of these are uninteresting and unpronounceable, like *xnzy*. The novel's character creates patented algorithms to determine names that are pronounceable and initially buys only those.

If you consider that domain names can be many characters in length, we should not run out of domain names soon. For example, there are about 6.27×10^{15} ten-character domain names and 3.94×10^{31} twenty-character domain names. Please note that these numbers are all approximations because the rules for domain name generation have additional complexities; for example, domain names cannot begin or end with a hyphen. Also, I have not considered the different domain name extensions, such as .net, .org, .com, and so on.

5.56 Where are the sea horses? Chamber *C*. To solve this problem theoretically, we assume that the sea horses swim randomly. In this sense, they behave like diffusing molecules in a gas. Therefore, the number of sea horses in each chamber is proportional to the volume of the chamber. The nature of the interconnecting tunnels should not matter if you give the diffusing sea horses sufficient time to come to an equilibrium state. In other words, Chamber *C* in figure 5.6 will have the most sea horses because it is the chamber with the largest volume. One assumption we make in this problem is that the large-area Chamber *C*, in this schematic slice through the aquarium, also represents a large volume if viewed in three dimensions.

5.57 Lotteries and superstition. Bill is wrong. Monica is just as likely to win as Bill is. In fact, she would be just as likely to win if she avoided 13, 14, and 15 and picked "16, 16, 16, 16, 16, 16." Any set of six numbers is just as likely to be picked as any other. The probability of winning is $1/4^6$ in every case.

5.58 Coin flips and the paranormal. I did a small survey of colleagues. Most said that they would begin to become suspicious after about 5 heads in a row. Most colleagues felt that a mathematician wouldn't think that the coin was biased until about 15 heads in a row.

Of course, a coin may be biased and not reveal its bias by always yielding a head. My friend has a set of dice that are weighted to roll 6s, but about 20 to 25 percent of the time, they still roll a non-6. To detect bias, most mathematicians probably wouldn't look for a consecutive number of heads but would rather like to see a sample set.

Other colleagues noted that if the initial results were somewhat random, and suddenly a string of heads resulted—say, 8 heads in a row—most people could accept that this might happen by chance and would consider that the coin was unbiased.

My friend Graham notes that if you just had to call a *single* coin toss of a biased coin, and the caller did not know that the bias existed, the procedure of calling the toss would still be fair. Consider the extreme case of a double-headed penny. It always comes down heads. However, if the caller takes a random guess of heads or tails, he or she still has a 50 percent chance of being correct. For a single toss, a biased coin is as "fair" as an unbiased one. It is similarly true that as long as I guess randomly each time, my expected proportion of correct guesses is still 50 percent no matter how biased the coin is and no matter how long the sequence.

In 2004, the mathematicians Persi Diaconis and Susan Holmes discovered that it is impossible to actually toss a coin into the air "fairly," because a coin is always more likely to land on the same face it started out on. In particular, their data indicate that a coin will land the same way it started 51 percent of the time, but a casual observer wouldn't notice this small bias. More alarming are the results of spinning a penny on its edge, in which case the penny will land on tails 80 percent of the time because the head side has slightly more mass. For further reading, see Erica Klarreich, "Toss Out the Toss-Up," *Science News* 165, no. 9 (February 28, 2004): 131–32.

As for the paranormal implications, Susan Blackmore, a researcher, uses the term *sheep* to refer to unskeptical people who believe in the paranormal. Blackmore has found that when sheep are given a string of random numbers, they overestimate the significance of ordinary repetitions in the sequence. Sheep are also more prone to generally misjudge probabilities. For example, if we flip a coin, sheep tend to underestimate or overestimate the odds of getting a head if a coin has gotten five heads in a row, because sheep assume that the previous tosses influence the next one. Finally, when sheep are asked to write down a sequence of random digits, sheep are more likely to avoid repeating digits than chance would dictate (John Horgan, *Rational Mysticism* [New York: Houghton Mifflin, 2003], p. 111).

5.59 Robot jaws. The top jaw has 100 possible combinations. The bottom jaw has 100 possible combinations. The probability of opening both locks on the first attempt is $(1/100) \times (1/100) = 1/10,000$.

If it takes you, on average, an hour to get the top combination right, how much time on average would it take you to get both combinations right on the first try?

5.60 Flies in a cube. It's possible to compute this average distance (known to physicists as the mean free path) by using the following formula:

$$L = 1/(4\pi\sqrt{2}\, ar^2)$$

Here, L is the average distance traveled before hitting another fly, a is the density of flies, and r is the radius of a fly. From the equation, you can see that as the radius of the fly increases or the density of the spheres increases, the distance traveled before hitting another fly (as you might expect) decreases. In our case, a is 1,000 flies/cubic mile.

Given all these hints, what value do you get for L? Does the answer you get seem reasonable? Are you surprised by the answer, which suggests that the average distance traveled between collisions is way over 500,000 miles?

5.61 Boomerang madness. The probability is 1 out of 20. The first boomerang merely indicates which tree the second boomerang must get caught in. It doesn't matter that they were thrown at the same time. If you throw one of them later, its odds of landing in any particular tree, and therefore the odds of hitting the tree with the other boomerang, are 1/20. I suspect that many people would falsely answer 1/400, the odds of both boomerangs landing in a particular tree.

5.62 Apes in a barrel. One out of 30—the same as if the puzzle were recast and you were to win if Tiffany emerged third or at any point in the sequence of apes leaving the barrel. It would be very easy to overanalyze this. But if you think about it, each ape has an equal chance of emerging last.

5.63 Dark hallway and vapor men. It turns out that an analysis of this problem is not too difficult. The chance of Man 1 vaporizing you is 50 percent. The probability of reaching the second man and being vaporized is less. Specifically, it is the probability that the first man didn't vaporize you times the probability that the second man did vaporize you, $0.5 \times 0.5 = 25$ percent. For the third man, we have $0.5^3 = 0.125$. This also means that in a large-enough sample, 1/2 of the people will be vaporized by Man 1, 1/4 by Man 2, 1/8 by Man 3, and so on. We can create a table that shows the chances of reaching and being vaporized by each man as you walk through the hallway:

Man Number	Percent Chance of Reaching and Being Vaporized by Man
1	50
2	25
3	12.5
4	6.25
5	3.125
6	1.5625
7	0.78125
8	0.390625
9	0.1953125
10	0.09765625

Your chances of reaching and being vaporized by the final man are slim indeed—only about 9/100ths of a percent. Therefore, most mathematicians whom I asked said that because they were limited to selecting one man, they would place their money on Man Number 1 as the fellow who will vaporize you. In other words, individuals would be most likely to be vaporized by this man.

How many attempts would it take to reach the end of the hallway without being vaporized along the way? Note that the chances of reaching each succeeding man are the same as for reaching and being vaporized by the previous one. Thus, the chances of passing by all the men safely are the same as reaching and being vaporized by Man 10. Therefore, the probability is 0.009765625; that is, 1,024 attempts would be required by a person, on average, to reach the end of the hallway safely.

My colleague J. Theodore Schuerzinger notes that because $1/2^{10}$ or $(1/1,024)$ people would make it all the way down the hallway without being vaporized by any of the mysterious men, this means that 1,023 out of 1,024 people would be vaporized. He notes that by using the formula $(1023/1024)^x = 1/2$, we can determine that out of the first x people to go down the hallway, there is a 50 percent chance that one person will make it down without being vaporized. The solution $x = 709.4$ satisfies the equation. Thus, I would bet that a person would make it all the way down on one of the first 710 attempts. In other words, after 710 attempts, the chance of someone succeeding exceeds 1/2.

5.64 Napoleon and Churchill. Churchill is much more likely to win than Napoleon is. In other words, THH is much more likely to appear before HHH does.

Most people reason that each triplet is equally likely, so that it is equally likely that

Napoleon and Churchill would win. However, such is not the case. HHH appears before THH only if the first three tosses come up heads. Any other result will allow THH to block HHH. Therefore, the probability that HHH appears before THH is 1/8, which is the probability that the first three tosses are HHH. If you don't see this, try writing down random strings of Hs and Ts on paper, and this logic will become apparent.

5.65 Triangles and spiders. To solve this problem, we must determine the area of the triangular target for which the x coordinate is less than the y coordinate and compare that area, A_1, to the area A_2 of the triangle in which the x coordinate is greater than the y coordinate. The two regions are separated by the line OP ($y = x$) (figure A5.3). Before we do any calculations, we can see that the meatball has less than a 50-50 chance of having the x-coordinate less than the y-coordinate, because the area of the region is smaller.

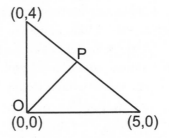

Figure A5.3   Triangles and spiders.

Let's calculate the area of the entire triangular target. Using the formula area = $(1/2) \times$ base $\times$ height, we find $A = 1/2 \times 5 \times 4 = 10$. Next, we compute A_1, the area of the smaller internal triangle. One way to do this is to first calculate the equation of the hypotenuse of the entire triangular target and then calculate the coordinates of point P.

The big hypotenuse connects points $(0,4)$ and $(5,0)$, so it has a slope of $-4/5$. The y-intercept is 4, so the equation of the hypotenuse line is $y = (-4/5)x + 4$. Point P must satisfy both $y = x$ and $y = (-4/5)x + 4$. Solving for y and x, we find that P is at $(2.22, 2.22)$.

Finally, we can compute the area of the smaller internal triangle using the usual $A = (1/2) \times b \times h$. Here we can visualize the base resting on the y axis so that $A_1 = 1/2 \times 4 \times 2.22 = 4.44$. Recall that the area of the entire triangle is 10. Of this, the area where the x coordinate is less than the y coordinate is 4.44. So the probability that the meatball lands on a point for which $x < y = 4.44/10$ or 44 percent.

5.66 Stylometry.

Stylometry is the science of measuring literary styles. In 2003, statistical tests, artificial intelligence methods, and other mathematical tools were used to determine that the fifteenth book in the Oz series, *The Royal Book of Oz*, was written by Ruth Plumly Thompson and not by Frank L. Baum. Jose Binongo, of the Collegiate School and Virginia Commonwealth University in Richmond, has verified the authorship by examining the 50 most frequently observed words in various Oz texts and exploring projections of a 50-dimensional space onto a plane. Researchers speculate that authors have certain habits ingrained in their neural pathways that favor particular word correlations. For further information, see Jose Binongo, "Who Wrote the 15th Book of Oz? An

Application of Multivariate Analysis to Authorship Attribution," *Chance* 16, no. 2 (2003): 9–18. Also see Erica Klarreich, "Bookish Math," *Science News* 164, no. 25 (December 20, 2003): 392–94.

5.67 Factors.

As I alluded to in chapter 2, the solution is $1/\xi(3)$, where

$$\xi(3) = 1/1^3 + 1/2^3 + 1/3^3 + 1/4^3 + \ldots = 1.2020569031595942853997381615114499907649862923 4049 \ldots$$

$\xi(3)$ is called Apery's constant. It's also the value of the Riemann Zeta function for $s = 3$:

$$\xi(3) = \sum_{n=1}^{\infty} \frac{1}{n^s}$$

Incidentally, the values of the Zeta function for $s = 2n$ are known to be fractions of π^{2n}. For example, Euler proved that

$$\xi(2) = \frac{\pi^2}{6}, \, \xi(4) = \frac{\pi^4}{90}, \, \xi(6) = \frac{\pi^6}{945}$$

5.68 Plato's choice.

The interval is 14 for the gods, and 13 for the mortals. How would your answer change if another god were inserted after Plato?

The mathematical puzzle involves the concept of *decimation*. Throughout history, the lawful penalty for mutiny on a ship was to execute one-tenth of the crew. First, the crew members were forced to randomly line up in a circle. It was then customary that the victims be selected by counting every tenth person from the circle, hence the term *deci*mation. Eventually, this term came to be applied to any depletion of a group by any fixed interval—

not just ten. (In real life, when we wish to dec-
imate a group of people, we don't usually line
them up in a circle because repeatedly going
around the circle kills everyone.)

This problem in combinatorics is an example of a class of problems that is sometimes
called Josephus problems: given a group of n
men arranged in a circle under the command
that every mth man will be executed going
around the circle until only one remains, find
the position $P(n,m)$ in which you should stand
in order to be the last survivor. For further
reading, see W. W. R. Ball and H. S. M. Coxeter, *Mathematical Recreations and Essays*,
13th ed. (New York: Dover, 1987), pp. 32–36.

6. Big Numbers and Infinity

6.1 Counting. It would take 31 years and 251
days. In reality, it would take you longer
because larger numbers would take more than
a second to pronounce.

6.2 Large law. This "law" refers to coincidences in everyday life. It essentially says that
many seemingly odd coincidences are likely to
happen with a large-enough sample. As Robert
Todd Carroll points out in *The Skeptic's Dictionary*, we might be amazed if a person wins the
lottery twice. However, the statisticians
Stephen Samuels and George McCabe of Pur-
due University calculated the odds of someone
winning the lottery twice to be about 1 in 30
for a four-month period—because players usually buy multiple tickets every week.

More than 17 million people on the planet
share your birthday. Carroll notes that at a
typical football game with 50,000 fans, most

fans are likely to share their birthdays with
about 135 others in attendance. In a random
selection of 23 people, there is a greater than
50 percent chance that at least 2 of them cele-
brate the same birthday.

Incidentally, in statistics and mathematics,
"the law of large numbers" has a different
meaning and refers to the rule that the average
of a large number of independent measurements of a random quantity tends toward the
theoretical average of that quantity.

6.3 The ultrahex. Its last digit is a 6, because 6
raised to any power produces a result that
ends in a 6.

6.4 The ubiquitous 7. You could essentially say
that *all* numbers have a 7 in them. Why? As
the numbers get larger, they contain more digits, increasing the probability that one of the
digits in them might be a 7. In fact, the probability that a 7 will *not* appear in a very long
number is vanishingly small. Moreover, you
can make a graph of the proportion of numbers with 7s in them as you scan larger and
larger numbers, and you will see it continuously rise and approach 1. We can also make
a table showing the percentage of numbers
with at least one 7 digit for the first X numbers. Here is a table, showing the actual number of numbers with sevens:

X	Percentage of Occurrence of 7
10	10
100	19
1,000	27
10,000	34

. . . rapidly approaching 100 percent

A formula describes this growth: $10^n - 9^n$, where n is the power of 10. Thus $(10^n - 9^n)/10^n$ gives you the percentage. According to the software package Mathematica, the limit as n approaches ∞ is 1. Or, even without Mathematica, the formula reduces to $1 - (9/10)^n$. The quantity $(9/10)^n$ approaches zero as n grows.

6.5 Knuth notation for ultra-big numbers.
No. $3\uparrow\uparrow4 = 3\uparrow3\uparrow3\uparrow3 = 3\uparrow3\uparrow27 = 3\uparrow7625597484987$. This number is huge, much larger than the number of atoms in the visible universe. (It's estimated that there are 10^{80} protons and the same number of electrons in the visible universe.) You can surely see that Knuth's notation is a compact way of representing large numbers.

One of the largest finite numbers ever contemplated by a mathematician is *Graham's number*, named after the mathematician Ronald L. Graham. As just the first step to understanding Graham's number, let $x_1 = 3\uparrow\uparrow\uparrow\uparrow3$, which is unimaginably large! Next, let $x_2 = 3$ (x_1 arrows) 3. In other words, x_2 has x_1 arrows. We continue the process, making the number of arrows equal to the number at the previous step, until you are 63 steps from x_1. The final value, beyond which any God could contemplate, is Graham's number, which is used in an area of mathematics called combinatorics and, more specifically, Ramsey theory.

6.6 The magnificent 7.
The chance is 100 percent. If I list all of the integers in numerical order, there is essentially a 100 percent chance of hitting a number with a 7, using a dart tossed randomly at the list. You learned from the last problem that as the numbers get larger, they contain more digits, increasing the probability that one of the digits in them might be a 7. In fact, the probability that a 7 will *not* appear in a very long number in my list is very low.

6.7 Computers and big numbers.
The computer that I am envisioning is one that is not an idealized or theoretical model, but one limited by known physical laws such as the speed of light, electrical resistance, RC (resistor-capacitor) delay, atomic dimensions, the world supply of silicon, the finiteness of programs and data, and other known phenomena. Can such a machine work with numbers so huge that they would require Donald Knuth's arrow notation to express them?

My colleague Dennis Gordon posed this big-number question to my discussion group. Most group members focused on the storage needed to hold such a number (with all its digits), which might depend on the amount of matter in the universe. If we limit ourselves to the visible universe, which contains about 10^{80} protons and the same number of electrons, then perhaps this provides an upper limit to the size of the binary number that we could conceivably store. If a quantum computer is to be considered, then all the matter in the multiverse should be considered. In addition to storage for the number, we would also need the computer itself, which would include a storage writer to write the number, even if we never wanted to read it. We would also need the engineer to design it.

My colleague Bill Gavin echoes this view when he notes that in order to determine the largest number that could be handled by a computer, we need to first imagine the ultimate computing machine, which would be the entire universe transformed into a single God-like computer. The largest number it could handle would then be the largest number that could be encoded in the universe itself. Making a simplified assumption that each particle may be encoded with one bit of information, we can determine the largest possible number that could be encoded in the universe by dividing the total mass of the universe (which is a constant) by the mass of the smallest particles that the universe could be broken up into.

But what is the smallest possible particle? We can get much smaller than a proton or an electron. The mass of any particle can be expressed as a function of its size and its energy. For the smallest possible particle, we assume the smallest possible size, a well-defined unit in physics known as the Planck length (very roughly 10^{-35} meters). The smallest possible particle would also have the lowest possible energy, which corresponds to the longest possible wavelength. The longest possible wavelength would be the "radius" of the universe itself, which is believed to be about 13 billion light years. With our smallest possible particle's "radius" and wavelength thus defined, its mass can be easily calculated. If we assume that the total mass of the universe is on the order of 10^{53} kg (the value favored by most physicists), the largest number of bits that could be encoded in the universe at its current size turns out to be approximately 10^{120}, which corresponds to a largest possible integer of $2^{10^{120}}$. How does this value change as the universe expands?

In 2004, Lawrence M. Krauss and Glenn D. Starkman, of Case Western Reserve University, submitted "Universal Limits on Computation" to *Physical Review Letters*. They show that the nature of the universe itself places limits on computation because it is not possible to transmit or receive information beyond the so-called global event-horizon in an accelerating, expanding universe. In particular, they determined how far an observer could travel in such a universe and still be able to transmit energy back to Earth. Their calculations show that the total number of computer bits that could be processed in the future would be less than 1.35×10^{120}. Their calculations also considered that the acceleration causes space to emit a form of energy known as de Sitter radiation, which would drain energy from regions of the universe and thus also diminish the computer resources that could be used for computation.

My solution concentrates on the largest number that can be stored, not worked with or yielded. To actually "work with a number," we need at least one other number. We also need space to hold intermediate work or at least the answer. Thus, if we want to multiply two 10-digit numbers (in any base), we need 20 digits for the two numbers and another 20 for the result. We also need some space for coding the multiplication algorithm!

6.8 Counting every rational number. Georg Cantor (1845–1918) boldly tamed the infinite and showed that there are different degrees of infinity. This subject had previously been left to pure speculation and mysticism. The rational numbers, for example, are *countably infinite*, which means it is possible to enumerate all of the rational numbers by means of an infinite list. In contrast, the real numbers are uncountable because we cannot enumerate them by means of an infinite list. A few examples will make this clear.

Cantor showed that it is possible, in theory, to count all rational (fractional) numbers if you arranged them in a pattern like this:

A	B	C	D	E	F	
1/1	2/1	3/1	4/1	5/1	6/1	. . .
1/2	2/2	3/2	4/2	5/2	. . .	. . .
1/3	2/3	3/3	4/3	. . .	. . .	. . .
1/4	2/4	3/4	. . .	. . .	. . .	. . .
1/5	2/5	. . .	. . .	. . .	. . .	. . .
1/6	. . .	. . .	. . .	. . .	. . .	. . .

Here is Cantor's method for listing all the rational numbers. Start your list with 1/1 at the top of column *A*. Next, go to 2/1 at the top of *B* and list all fractions down and to the left: 2/1, 1/2. Next, go to the top of *C* and list rational numbers along the diagonal: 3/1, 2/2, 1/3. As you continue this process, you can check whether a number has been counted already, such as when you get to 2/4, which is the same as 1/2. You can eliminate repeats. You can also reduce fractions to the lowest terms; for example, 2/1 = 2. We can continue this process to create a list of rational numbers (which includes the integers):

$$1, 2, 1/2, 3, 1/3, 4, 3/2, 2/3, 1/4, 5, \ldots$$

Any positive rational number will be reached sooner or later. On the other hand, Cantor showed that for every given infinite sequence of real numbers it is possible to construct a *real* number that is not on that list. Consequently, it is impossible to enumerate the real numbers; they are uncountable and represent a "greater" infinity than the infinity of rational numbers. Here's how he showed this.

Let's try to make a list of all the positive real numbers, just as we did for the positive rational numbers. Such a list would include numbers like pi, which can't be expressed as a fraction. For simplicity, let's imagine a list of all the real numbers between 0 and 1. The list might, in no special order, look something like this, each number with an infinite number of digits:

$N_1 = 0.398471 \ldots$

$N_2 = 0.281910 \ldots$

$N_3 = 0.538567 \ldots$

$N_4 = 0.790193 \ldots$

. . .

Is it possible to find a real number that is not on this infinite list? The answer is yes. To construct this number, add one to each underlined digit:

$N_1 = 0.\underline{3}88471 \ldots$

$N_2 = 0.2\underline{8}1510 \ldots$

$N_3 = 0.53\underline{8}567 \ldots$

$N_4 = 0.740\underline{1}83 \ldots$

We continue with every digit and every number *N* to produce a new number: $M = 0.4982$ . . . Notice that this mystery real number, *M*, is different from N_1 in the first digit, different from N_2 in the second digit, and so forth. Where does *M* itself reside on the list of real numbers? Nowhere! It differs from every number on the list. Thus, it is not possible to "enumerate" every real number, as it is for the rational numbers. (Or, perhaps more accurately, I should say that we have shown that the list of real numbers must have more elements than the set of integers does.) The exact values for numbers in our $N_1, N_2, N_3, \ldots$ list do not affect the basis of our argument.

I have simplified our discussion of Cantor's diagonal proof somewhat, and additional complexities are discussed in "Cantor's diagonal proof," www.mathpages.com/home/kmath371.htm.

6.9 1597 problem. Yes. The reason it would take even your computer so long to find the infinite number of solutions is the fact that the smallest integer value for y is

$$y = 5197115277554630962242663853756384499943026746249$$

for an x value of

$$x = 1300498608879077225030950464390867 1520836229100$$

6.10 Infinite surface. There are many answers to this question, but my favorite involves the Funnel of Zeus, a hornlike object created by revolving $f(x) = 1/x$ for $x \in$ of $[1, \infty)$ about the x axis. Figure A6.1 shows a cross-section of the funnel, which can be created by revolving the curves about the x axis. Standard calculus methods can be used to demonstrate that the Funnel of Zeus has finite volume but infinite surface area!

In *777 Mathematical Conversation Starters*, John de Pillis explains that mathematically speaking, pouring red paint into the Funnel of Zeus could fill the funnel and, in so doing, you could paint the entire inside, an infinite surface—even though you have a

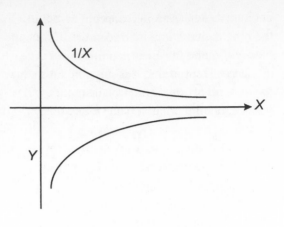

Figure A6.1   Funnel of Zeus, Gabriel's horn, or Torricelli's trumpet.

finite number of paint molecules. This seeming paradox can be partly resolved by remembering that the Funnel of Zeus is actually a mathematical construct, and our finite number of paint molecules that "fills" the horn is an approximation to the actual finite volume of the horn.

For what values of a does $f(x) = 1/x^a$ produce a horn with finite volume and infinite area? This is something for you to ponder as you drift off to sleep tonight.

The Funnel of Zeus is sometimes called Gabriel's horn or Torricelli's trumpet, after Evangelista Torricelli (1608–1647), a mathematics teacher who studied this object. He was astounded by this object that seemed to be an infinitely long solid with an infinite-area surface and a finite volume. He thought it was a paradox and, unfortunately, did not have the tools of calculus to fully appreciate and understand the object. Today, Torricelli is remembered for the telescopic astronomy he did with Galileo.

6.11 Infinity keyboard. The concept of infinity has challenged humans for centuries. For example, Zeno (an Eleatic philosopher living in the fifth century B.C.) posed a famous paradox that involved infinity. The paradox seemed to imply that you can never leave the room you are in. As Zeno reasoned, in order to reach the door, you must first travel half the distance there. Once you get to the halfway point, you must still traverse the remaining distance. You need to continue to half the remaining distance. If you were to jump 1/2 the distance, then 1/4 the distance, then 1/8 of the distance, and so on, would you reach the door? Not in a finite number of jumps! In fact, if you kept jumping forever at a rate of 1 jump per second until you were out the door, you would jump forever. Mathematically, one can represent this limit of an infinite sequence of actions as the sum of the series (1/2 +1/4 + 1/8 + . . .). The modern tendency is to resolve Zeno's paradox by insisting that the sum of this infinite series 1/2 + 1/4 + 1/8 is *equal* to 1. Since each step is done in half as much time, the actual time to complete the infinite series is no different than the real time required to leave the room.

As for typing the two H and J keys, the answer is that the key value is indeterminate or undefined. My friend Nick Hobson notes that this is perhaps not too surprising when you reflect on the fact that the keys have been pressed an infinite number of times! Infinity is neither even nor odd.

Nick agrees that the typing experiment does end, since the sum to infinity of 1 + 1/2 + 1/4 + . . . is well-defined and finite. "On the other hand, the sum to infinity of 1 − 1 + 1 − 1 + . . . is neither of these. It is a monstrosity

of which we cannot speak! It is this sum that corresponds to the final key struck. The sum is not well-defined; it is neither 1 nor 0 (nor is it 1/2); therefore it is not possible to say that the final key struck is either J or H."

My friend Chuck says, "When using a real keyboard, the final state of the keyboard is obviously . . . broken."

For an analogous situation, consider a husky dog taking a walk in the snow. The dog walks one mile north, instantly turns around and walks half a mile south, instantly turns around and walks a quarter of a mile north. . . . When the dog stops, we know it is 2/3 mile north of where it started, but we can't say which direction it is facing! According to Nick Hobson, the dog's final position is 1 − 1/2 + 1/4 − 1/8 + . . . miles north of its starting point—considering the direction north as positive and south as negative. The sum to infinity of a geometric series with initial term a and common ratio r is $a/(1 − r)$ if $|r| < 1$. (If $|r| \geq 1$, the sum to infinity is not defined.) Our dog series has $a = 1$ and $r = −1/2$. So its sum is $1/(1 + 1/2) = 2/3$.

6.12 Infinite gift. The volumes of the cubes that form the gift are part of this series:

$$1 + \frac{1}{2\sqrt{2}} + \frac{1}{3\sqrt{3}} + \frac{1}{4\sqrt{4}} + \ldots + \frac{1}{n\sqrt{n}} + \ldots$$

For example, if our units of measurement are feet, the first box would have a volume of one cubic foot, and the next box would have a volume of about 0.35 cubic feet. This series converges. The total volume of the gift is finite, but the surface area is infinite! Of course, an infinite object such as this cannot truly be constructed because the boxes would eventually become smaller than an atom, but this is a wonderful example of a wide class of mathematical objects the members of which have finite volumes but infinite surface areas. By summing the series on a computer, we find that the series converges to a value of around 2.61.

6.13 Large number contest.
Here is the answer for the first part of the contest:

$$.3^{-(.2^{-(.1^{-4})})} = 3.33^{(5^{10000})} \quad \text{or} \quad 3.3^{5 \times 10^{6989}}$$

This number is roughly 3 raised to the nth power, where n has about 6,990 digits. Recall that a number raised to a negative power is simply 1 over the number raised to the positive value of the power. To determine the number of digits in a number, you simply take the log of the number and add 1.

The number of cubic inches in the whole volume of space comprising the observable universe is almost negligible compared to this quantity.

I am not certain about the answer to the second part of the contest. Here are some attempts by friends who answered the question that I posed in my discussion group:

$$10^{10^{10^{10^{10}}}} \quad \text{which has} \quad 10^{10^{10^{10}}} + 1 \text{ digits (Jon Anderson)}$$

$$10^{10^{10 \cdot .1^{-100}}} = 10^{10^{10^{10^{100}}}}, \text{ which has } 10^{10^{10^{100}}} + 1 \text{ digits (Bill Olsen)}$$

$$.1^{-.1^{-.1^{-.1^{-100000}}}}, \text{ which has an amazing } 10^{10^{10^{10^{100000}}}} + 1 \text{ digits (Bill Olsen)}$$

References

I've compiled the following list, which identifies much of the material I used to research and write this book. It includes information culled from books, journals, and Web sites. As many readers are aware, Internet Web sites come and go. Sometimes they change addresses or completely disappear. The Web site addresses listed here provided valuable background information when this book was written. You can, of course, find numerous other Web sites that relate to the mathematical curiosities by using Web search tools, such as the ones provided at www.google.com.

If I have overlooked an interesting mathematical puzzle, person, reference, or factoid that you feel has never been fully appreciated, please let me know about it. Visit my Web site, www.pickover.com, and send me an e-mail explaining the idea and how you feel it influenced the mathematical world.

Beeler, Michael, William Gosper, and Rich Schroeppel. *HAKMEM*. MIT AI Memo 239, February 29, 1972. See also www.inwap.com/pdp10/hbaker/hakmem/hackmem.html. Work reported in this document was conducted at the Artificial Intelligence Laboratory, a Massachusetts Institute of Technology research program supported in part by the Advanced Research Projects Agency of the Department of Defense and monitored by the Office of Naval Research.

Berndt, Bruce. *Ramanujan's Notebooks*. 2 vols. New York: Springer, 1985.

Borwein, J., and P. Borwein. "Strange Series and High Precision Fraud." *The American Mathematical Monthly* 99, no. 7 (August–September 1992): 622–40.

Borwein, J., P. Borwein, and D. Bailey. "Ramanujan, Modular Equations, and Approximations to Pi, or How to Compute One Billion Digits of Pi." *The American Mathematical Monthly* 96, no. 3 (March 1989): 201–19. Also available at www.cecm.sfu.ca/personal/pborwein/.

Clawson, Calvin. *Mathematical Mysteries*. New York: Plenum, 1996.

de Pillis, John. *777 Mathematical Conversation Starters*. Washington, D.C.: The Mathematical Association of America, 2002.

Gardner, Martin (editor). *Mathematical Puzzles of Sam Loyd*. New York: Dover, 1959.

——— (editor). *More Mathematical Puzzles of Sam Loyd*. New York: Dover, 1960.

Gulberg, Jan. *Mathematics from the Birth of Numbers*. New York: Norton, 1997.

Kanigel, Robert. *The Man Who Knew Infinity*. New York: Scribner, 1991.

Klarreich, Erica. "Math Lab: Computer Experiments Are Transforming Mathematics." *Science News* 165, no. 17 (April 24, 2004): 266–68.

Pappas, Theoni. *Mathematical Scandals*. San Carlos, Calif.: Wide World Publishing/Tetra, 1997.

Ribenboim, Paulo. *The Little Book of Big Primes*. New York: Springer, 1991.

———. *The Book of Prime Number Records*, 2nd ed. New York: Springer, 1989.

Poundstone, William. *How Would You Move Mount Fuji?* New York: Little Brown, 2003.

Pickover, Clifford. *The Loom of God*. New York: Plenum, 1997.

———. *Computers and the Imagination*. New York: St. Martin's Press, 1991.

———. *Wonders of Numbers*. New York: Oxford University Press, 2000.

Room, Adrian. *The Guinness Book of Numbers*. Middlesex, England: Guinness Publishing, 1989.

Sardar, Ziauddin, Jerry Ravetz, and Borin Van Loon. *Introducing Mathematics*. New York: Totem Books, 1999. A great cartoon guide to mathematics.

Schimmel, Annemarie. *The Mystery of Numbers*. New York: Oxford University Press, 1993.

Vaderlind, Paul, Richard Guy, and Loren Larson. *The Inquisitive Problem Solver*. Washington, D.C.: The Mathematical Association of America, 2002.

vos Savant, Marilyn. *More Marilyn*. New York: St. Martin's Press, 1994.

Wells, David. *The Penguin Dictionary of Curious and Interesting Numbers*. Middlesex, England: Penguin Books, 1986.

———. *The Penguin Dictionary of Curious and Interesting Geometry*. Middlesex, England: Penguin Books, 1991.

———. *The Penguin Dictionary of Curious and Interesting Puzzles*. Middlesex, England: Penguin Books, 1992.

Web Pages

Dr. Math. "Ask Doctor Math," mathforum.org/dr.math/.

Elmosaly, Mohamed. "Newton's Approximation of Pi," members.tripod.com/egyptonline/newton.htm.

Finch, Steve. "Mathematical Constants," pauillac.inria.fr/algo/bsolve/constant/constant.html.

Hobson, Nick. "Nick's Mathematical Puzzles," www.qbyte.org/puzzles/.

Imaeda, Mari. "Octonions and Sedenions," www.geocities.com/zerodivisor/.

Kosinov, Nikolay. "Connection of Three Major Constants: Fine Structure Constant, Number Pi, and Golden Ratio," www.laboratory.ru/articl/hypo/ax140e.htm.

Mayer, Steve. "Mathematical Constants," dspace.dial.pipex.com/town/way/po28/maths/constant.htm.

Munafo, Robert. "The Published Data of Robert Munafo," www.mrob.com/pub/index.html. In particular, see Robert's pages on "Numbers and Large Numbers." Also, see his sites, such as home.earthlink.net/~mrob/pub/math/numbers-10.html and home.earthlink.net/~mrob/pub/math/numbers.html.

O'Connor, John J. and Edmund F. Robertson. "The MacTutor History of Mathematics Archive," University of St. Andrews, Scotland, www-gap.dcs.st-and.ac.uk/~history/ or www-history.mcs.st-and.ac.uk/~history/index.html.

Pe, Joseph. "JLPE's Number Recreations Page," www.geocities.com/windmill96/numrecreations.html.

Peterson, Ivars. "MathTrek Cattle Problem," www.maa.org/mathland/mathtrek_4_20_98.html.

Schneider, Walter. "Recreational Mathematics," www.wschnei.de/index.html

Syvum Technologies, Inc. "Brain Teasers," www.syvum.com/teasers/.

Weisstein, Eric W. *CRC Concise Encyclopedia of Mathematics*, 2nd ed. New York: CRC Press, 2002. Also see Eric W. Weisstein, "MathWorld," mathworld.wolfram.com/.

Weisstein, Eric W. "Pi Formulas from MathWorld," mathworld.wolfram.com/PiFormulas.html.

Wikipedia Encyclopedia. www.wikipedia.org/.

Wikipedia Encyclopedia. "Pi," www.wikipedia.org/wiki/Pi.

Index

Note: Many special kinds of numbers, such as rational, prime, or perfect, can be found under the index entry "numbers." Special constants, such as the golden ratio (ϕ), Euler's number (e), or pi (π), can be found under the index entry "constants."

Clifford A. Pickover has a passion for mathematics. He received his Ph.D. from Yale University's Department of Molecular Biophysics and Biochemistry. He graduated first in his class from Franklin and Marshall College, after completing the four-year undergraduate program in three years. His many books have been translated into Italian, French, Greek, Spanish, German, Japanese, Chinese, Korean, Portuguese, Polish, and more. One of the most prolific and eclectic authors of our time, Pickover is the author of these popular books: *Sex, Drugs, Einstein, and Elves, Calculus and Pizza, The Mathematics of Oz, Zen of Magic Squares, Circles, and Stars, The Paradox of God and the Science of Omniscience, The Stars of Heaven, Dreaming the Future, Wonders of Numbers, The Girl Who Gave Birth to Rabbits, Surfing through Hyperspace, The Science of Aliens, Time: A Traveler's Guide, Strange Brains and Genius: The Secret Lives of Eccentric Scientists and Madmen, The Alien IQ Test, The Loom of God, Black Holes—A Traveler's Guide*, and *Keys to Infinity*. He is also the author of numerous other highly acclaimed books, including *Chaos in Wonderland: Visual Adventures in a Fractal World, Mazes for the Mind: Computers and the Unexpected, Computers and the Imagination*, and *Computers, Pattern, Chaos, and Beauty*, all published by St. Martin's Press—as well as the author of more than 200 articles concerning topics in science, art, and mathematics. He is the coauthor, with Piers Anthony, of *Spider Legs*, a science-fiction novel, as well as the author of the popular "Neoreality" science-fiction series (*Liquid Earth, Sushi Never Sleeps, The Lobotomy Club*, and *Egg Drop Soup*) in which characters explore strange realities. Pickover is currently an associate editor for the scientific journal *Computers and Graphics* and is an editorial board member for *Odyssey, Leonardo*, and *YLEM*.

Dr. Pickover's primary interest is finding new ways to continually expand creativity by melding art, science, mathematics, and other seemingly-disparate areas of human endeavor. The *Los Angeles Times* recently proclaimed, "Pickover has published nearly a book a year in which he stretches the limits of computers, art and thought." *OMNI* magazine described him as "Van Leeuwenhoek's twentieth century equivalent." *Scientific American* several times featured his graphic work, calling it "strange and beautiful, stunningly realistic." *Wired* magazine wrote, "Bucky Fuller thought big, Arthur C. Clarke thinks big, but Cliff Pickover outdoes them both." Among his many patents, Pickover has received U.S. Patent 5,095,302 for a 3-D computer mouse; 5,564,004 for strange computer icons; and 5,682,486 for black-hole transporter interfaces to computers.

Dr. Pickover is currently a research staff member at the IBM T. J. Watson Research Center, where he has received forty invention achievement awards and three research division awards. For many years, Dr. Pickover was the lead columnist for *Discover* magazine's "Brain-Boggler" column, and he currently writes the "Brain-Strain" column for *Odyssey*.

Dr. Pickover's hobbies include the practice of Ch'ang-Shih Tai-Chi Ch'uan and Shaolin Kung Fu, raising golden and green severums (large Amazonian fish), and piano playing (mostly jazz). He is also a member of the SETI League, a group of signal-processing enthusiasts who systematically search the sky for intelligent extraterrestrial life. Visit his Web site, which has received over a million visits: www.pickover.com. He can be reached at P.O. Box 549, Millwood, NY 10546-0549, USA.